Statistical Theory and Inference in Research

STATISTICS: Textbooks and Monographs

A SERIES EDITED BY

D. B. OWEN, Coordinating Editor
Department of Statistics
Southern Methodist University
Dallas, Texas

Volume 1: The Generalized Jackknife Statistic, *H. L. Gray and W. R. Schucany*

Volume 2: Multivariate Analysis, *Anant M. Kshirsagar*

Volume 3: Statistics and Society, *Walter T. Federer*

Volume 4: Multivariate Analysis: A Selected and Abstracted Bibliography, 1957-1972, *Kocherlakota Subrahmaniam and Kathleen Subrahmaniam* (out of print)

Volume 5: Design of Experiments: A Realistic Approach, *Virgil L. Anderson and Robert A. McLean*

Volume 6: Statistical and Mathematical Aspects of Pollution Problems, *John W. Pratt*

Volume 7: Introduction to Probability and Statistics (in two parts)
Part I: Probability; Part II: Statistics, *Narayan C. Giri*

Volume 8: Statistical Theory of the Analysis of Experimental Designs, *J. Ogawa*

Volume 9: Statistical Techniques in Simulation (in two parts), *Jack P. C. Kleijnen*

Volume 10: Data Quality Control and Editing, *Joseph I. Naus*

Volume 11: Cost of Living Index Numbers: Practice, Precision, and Theory, *Kali S. Banerjee*

Volume 12: Weighing Designs: For Chemistry, Medicine, Economics, Operations Research, Statistics, *Kali S. Banerjee*

Volume 13: The Search for Oil: Some Statistical Methods and Techniques, *edited by D. B. Owen*

Volume 14: Sample Size Choice: Charts for Experiments with Linear Models, *Robert E. Odeh and Martin Fox*

Volume 15: Statistical Methods for Engineers and Scientists, *Robert M. Bethea, Benjamin S. Duran, and Thomas L. Boullion*

Volume 16: Statistical Quality Control Methods, *Irving W. Burr*

Volume 17: On the History of Statistics and Probability, *edited by D. B. Owen*

Volume 18: Econometrics, *Peter Schmidt*

Volume 19: Sufficient Statistics: Selected Contributions, *Vasant S. Huzurbazar (edited by Anant M. Kshirsagar)*

Volume 20: Handbook of Statistical Distributions, *Jagdish K. Patel, C. H. Kapadia, and D. B. Owen*

Volume 21: Case Studies in Sample Design, *A. C. Rosander*

Volume 22: Pocket Book of Statistical Tables, *compiled by R. E. Odeh, D. B. Owen, Z. W. Birnbaum, and L. Fisher*

Volume 23: The Information in Contingency Tables, *D. V. Gokhale and Solomon Kullback*

Volume 24: Statistical Analysis of Reliability and Life-Testing Models: Theory and Methods, *Lee J. Bain*

Volume 25: Elementary Statistical Quality Control, *Irving W. Burr*

Volume 26: An Introduction to Probability and Statistics Using BASIC, *Richard A. Groeneveld*

Volume 27: Basic Applied Statistics, *B. L. Raktoe and J. J. Hubert*

Volume 28: A Primer in Probability, *Kathleen Subrahmaniam*

Volume 29: Random Processes: A First Look, *R. Syski*

Volume 30: Regression Methods: A Tool for Data Analysis, *Rudolf J. Freund and Paul D. Minton*

Volume 31: Randomization Tests, *Eugene S. Edgington*

Volume 32: Tables for Normal Tolerance Limits, Sampling Plans, and Screening, *Robert E. Odeh and D. B. Owen*

Volume 33: Statistical Computing, *William J. Kennedy, Jr. and James E. Gentle*

Volume 34: Regression Analysis and Its Application: A Data-Oriented Approach, *Richard F. Gunst and Robert L. Mason*

Volume 35: Scientific Strategies to Save Your Life, *I. D. J. Bross*

Volume 36: Statistics in the Pharmaceutical Industry, *edited by C. Ralph Buncher and Jia-Yeong Tsay*

Volume 37: Sampling from a Finite Population, *J. Hájek*

Volume 38: Statistical Modeling Techniques, *S. S. Shapiro and A. J. Gross*

Volume 39: Statistical Theory and Inference in Research, *T. A. Bancroft and C.-P. Han*

OTHER VOLUMES IN PREPARATION

Statistical Theory and Inference in Research

T. A. BANCROFT CHIEN-PAI HAN
Iowa State University
Ames, Iowa

MARCEL DEKKER, INC. New York and Basel

Library of Congress Cataloging in Publication Data

Bancroft, Theodore Alfonso, [date]
Statistical theory and inference in research.

(Statistics, textbooks and monographs ; v. 39)
Rev. and expanded version of the 1st pt. of the 1-vol. Statistical theory in research / by R.L. Anderson and T.A. Bancroft. 1952.
Includes indexes.
1. Mathematical statistics. 2. Research--Methodology. I. Han, Chien-Pai, [date]. II. Anderson, R. L. (Richard Loree), [date]. Statistical theory in research. III. Title. IV. Series.
QA276.B2718 1981 519.5 81-15205
ISBN 0-8247-1400-8 AACR2

MARCEL DEKKER, INC.
270 Madison Avenue, New York, New York 10016

Current printing (last digit):
10 9 8 7 6 5 4 3 2 1

PRINTED IN THE UNITED STATES OF AMERICA

To Lenore and Maria

PREFACE

Modern statistics is primarily concerned with the development and use of an effective methodology for making inferences (estimation, prediction, tests of hypotheses) to a population (or populations) of possible observations from usually a small sample (or samples) of such observations. Such a sample, small or large, may, for example, result from a sample survey taken by a social scientist (economist, psychologist, or sociologist) or an experiment conducted by a natural scientist (biologist, physical scientist, or engineer). It is clear, then, that modern statistics is an important part of scientific methodology, that is, an investigative procedure for obtaining new knowledge in a number of scientific and technological areas.

In view of the above, the discipline of statistics has a dual nature: (1) it is one of the mathematical sciences as regards the development and understanding of the "know why" of statistical theory and methodology; and (2) it is an important part of scientific methodology as regards the "know how" of statistical methods and techniques.

In the following, the authors, after setting out certain principles, have attempted to delineate the scope of this book, which could be used as a text for advanced undergraduate and graduate courses and as a reference, in particular, in those courses presenting the "know why" of statistical theory and methodology without excessive mathematical background.

Based on many years of teaching, consulting, and research at a university statistical center, the firm conviction of the authors is that the usual text or treatise on statistical methods or applied statistics, emphasizing statistical techniques or the "know how" of statistics, is not enough background for the serious and discriminating use of such procedures in substantive area investigations. In particular, in addition to texts and treatises prepared primarily for advanced undergraduate and graduate students in statistics per se, teaching and self-study materials concerned with the "know why" of statistical methodology, appropriately specialized, should be available for advanced undergraduate and graduate students in substantive areas that use statistical methodology, such as the physical, engineering, social, and biological sciences. Instead of the standardized highly theoretical text, usually requiring mathematics through advanced calculus (or at least taken concurrently), a text or treatise especially prepared for advanced undergraduates and graduate students in substantive areas other than statistics should require mathematics through undergraduate integral and differential calculus. Again the statistical topics covered should be concerned with the theory or "know why" of importance in substantive area application. In presenting the material, emphasis should be placed on clarity and understanding rather than the level of mathematical rigor that should be required of a statistics or mathematics major.

A book containing such materials should also be appropriate as a text for graduate students majoring in applied statistics and graduate students who wish a joint major program between statistics and some substantive discipline which uses statistics as an important part of its research methodology.

The authors have attempted to follow the above principles in preparing the material for this text, designed especially for advanced undergraduate or graduate students in various substantive areas or in applied statistics including joint majors. The publication of material for such a course is not new; actually Bancroft, one of the authors of this text, had the primary responsibility of writing Part I, Basic Statistical Theory, of *Statistical Theory in Research* (1952) by R. L. Anderson and T. A. Bancroft. The material of *Basic Statistical Theory* has been used in a two-quarter course offered by the Department of Statistics at Iowa State University twice a year as a service course, primarily for advanced undergraduate and graduate students in substantive fields, since 1950. However, this earlier text is now out of print. For the past several years work was undertaken to update and rewrite the material of *Basic Statistical Theory*, which has resulted in this text, *Statistical Theory and Inference in Research*.

In the above connection, we are of the opinion that textbooks, in particular those designed to be used in service courses and for self-study, should have the intended users as recognized co-authors. Since it was expected that the users of *Statistical Theory and Inference in Research* would be (1) advanced undergraduate and graduate students in substantive fields, joint majors, and majors in applied statistics; (2) substantive area scientists for reference and self-study; and, of course, (3) teachers of courses designed for students described in (1), plans were made to obtain the opinion and advice of these potential users in making improvements.

To test the effectiveness of a first draft of the updated material, Bancroft taught the material during two quarters in 1976-1977, and a subsequent revision in the offset printing form was used by Han in a second effectiveness test in teaching the two-quarter sequence from 1977 to 1981. This text is based on a final revision, taking into account the feedback information obtained from these teaching experiences and the advice of users.

This textbook is intended for a one-year, three-credit course. It may be used for a one-semester or two-quarter course consisting of the following suggested material: Sections 1.1, 2.1-2.7, 3.1-3.4, 3.7-3.10, 3.13, 4.1-4.6, 5.1-5.8, 6.1, 6.2, 6.5-6.12, 7.1-7.8, 7.11-7.13, 9.1-9.4, 10.1-10.5, 10.7-10.8, 11.4-11.6, 13.1-13.7, 14.1-14.3, 14.5.

Both authors are indebted to the late Professor William G. Cochran for his contributions in research and teaching in statistical theory and inference in providing research methods and techniques.

We wish to thank Mrs. Marylou Nelson and Mrs. Janice Peters for the typing of the draft of the manuscript. Thanks are also due to

Mrs. Marlene Sposito for the final typing in accordance with special instructions of the publisher.

Ames, Iowa

T. A. Bancroft
C.-P. Han

CONTENTS

Statistical Theory and Inference in Research

1

INTRODUCTION

1.1 STATISTICS: In our daily lives we have knowingly or unknowingly encountered descriptive quantities which are determined by the use of statistics. To name a few, the following terms are repeatedly heard: Dow Jones average of 30 industrial stocks, median income, batting averages of baseball players, average life of light bulbs, unemployment rate, pollution standards, percentage of votes candidate A will receive, etc. Statistics is widely used not only to describe present conditions but also to predict future conditions and to make decisions. It is commonplace for experimenters or investigators to collect data from experiments and surveys. Based on such data, inferences are made concerning the population which generated the data. In order to insure that description, prediction, and decision making be correct with high precision, statistical methodology is derived using sound statistical theory established by means of inductive reasoning based on the mathematics of probability. Hence, a definition of statistics may be stated as follows:

Statistics is the science and art of the development and application of the most effective methods of collecting, tabulating, and interpreting data in such a manner that the fallibility of conclusions may be assessed by means of inductive reasoning based on the mathematics of probability.

A definition and theory of probability form the basis of a statistical methodology. Since there are several definitions and interpretations of probability, the result is that there are several resulting theories and accompanying methodologies as regards modern statistical theory as a basis for applied statistics. Two of these are the *frequency definition of probability* and the *subjective definition of probability* and their respective resulting theories and methodologies.

Using statistical procedures based on the frequency definition of probability (most often used in applied statistics), we may gain some insight into the nature of statistics from the steps taken in analyzing the data from a simple yet representative investigation.

1.2 A REPRESENTATIVE INVESTIGATION: The problem: Is the newly developed drug A more effective than standard drug B in the treatment of a certain swine disease?

Step 1. A hypothesis, sometimes referred to as a null hypothesis, is formulated. In this case a pertinent hypothesis for which simple techniques for testing are available is: Drug A is equally effective as drug B in the treatment of the swine disease.

Step 2. An experiment is designed to test the hypothesis. Twenty diseased swine of the same intensity of infection are divided at random into two groups of 10 each. Treatment A is assigned at random to one group and B to the other. Except for the different drugs all 20 swine were treated alike for the duration of the experiment. At the end of a predetermined treatment period an objective determination of improvement or no improvement was available for each animal.

Step 3. Pertinent data are collected and tabulated. The number of diseased animals showing improvement and those showing no improvement are given in Table 1.1.

TABLE 1.1

	Diseased Swine		
Drug	Improved	Unimproved	Total
A	a = 9	b = 1	10
B	c = 3	d = 7	10
Total	12	8	20

Now, if A is actually more effective than B, we should expect the experiment to provide evidence of this; that is, the sample improvement rate of A would be expected to be greater than that for B, which is in fact the case since 9/10 is greater than 3/10. A critical question is now in order: Does this high value for A, from this one experiment, indicate that A is truly more effective than B, or is there a high probability that one could obtain a value of 9/10 for A by chance when A was actually no more effective than B?

Step 4. The distribution of the data on the assumption that the null hypothesis is true is obtained, that is, assuming that any treatment difference observed in this one experiment may be attributed to chance. In order to answer the critical question posed in step 3, we assume that the null hypothesis is true, that is, that A and B are equally effective in treating the swine disease, and find the probability α' that such a high value (or a higher one) as 9/10 for treatment A could have been obtained under this assumption. If this probability is low, we reject the null hypothesis, knowing that there is a very small probability (measured by α) that the null hypothesis is correct.

In order to accomplish the above, we make use of the exact treatment of 2 × 2 contingency tables due to R. A. Fisher (1970). If the entries in the cells of a 2 × 2 contingency table, such as those in Table 1.1, be designated a, b, c, and d and if a + b + c + d = n, Fisher shows by simple mathematics that the exact probability of any observed set of entries is given by

$$\frac{(a + b)!(c + d)!(a + c)!(b + d)!}{n!a!b!c!d!}$$

where the marginal frequencies are assumed to be fixed in repeated sampling.

Using the above formula and the observed cell entries from Table 1.1, we find

$$P\left(A = \frac{9}{10}, B = \frac{3}{10}\right) = \frac{10!10!12!8!}{20!9!1!7!3!}$$
$$= \frac{40}{4199}$$

Now, as stated in step 4, we need to increase this by $P(A = 10/10, B = 2/10) = 3/8398$. Hence, the probability for treatment A to do as well or better, on the assumption that the null hypothesis is true, is 83/8398 or approximately .01.

It follows that there is only 1 chance out of 100 that treatment A could be as effective or better than it was in the performed experiment. Or stated another way, if A and B were equally effective, we could expect as good or better showing by A in only 1 out of 100 experiments on the average.

Step 5. A test of significance is performed. Usually we decide on an acceptable probability level, α, before the experiment is conducted, and if the results give a probability $\alpha' \leq \alpha$, we state that the null hypothesis is rejected at the α significance level. Two commonly used values of α are .05 and .01. If the null hypothesis is rejected at the $\alpha = .05$ level, the results are said to be *significant*; if at the $\alpha = .01$ level, *highly significant*.

In our experiment $\alpha' = .01$. Hence we may well reject the null hypothesis at the $\alpha = .01$ significance level. In this case, if the null hypothesis was not rejected, we should be forced to accept the occurrence of an improbable event.

Instead of using Fisher's exact treatment of 2×2 contingency tables, use may be made of the chi-square (χ^2) probability distribution to obtain a satisfactory test of our null hypothesis with computational ease even for large cell entries. However, this latter test requires the development of a more complex theoretical background and the provision of special tables.

The representative investigation described above is concerned with an experiment. Another large class of investigations involving the use of statistical methodologies is that of the sample survey. One may wish to use the data from a sample survey to make tests of significance in making comparisons; however, many sample surveys are concerned with estimating means and totals and accompanying measures of precision.

It should be noted that, for both an experiment and a sample survey, the investigator is concerned with obtaining information regarding comparisons and estimates in an existing or conceptual population of possible observations from the particular observations (usually a small number) taken for a particular investigation. In addition to hypothesis testing and estimation, the investigator may be concerned with establishing a prediction equation from the data from an experiment or a survey. Again, instead of drawing conclusions from observed data, the investigator may wish to make a decision. Since information pertaining to population is to be obtained from a part of the population, we may classify hypothesis testing, estimation, prediction, and decision making as statistical inferences.

Now, workers engaged in the various fields of scientific inquiry are concerned with making inferences which may or may not involve the use of statistical methodology. What essential characteristic, then, is peculiar to statistical inference procedures? As indicated in the definition of *statistics*, this essential characteristic is the use of the mathematics of probability to calculate from the observations themselves a measure of the fallibility of the respective inferences drawn from the sample observation to the population.

In order that the valid and efficient measures of the fallibility of inferences in terms of probability statements be possible and useful, steps in the investigation must be taken with this end product in mind. Hence, in using statistical inference procedures, the investigator may find himself vitally concerned with matters that are necessary although indirectly related to the particular inference procedure technique, such as: design of experiments and sample surveys, questionnaire construction and training and supervision of enumerators, experimental techniques, collection and tabulation of data, model specification, and interpretation of results.

EXERCISES

1.1 Suppose that the cell frequencies in Table 1.1 were a = 8, b = 2, c = 4, d = 6. Now test the hypotheses that drug A and drug B are equally effective in the treatment of the swine disease using the procedures of Section 1.2. Interpret this result, if α is chosen to be .05.

1.2 A random sample of 22 married men, all retired, were classified according to whether or not they had graduated from a four-year undergraduate college and had "3 or less" or "over 3" children. Use the data below and Fisher's exact method to test the hypothesis that the fathers who have graduated from college have the same percentage of family size of over 3 children as fathers with no college education in the senior population.

	Number of Children		
Education	3 or Less	Over 3	Total
College	5	1	6
No college	2	14	16
Total:	7	15	22

REFERENCE

Fisher, R. A. (1970). *Statistical Methods for Research Workers*, 14th ed. Hafner, New York.

2

PROBABILITY

2.1 INTRODUCTION: It was stated earlier that statistics concerns itself with inductive reasoning based on the mathematics of probability. In developing statistical methodology the statistician makes use of the definitions, postulates, and theorems of mathematical probability. What can be said of this framework, these "bones," of statistics? The theory of probability had its genesis in the application of mathematics to determining the odds in various games of chance: dice, cards, spun wheels, etc. In particular, the foundations of the science of probability were laid by two seventeenth-century mathematicians, Pascal and Fermat, in their private correspondence concerning questions raised regarding the gambling observations of the French nobleman Chevalier de Méré. Game theory has become a branch of study in statistics.

Statistics, then, is no dry-as-dust subject concerning itself with the compilation of innumerable tables and charts. On the contrary, it deals with the development and application of an important methodology based on the fascinating subject of probability. Since the turn of the century, theory and methods of statistics have been steadily developing. This methodology has become of great importance as a research tool in the physical, biological, engineering, and social sciences.

2.2 SOME INTERPRETATIONS OF PROBABILITY: Since statistics uses inductive reasoning based on the mathematics of probability, statistics is a branch of applied mathematics whose methodologies stem from the axioms and theorems of probability, which in turn is a branch of pure mathematics. A definition of the probability of the occurrence of an event would appear in order.

In order to arrive at a proper definition we will first define an event. An *event* is a collection or a set of elementary events. An *elementary event* is one of many ways an experiment can terminate. For example, in a lottery an urn contains 20 balls marked 1 to 20. One ball is drawn at random for a prize. Then there are 20 elementary events. An event may be defined as that the ball drawn is either numbered 5 or 17. One may ask, "What is the probability of this event?" Since there are 2 chances out of 20 that the event will occur, we might reasonably set the probability as equal to 1/10. An interpretation of this kind of probability is that if one continues the drawing (with replacement) a large number of times under similar conditions, the relative frequency of the occurrence of 5 and 17 tends to the limit 1/10. This interpretation depends on the *relative frequency* of the occurrence of the event.

The conditions for the frequency interpretation of probability are not definite enough to serve as a definition of probability for scientific purposes. For example, how large a drawing is "large?" Would 1000 suffice, or would one need 10,000? Since it is not physically possible to make an infinite number of drawing, this answer must be ruled out. Again, in practical applications, it may not be possible to guarantee "similar conditions."

A second interpretation of probability leads to the *classical* definition of probability:

> If an experiment can terminate in N equally likely and mutually exclusive (the occurrence of one precludes the occurrence of any other) ways and if A occurs in n of these ways, th en the probability of the occurrence of A is n/N.

This definition of probability assumes that it is possible logically to determine, before trials are made, all the equally likely and mutually exclusive ways that an event may happen and to assign n of these ways to the occurrence of attribute A.

EXAMPLE 2.1 What is the probability of obtaining a head with a penny on a single toss? Assuming the coin is a "true" coin, we reason that it may fall 2 equally likely ways and that 1 of them must be heads; hence, the probability is 1/2.

Notice that the classical definition of probability, in using the words "equally likely," assumes a knowledge of probability in order to define the term. Logically, of course, this is certainly undesirable.

It should be noted that both the *frequency* and the *classical* interpretations of probability are objective in nature.

Let us now turn to a third interpretation of probability. Consider the following example. A store sells bicycles. For inventory purposes or otherwise, the question "What is the probability that the store will sell 50 or more bicycles next month?" may be of interest. The store manager, based on his experience, may assess that the probability is .9. Others may give other assessments; e.g., the Chamber of Commerce may assess the probability to be .8. These probabilities cannot be interpreted by relative frequency. There is no actual repetition of the event. The probability given by the store manager is based on his previous experience or his personal belief. This interpretation depends on personal assessment of *degree of belief*. Probability interpreted this way is called *subjective probability*. Like classical probability this is also *a priori probability*, that is, the assessment is made before the experiment. If data are collected after the experiment, this a priori subjective probability can be revised by the data. This is done by using Bayes' formula. Bayes' formula will be discussed in Section 2.6. The revised probability is called *a posteriori probability*.

There are, of course, difficulties in using subjective probability as a basis for statistical inference: (1) a person's belief as to the relative likelihood of an infinite number of events must be completely consistent and free from contradictions; and (2) a subjective interpretation provides no "objective" basis for scientists, working in some scientific area of common interest, to arrive at a common evaluation of the state of knowledge of the area.

As mentioned above, Bayes' formula is used to combine a priori subjective probability and the information supplied by data, from which a posteriori probability is obtained. Statistical inference based on such a procedure is called Bayesian inference. Bayesian inference will be treated later.

We have discussed three useful interpretations of probability. As to which one to use, this should be made to depend on the actual problem in hand. It is our opinion that if the experimenter has previous experience or information, such should be used either at the model specification stage using an objective analysis of the data or incorporated with the data in case a subjective approach is used. If he or she does not have any such experience or information, the experimenter should analyze the data objectively.

2.3 SET OPERATIONS: In order to enumerate the probability of an event it is better to use set operations. A set is a collection of objects, for example, a set of furniture, a set of tools, etc. This can be generalized, in our context, to state that a set is a collection of elementary events. These elementary events are elements of the set. For example, if a die is tossed, a set may consist of all even numbers, i.e., $\{2,4,6\}$. Another set may consist of all possible outcomes, which is $\{1,2,3,4,5,6\}$. Sets will be denoted by A, B, ..., so we may let $B = \{2,4,6\}$, $A = \{1,2,3,4,5,6\}$. The set consisting of all possible outcomes is called the *universal set*. The set consisting of no outcome is called the *empty set* or *null set*. In statistics, the universal set is called the *sample space* and the elementary events are called the *sample points*.

Some important rules of operation on sets are:

1. The union of two sets A and B is the set consisting of the elements either in A or in B or in both. The union is denoted by $A \cup B$ or $A + B$.

2. The intersection of two sets A and B is the set consisting of the elements both in A and in B. The intersection is denoted by $A \cap B$ or AB.

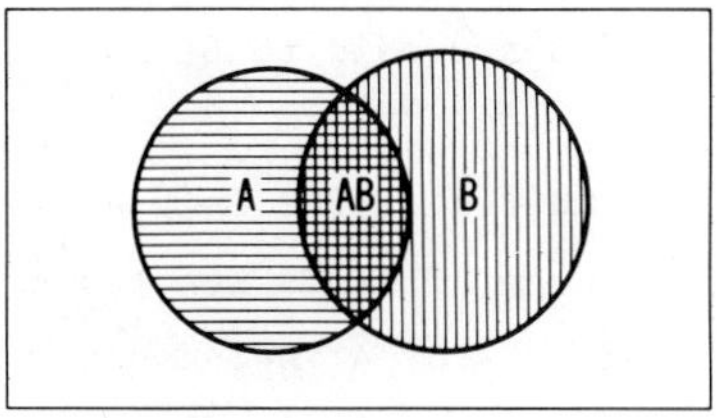

FIG. 2.1 Venn diagram.

3. The difference of two sets, denoted by A - B, is the set consisting of the elements in A but not in B.

4. The complement of a set A, denoted by $\bar{A}$, is the set consisting of the elements not in A.

These operations can be illustrated by means of a "Venn diagram." In Fig. 2.1 let the rectangle represent the universal set, the two circles represent any two sets. The union $A \cup B$ is the area shaded by *either* horizontal *or* vertical lines. The intersection $A \cap B$ is the area shaded by *both* horizontal and vertical lines. The difference A - B is the area shaded *only* by horizontal lines but not by vertical lines. The complement $\bar{A}$ is the area outside the circle A.

Some laws similar to those of ordinary addition and multiplication are as follows:

1. *Commutative law:*

$$A \cup B = B \cup A \quad \text{and} \quad A \cap B = B \cap A$$

2. *Associative law:*

$$A \cup (B \cup C) = (A \cup B) \cup C \quad \text{and} \quad A \cap (B \cap C) = (A \cap B) \cap C$$

3. *Distributive law:*

$$A \cup (B \cap C) = (A \cup B) \cap (A \cup C) \quad \text{and} \quad A \cap (B \cup C) = (A \cap B) \cup (A \cap C)$$

The reader should satisfy himself of the validity of these laws by using a Venn diagram. Further, it is easy to see that the complement of $\bar{A}$ is A. $A \cap \bar{A}$ is the empty set and $A \cup \bar{A}$ is the universal set. If $A \cap B$ is the empty set, then A and B are said to be *mutually exclusive or disjoint*. Also $A = AB \cup A\bar{B}$.

Occasionally subsets of a set are considered. If B is a subset of A, then all the elements in B are also in A. This is denoted by $B \subset A$ or $A \supset B$, which reads as B is included in A or A includes B. For example, the set $B = \{2,4,6\}$ is a subset of $A = \{1,2,3,4,5,6\}$. Since the universal set consists of all the elements, any set is a subset of the universal set. A proper subset is a subset other than the set itself. The subset B above is a proper subset of A.

EXAMPLE 2.2 Suppose there are 9 candidates on a ballot and they are numbered from 1 to 9. Candidates 1 to 3 are Democrats and 4 to 9 are Republicans. The candidates were asked their stands on the issue of imposing wage and price control to curb inflation. Candidates 1, 6, 7, and 9 oppose imposing the control and the others are for the control. Let A be the set of Democratic candidates and B the set of candidates who oppose the control. Then $A = \{1,2,3\}$ and $B = \{1,6,7,9\}$. It is easily seen from Fig. 2.2 that

$$\begin{aligned}
\bar{A} &= \{4,5,6,7,8,9\} \\
\bar{B} &= \{2,3,4,5,8\} \\
A \cap B &= \{1\} \\
A \cap \bar{B} &= \{2,3\} \\
\bar{A} \cap \bar{B} &= \{4,5,8\}
\end{aligned}$$

The set $A \cap B$ is the set of Democrats who oppose the control; $A \cap \bar{B}$ is the set of Democrats who agree with the control and $\bar{A} \cap \bar{B}$ is the set of Republicans who agree with the control.

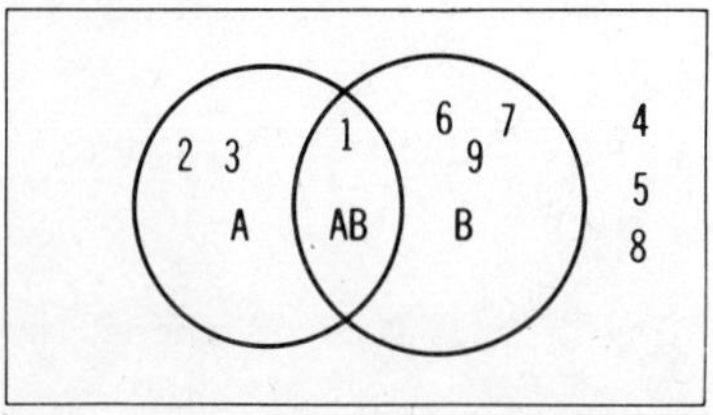

FIG. 2.2 Venn diagram for the candidates.

2.4 MATHEMATICAL THEORY OF PROBABILITY: The development and use of probability, based on intuitive and "reasonable" interpretations, did not wait for the development of a rigorous mathematical theory of probability. The Russian mathematician Kolmogorov (1933) gave a definitive and complete formulation of the mathematical theory of probability. Essentially Kolmogorov's formulation may be stated in the following manner. For an experiment E we need to assign to each event A_i in the sample space S a number $P(A_i)$ which indicates the probability that A_i will occur. If the number assigned $P(A_i)$ satisfies three axioms, then it will satisfy the mathematical definition of probability. The axioms require that $P(A_i)$ possesses certain properties which we intuitively and "reasonably" expect a probability to possess under any of the three interpretations described in Section 2.2.

Stated in words and symbolically the axioms are:

Axiom 1 The probability of every event must be nonnegative, i.e., $P(A_i) \geq 0$.

Axiom 2 If any event is certain to occur, then its probability is 1, i.e., $P(S) = 1$.

Axiom 3 For any infinite sequence of disjoint* events A_1, A_2, ..., the probability of the union or sum is the sum of the probabilities, i.e.,

$$P\left(\bigcup_{i=1}^{\infty} A_i\right) = \sum_{i=1}^{\infty} P(A_i)$$

As a consequence of the above, we are now able to give the mathematical definition of probability as follows:

A probability on a sample space S is a specification of numbers $P(A_i)$ which satisfy Axioms 1, 2, and 3.

---

*If two events, say A and B, are disjoint, then their Venn diagrams (see Fig. 2.1) would not intersect, i.e., A ∪ B would consist of all the elements in A and all the elements in B. Hence, it is logical to assume that the probability that either occur is the sum of their individual probabilities. Also assume true for any finite or infinite sequence of disjoint events.

Now, with these axioms we may prove fundamental theorems providing further properties of probability. Continuing in this manner it is possible to prove the fundamental theorems and laws of probability, many of which had already been stated and used in the early development of probability and statistics from an intuitive and "reasonable" approach. Since our purpose is to make use of needed results from probability in the development of statistical inference, we will forego a rigorous approach in the proof of needed theorems or rules in order to concentrate on their uses in the development of basic statistical theory.

In order to calculate the mathematics of probability of the occurrence of an event, we must know how to compute the number of ways that an event may occur. A fundamental tool is to consider permutations and combinations. This is given in the next section.

2.5 PERMUTATIONS AND COMBINATIONS: The number of ways in which an event may occur may be determined by the enumeration of possible outcomes in the set or by the use of some simple rules from college algebra. The latter method is simpler for more complicated problems. Two fundamental rules are:

RULE 2.1 If A can happen in m ways and B in n ways independent (the occurrence of one does not affect the chance of the occurrence of the other) of m, then both A and B can happen in mn ways.

EXAMPLE 2.3 If two ordinary dice numbered I and II are tossed, one may appear face up in 6 ways which are independent of the 6 ways in which the second may appear. Hence, both may appear face up together in 36 different ways.

RULE 2.2 If A can happen in m ways and B in n ways mutually exclusive of m, then either A or B can happen in m + n ways.

EXAMPLE 2.4 Either an ace or a king (1 card) may be drawn from an ordinary deck of cards in 4 + 4 = 8 ways.

For multiple arrangements, a rule for *permutations* can be applied. If it is desired to arrange n different objects into sets of r objects per set, the number of such arrangements is called "the number of permutations of n objects taken r at a time" and is indicated by P(n,r). The first of the r positions can be filled in n ways, the second in (n - 1) ways since one object will have been used in the first position, the third in (n - 2) ways, etc. Hence, by Rule 2.1:

RULE 2.3 $P(n,r) = n(n-1)(n-2)\cdots(n-r+1) = n!/(n-r)!$, *where* $n! = n(n-1)\cdots 2\cdot 1$.

It should be noted that when r = n, P(n,n) = n!, which implies that 0! = 1.

EXAMPLE 2.5 The number of different ways of selecting a president, vice-president, and secretary from a suggested slate of 6 is

$$P(6,3) = 6\cdot 5\cdot 4 = 120$$

Suppose that all n objects in the arrangement are used, but certain groups n_1, n_2, etc. are alike. Any rearrangement of the objects of any n_i group will not change any particular arrangement; hence, the total number of arrangements will be less than if all the objects were different from one another. Now, any group of n_i alike objects can be arranged $n_i!$ ways, and since these $n_i!$ arrangements are alike for every arrangement of the other objects, the total number of different arrangements will be given as below:

RULE 2.4 $P(n;n_1,n_2,n_3,\ldots) = n!/n_1!n_2!n_3!\ldots$, where $P(n;n_1,n_2,n_3,\ldots)$ represents the total number of permutations, given that n_1 are alike, n_2 alike but different from the first group, etc., and $\Sigma n_i = n$.

EXAMPLE 2.6 How many different 6-flag signals may be made if 3 are red, 2 blue, and 1 yellow? $P(6;3,2,1) = 6!/3!2!1! = 60$.

If interest lies only in groups of objects and not in the arrangements within the groups, then combinatorial rules apply. The total number of combinations of n objects taken r at a time is denoted symbolically as $\binom{n}{r}$ or C(n,r). It is easily seen that

$$P(n,r) = \binom{n}{r} P(r,r)$$

since each combination of r objects may be permuted P(r,r) times. The following rule is now derived:

RULE 2.5 $\binom{n}{r} = \frac{P(n,r)}{P(r,r)} = \frac{n!}{(n - r)!r!}$

EXAMPLE 2.7 The total number of different bridge hands of 13 cards which can be dealt from a deck of 52 cards is $\binom{52}{13} = 52!/13!39!$. Again the total number of sets of 4 bridge hands is

$$\binom{52}{13} \cdot \binom{39}{13} \cdot \binom{26}{13} \cdot \binom{13}{13} = \frac{52!}{(13!)^4}$$

The use of the rules of permutations and combinations involves factorials, some with quite large values of n. Stirling's formula,

$$n! = \sqrt{2\pi n}\, n^n e^{-n} \left(1 + \frac{1}{12n} + \frac{1}{288n^2} + \cdots\right)$$

may be used to obtain quickly an approximation to n!. The first term,

$$n! \doteq \sqrt{2\pi n}\, n^n e^{-n} \tag{2.1}$$

gives a suitable approximation in many cases, where $\doteq$ indicates "approximately equal to." When n is reasonably small, n! may be obtained from a calculator or on computer.

EXAMPLE 2.8 Evaluate 13! by the use of Stirling's formula:

$$13! \doteq \sqrt{2\pi(13)} \left\{13^{13} e^{-13} \left[1 + \frac{1}{(12)(13)}\right]\right\}$$

$$\log 13! \doteq \frac{1}{2}(\log 26 + \log \pi) + 13 \log 13 - 13 \log e + \log \frac{157}{156}$$

$$13! \doteq 6.2271 \times 10^9$$

using a five-place logarithm table.

After obtaining the total number of mutually exclusive and equally likely ways and those which possess attribute A by the use of the rules of permutations and combinations, it is then possible to write the required probability by applying the classical definition

$$p = \frac{\text{number of ways that possess attribute A}}{\text{total number of ways}} \qquad (2.2)$$

EXAMPLE 2.9 A bag contains 4 red and 3 white balls. What is the probability of obtaining exactly 3 red balls when 3 balls are drawn?

$$p = \frac{\binom{4}{3}}{\binom{7}{3}} = \frac{4}{35}$$

2.6 *FUNDAMENTAL LAWS OF PROBABILITY:*

LAW 1 If A and B are two mutually exclusive events, then the probability of either of them happening is the sum of the respective probabilities. Symbolically, $P(A \cup B) = P(A) + P(B)$, where $P(A)$ denotes the probability of A occurring.

EXAMPLE 2.10 The probability of throwing either a 7 or an 8 with two dice is $6/36 + 5/36 = 11/36$.

LAW 2 If A and B are two independent events, so that the occurrence of one does not affect the chance of the occurrence of the other, the probability that both happen is the product of their respective probabilities. Symbolically, $P(A \cap B) = P(A) \cdot P(B)$.

This law may be extended to n events. The events $A_1, A_2, \ldots, A_n$ are mutually independent if $P(A_1 \cap A_2 \cap \cdots \cap A_n) = P(A_1) \cdot P(A_2) \cdots P(A_n)$.

EXAMPLE 2.11 The probability of getting 2 red balls in drawing 1 ball from each of two urns containing 6 red balls and 4 black balls is $(6/10)(6/10) = 9/25$

If A and B are not independent of one another so that the occurrence of one affects the probability of the occurrence of the other, then a definition is needed for the *conditional probability* of A given that B has occurred, which is denoted $P(A|B)$. Similarly, the conditional probability of B given that A has happened is $P(B|A)$. In such cases fundamental Law 2 becomes:

LAW 3 $P(A \cap B) = P(A) \cdot P(B|A) = P(B) \cdot P(A|B)$. If A and B are independent, $P(B|A) = P(B)$ and $P(A|B) = P(A)$.

EXAMPLE 2.12 If both balls were drawn in succession from one of the urns in Example 2.10 without replacement, i.e., after the first ball is drawn, it is not returned before the second drawing, then the probability of obtaining 2 red balls is $(6/10)(5/9) = 1/3$.

LAW 4 If two events are not mutually exclusive, then the probability of at least one of them occurring is $P(A \cup B) = P(A) + P(B) - P(A \cap B)$.

Proof: Let $\bar{A}$ and $\bar{B}$ represent the nonoccurrence of A and B, respectively. Then,

$$P(A) + P(B) = P(A \cap B) + P(A \cap \bar{B}) + P(A \cap B) + P(B \cap \bar{A})$$
$$= P(A \cap B) + P(A \cup B)$$

Therefore $$P(A \cup B) = P(A) + P(B) - P(A \cap B)$$

where $P(A \cap \bar{B})$ is the probability of A occurring and B not occurring, and similarly for $P(B \cap \bar{A})$.

Law 4 is illustrated in Fig. 2.3, where the outcomes of a chance event are represented by points in a plane. Then the outcomes belonging to either A or B may be represented by the points in A and B less the points common to region AB since AB would be counted twice.

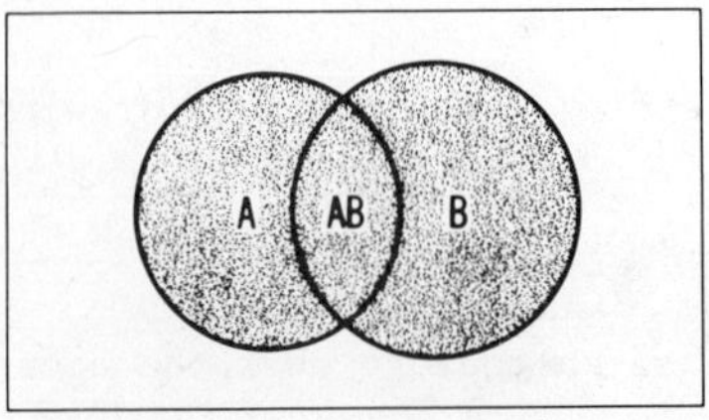

Fig. 2.3 Venn diagram.

This law may be extended, for example,

$$P(A \cup B \cup C) = P(A) + P(B) + P(C) - P(A \cap B) - P(A \cap C) - P(B \cap C) + P(A \cap B \cap C)$$

EXAMPLE 2.13 In Example 2.10 the probability of obtaining at least 1 red ball is

$$P(A \cup B) = P(A) + P(B) - P(A \cap B) = \frac{6}{10} + \frac{6}{10} - \frac{6}{10} \cdot \frac{6}{10} = \frac{21}{25}$$

LAW 5 If the probability of an event occurring in a single trial is p, the probability of its occurring r times out of n independent trials is given by

$$\binom{n}{r} p^r (1 - p)^{n-r} = \binom{n}{r} p^r q^{n-r}$$

where $1 - p = q$. This is the $(r + 1)$st term of $(q + p)^n$.

Proof: If the event occurs r times out of n trials, it will fail to occur $n - r$ times; hence, the probability of the occurrence of any sequence of r successes and $n - r$ failures is $p^r(1 - p)^{n-r}$. But the number of possible sequence is given by $\binom{n}{r}$.

EXAMPLE 2.14 The probability of obtaining exactly 3 heads on single toss of 5 coins is $\binom{5}{3}$ $(1/2)^3(1/2)^2 = 5/16$.

2.7 BAYES' FORMULA: In the previous sections it was assumed that the causal system was known a priori, hence, exact probabilities of various results were readily calculated. In tossing a die it was assumed that all 6 faces appeared "equally likely" and that a random toss of the die was made. In such cases the probability of obtaining any number from 1 to 6 was easily seen to be 1/6. With these same assumptions and use of the fundamental laws of probability it was also easy to state the probability of obtaining a 7 on a single throw, say, with two dice.

In statistics, however, one is often faced with exactly the reverse of this situation. A batch of data resulting from some experiment is at hand, and we wish to state the probability that such data could have been produced by a given causal system. For example, it is noted that 200 7s were obtained in tossing two dice 1000 times. We now wish to know the probability that such a result could have been produced with unbiased dice. Or we may wish to state, on the basis of these results, the expected number of 7s to be obtained on the next 100 tosses of two dice. These problems concern a posteriori probability based on previous occurrences.

A posteriori probabilities, under certain conditions, may be obtained by the use of Bayes' formula. Let $B_1, B_2, \ldots, B_n$ be n mutually exclusive random events, of which one is certain to occur. Let $P(B_i)$ be the probability of the occurrence of B_i. Let E be an event which can occur only if one of the set of B_i values occurs. Let $P(E|B_i)$ be the conditional probability for E to occur, assuming the occurrence of B_i. We wish to know how the probability of B_i changes with the added information that E has actually happened. In other words, we wish the conditional probability $P(B_i|E)$.

Using Law 3,

$$P(E \cap B_i) = P(B_i)P(E|B_i)$$

and

$$P(E \cap B_i) = P(E)P(B_i|E)$$

Equating the right-hand sides of these two equations and solving for $P(B_i|E)$, we obtain

$$P(B_i|E) = \frac{P(B_i)P(E|B_i)}{P(E)} \qquad P(E) \neq 0$$

Since E may occur with any of the B_i mutually exclusive events,

$$\begin{aligned} P(E) &= P(E \cap B_1) + P(E \cap B_2) + \cdots + P(E \cap B_n) \\ &= P(B_1)P(E|B_1) + P(B_2)P(E|B_2) + \cdots + P(B_n)P(E|B_n) \end{aligned}$$

Upon substituting this last result in the denominator of the preceding equation, we obtain *Bayes' formula*:

$$P(B_i|E) = \frac{P(B_i)P(E|B_i)}{P(B_1)P(E|B_1) + P(B_2)P(E|B_2) + \cdots + P(B_n)P(E|B_n)} \quad (2.3)$$

If we consider the events B_1, B_2, ..., B_n as hypotheses to account for the occurrence of E, then Bayes' formula provides a means of calculating probabilities of hypotheses. In this case $P(B_1)$, $P(B_2)$, ..., $P(B_n)$ are called a priori probabilities of the hypotheses B_1, B_2, ..., B_n, and $P(B_1|E)$, $P(B_2|E)$, ..., $P(B_n|E)$ are called a posteriori probabilities of the same hypotheses.

EXAMPLE 2.15 Urn I contains 2 white, 1 black, and 3 red balls. Urn II contains 3 white, 2 black, and 4 red balls. Urn III contains 4 white, 3 black and 2 red balls. One urn is chosen at random, and 2 balls are drawn. They happen to be red and black. What is the probability that both balls came from Urn I? Urn II? Urn III?

We identify E as the event that the 2 balls were, respectively, red and black. To explain this occurrence, we have three hypotheses: the urn was I, II, III. We identify these hypotheses with B_1, B_2, B_3.

Then $P(B_1) = P(B_2) = P(B_3) = 1/3$, and $P(E|B_1) = (3/6) \cdot (1/5) + (1/6) \cdot (3/5) = 1/5$. Similarly, $P(E|B_2) = 2/9$, and $P(E|B_3) = 1/6$. Substituting in Bayes' formula,

$$P(B_1|E) = \frac{\frac{1}{3} \cdot \frac{1}{5}}{\frac{1}{3} \cdot \frac{1}{5} + \frac{1}{3} \cdot \frac{2}{9} + \frac{1}{3} \cdot \frac{1}{6}} = \frac{18}{53}$$

Similarly,

$$P(B_2|E) = \frac{\frac{1}{3} \cdot \frac{2}{9}}{\frac{1}{3} \cdot \frac{1}{5} + \frac{1}{3} \cdot \frac{2}{9} + \frac{1}{3} \cdot \frac{1}{6}} = \frac{20}{53}$$

$$P(B_3|E) = \frac{\frac{1}{3} \cdot \frac{1}{6}}{\frac{1}{3} \cdot \frac{1}{5} + \frac{1}{3} \cdot \frac{2}{9} + \frac{1}{3} \cdot \frac{1}{6}} = \frac{15}{53}$$

Note that the sum, as we should expect, is 1.

EXAMPLE 2.16 (due to Neyman) Consider the cross of two hybrids x_1 and x_2 which are heterozygous (Aa) and its progeny y_1 having the appearance of a dominant, that is, either doubly dominant (AA) or heterozygous (Aa). Suppose that y_1 is crossed with a recessive (aa), designated y_2, resulting in n offspring: $z_1, z_2, \ldots, z_n$. Suppose that not one of these offspring is a recessive, that is, either a dominant or a hybrid. It is proposed to find the probability that y_1 is (Aa).

Let E be the event that n offspring have the appearance of dominants; B_1 may be identified with the event that y_1 is (Aa) and B_2 with the event that y_1 is (AA). Then, $P(B_1) = P(y_1 = Aa) = 2/3$, and

$$P(E|B_1) = P(E|y_1 = Aa) = \frac{1}{2^n}$$

and $P(E|B_2) = P(E|y_1 = AA) = 1$. Substituting in Bayes' formula,

$$P(B_1|E) = P(y_1 = Aa|E) = \frac{(2/3)(1/2^n)}{(2/3)(1/2^n) + (1/3)1} = \frac{1}{1 + 2^{n-1}}$$

Giving n the values 1, 2, 3, 4, 5, we obtain Table 2.1.

Suppose, however, that we do not know anything about the origin of y_1. In that case the a priori probabilities $P(B_1) = P(y_1 = Aa)$ and $P(B_2) = P(y_1 = AA)$ would be unknown, and we would not have sufficient information to evaluate the right side of Bayes' formula. In

TABLE 2.1 A Posteriori Probabilities

n	$P(y_1 = Aa\|E)$
1	.500
2	.333
3	.200
4	.111
5	.059

the past it has been suggested that since $P(B_1)$ and $P(B_2)$ are unknown, and we have no reason to favor one more than another, we should assign 1/2 to each. Using an objective approach, modern statistics provides other ways of attacking such problems which seem more reasonable in the absence of a priori probabilities. These methods will be discussed later.

EXERCISES

2.1 Suppose airline A serves the cities in set A = {Boston, New York, Washington, D.C., Atlanta, Cleveland, Chicago, St. Louis, Minneapolis, Kansas City}; airline B serves the cities in set B = {Seattle, Portand, San Francisco, Los Angeles, Salt Lake City, Denver, Minneapolis, Kansas City, Chicago, St. Louis}. Find (a) $A \cup B$; (b) $A \cap B$; (c) $A \cap \bar{B}$; (d) $\bar{A} \cap B$. Interpret each of these sets in words and also express the set $\bar{A}$ in words.

2.2 Use Venn diagrams to show that

$$P(AB) \leq P(A) \leq P(A \cup B) \leq P(A) + P(B)$$

Under what condition does each equality hold?

2.3 Let A, B, C be three sets and $A = \{0 \leq x \leq 5\}$; $B = \{2 \leq x \leq 10\}$; $C = \{-1 \leq x \leq 3\}$. Find (a) $A \cap B \cap C$; (b) $A \cup B \cup C$; (c) $A \cap B \cap \bar{C}$; (d) $A \cap B \cup \bar{C}$.

2.4 An agronomist is designing an experiment involving the use of 4 varieties, 3 fertilizers, and 3 spacings. How many different treatment combinations, using one from each of the three kinds of treatments, does he have?

2.5 In how many different ways may a Jersey or a Holstein be drawn from a mixed herd of 5 Jerseys, 7 Holsteins, 10 Guernseys, and 6 Brahmans?

2.6 How many different ways may a horticulturist arrange 5 different potted plants along a line on a greenhouse bench?

2.7 How many different ways may a student select a major and a minor from 5 possible fields?

2.8 How many different arrangements can be made using the 10 letters from the word *statistics*?

2.9 A clerk in a bookstore has 3 identical statistics books, 2 identical mathematics books, 1 physics book, and 1 chemistry book. (a) How many different ways can the clerk arrange all the books on a shelf? (b) If the physics book and the chemistry book must be next to each other, how many different arrangements are possible?

2.10 How many signals can be made by hoisting 6 flags of different colors if there are 6 significant positions on the flagpole? Any number of the flags may be hoisted at a time.

2.11 An organism has the possibility of having 1, 2, 3, 4, or 5 out of a total of 15 characters. What are the total possible combinations?

2.12 An industrial engineer is designing an experiment arranged to measure sources of variation from 4 factors (runs, journeys, cylinders, and pots). If we let R, J, C, and P represent the respective factors, how many 2-factor interactions of the type RJ etc. are there? How many 3-factor interactions? How many 4-factor interactions?

2.13 Using the relationship

$$(1 + x)^n = 1 + \binom{n}{1}x + \binom{n}{2}x^2 + \cdots + \binom{n}{n-1}x^{n-1} + \binom{n}{n}x^n$$

show that

$$2^n - 1 = \binom{n}{1} + \binom{n}{2} + \cdots + \binom{n}{n-1} + \binom{n}{n}$$

How many ways can we make a selection of 5 breeds of chickens, taking some or all?

2.14 Show that $\binom{n}{r} = \binom{n}{n-r}$

2.15 If $\binom{n}{10} = \binom{n}{6}$, find $\binom{n}{3}$.

2.16 If $\binom{16}{r} = \binom{16}{r-2}$, find r.

2.17 If P(56, r + 6)/P(54, r + 3) = 30,800, find r.

2.18 A random sample of size n from a finite population of N sampling units is one in which every possible combination of size n has an equal chance of being chosen. How many different samples of size 10 may be drawn from a list of 100 names? Use Stirling's approximation to evaluate the factorials.

2.19 Suppose that in selecting the sample of size 10 in Exercise 2.18 we draw a number from 1 to 10 at random, say 6, and select every 10th name on our list thereafter, that is, 16, 26, etc. Is this method of selection equivalent to random sampling? Why?

2.20 From a pack of 52 cards, 2 are drawn at random; find the probability that one is a queen and the other a king.

2.21 There are three events, A, B, C, one of which must, and only one can, happen; the probability of A not happening is 8/11, and the probability of B not happening is 5/7. Find the probability of C happening.

2.22 A person has 6 keys of which only 1 fits the door. If the keys are tried successively without replacement, it may require 1, 2, 3, 4, 5, or 6 trials until the correct key is found. Show that each of these 6 outcomes has probability 1/6.

2.23 The probability of A solving a certain problem is 3/7, and the probability of B solving the same problem is 5/12. What is the probability that the problem will be solved if both try?

2.24 In a family with 6 children, what is the probability that (a) all children will be girls; (b) all children will be of the same sex; (c) the first 5 children will be boys and the sixth a girl; (d) 3 of the children will be boys? Assume the sex ratio is 1/2.

2.25 Show that under the conditions of Law 5 the probability that an event happens at least r times in n trials is

$$\binom{n}{r}p^r q^{n-r} + \binom{n}{r+1}p^{r+1}q^{n-r-1} + \cdots + \binom{n}{n-1}p^{n-1}q + p^n$$

or the sum of the last $n - r + 1$ terms of the expansion of $(q + p)^n$.

2.26 Show that for any positive integer n and positive integer $k \leq n$,

$$\binom{n}{k-1} + \binom{n}{k} = \binom{n+1}{k}$$

2.27 (Fisher's tea-tasting experiment) A lady declares that by tasting a cup of tea made with milk she can discriminate whether the milk or the tea infusion was first added to the cup. Eight cups of tea were mixed, 4 in one way and 4 in the other, and the lady was so informed. The cups were then presented, in random order, to her for judgment. She was asked to divide the 8 cups into two sets of 4, agreeing, if possible, with the treatments received. The lady selected 3 right and 1 wrong in each set of the same treatment. On the assumption that the lady cannot discriminate between the two methods, show that the probability of her doing as well or better by chance is 17/70.

2.28 An urn contains 6 balls which are known to be all red or 4 red and 2 black. A ball is drawn and found to be red. What is the probability that all the balls are red?

2.29 A male rat is either doubly dominant (AA) or heterozygous (Aa), and owing to Mendelian properties, the probabilities of either being true is 1/2. The male rat is bred to a doubly recessive (aa) female. If the male rat is doubly dominant, the offspring will exhibit the dominant characteristic; if heterozygous, the offspring will exhibit the dominant characteristic 1/2 of the time and the recessive characteristic 1/2 of the time. Supposing all of 3 offspring exhibit the dominant characteristic, what is the probability that the male is doubly dominant?

2.30 Chevalier de Méré's problem was concerned with a certain game of dice. Twenty-four throws of a pair of dice are to be allowed, and the player is permitted to bet even money either on the occurrence of at least one "double 6" in the course of the 24 throws or against it. Certain theoretical considerations let de Méré to believe that betting on the double 6 is advantageous. On the other hand, his empirical trials appeared to contradict this conclusion. Pascal's solution stated that if the dice are "fair," the probability of obtaining at least one double 6 in 24 throws is .491. Check Pascal's results.

REFERENCE

Kolmogorov, A. N. (1933). Grundbegriffe der Wahrscheinlichkeitsrechnung. Ergeb. Math. (Berlin) 3.

3

UNIVARIATE PARENT POPULATION DISTRIBUTIONS

3.1 SPECIFICATION: In Chapter 2, we were interested in obtaining the probability of the occurrence of a single chance event. In scientific research using statistical methodology, it is often required to obtain the probabilities of the occurrence of all possible values of a chance event. This usually happens when statistical assumptions have to be specified for a particular experiment or investigation. A table of the possible values which a chance event may assume with a corresponding probability for each value is called a *probability distribution* for the parent population. Table 3.1 gives the probability distribution of the sum of two unbiased dice. In this table, all possible values of the variable X are given; the variable X representing the chance event of the sum of two dice is called a *random variable* or *variate*. A random variable is usually denoted by the capital letters X, Y, Z, etc. and the value of the random variable is denoted by the lower case letters x, y, z, etc.

Ordinarily, in statistics, instead of a table of values such as Table 3.1, specification is accomplished by selecting a mathematical function, for example, the normal, binomial, or Poisson, on the basis

TABLE 3.1 Probability Distribution of Sum of Two Dice

x	2	3	4	5	6	7	8	9	10	11	12
f(x)	$\frac{1}{36}$	$\frac{2}{36}$	$\frac{3}{36}$	$\frac{4}{36}$	$\frac{5}{36}$	$\frac{6}{36}$	$\frac{5}{36}$	$\frac{4}{36}$	$\frac{3}{36}$	$\frac{2}{36}$	$\frac{1}{36}$

of theoretical or empirical evidence and stating that the observations form a sample of all possible values of the variate. Quoting from R. A. Fisher (1922): "We may know by experience what forms are likely to be suitable, and the adequacy of our choice may be tested *a posteriori*. We must confine ourselves to those forms which we know how to handle, or for which any tables which may be necessary have been constructed."

3.2 DISCRETE DISTRIBUTIONS: Functions like f(x) in Table 3.1 are called *discrete probability density functions*, to distinguish distributions of this type from *continuous probability density functions*, to be discussed later. The various values of f(x) may be thought of as giving the *relative* frequencies of occurrence corresponding to the particular values of X. Since some one of the 11 events must occur on any one trial, the sum of all the probabilities is 1 or symbolically

$$\sum_{x=2}^{12} f(x) = 1 \qquad (3.1)$$

The distinguishing characteristic of the discrete distribution is that the variate X can take only isolated values, that is, in Table 3.1 only the *whole* numbers 2 through 12. Since f(x) is the probability that X = x, f(x) must be nonnegative.

3.3 BINOMIAL DISTRIBUTION: The discrete distribution called the *binomial distribution* is the distribution of successes X in n repeated independent trials, in which the probability of success on any trial is a constant p. It has been named the binomial distribution because the successive probabilities are given by the respective terms of the expansion of the binomial $(q + p)^n$, where q = 1 - p, which is the probability of failure on any trial. One property of the binomial theorem is that the (x + 1)st term of the expansion is

$$f(x) = \binom{n}{x} p^x q^{n-x} \qquad x = 0, 1, \ldots, n \qquad (3.2)$$

which by the methods of Chapter 2 also gives the probability of exactly x successes in n independent trials. On the right side of (3.2), x is the value of the variate, and p and n are parameters, that is, for any particular member of this family of distributions special values of p and n must be specified. It should be noted that parameters are values which characterize the distribution of a random variable. Since

$$\sum_{x=0}^{n} f(x) = (q + p)^n = 1$$

this distribution fulfills the requirement that the sum of the probabilities is 1. When n = 1 there is only a single trial; this distribution is called the *Bernoulli distribution* and the random variable is called the *Bernoulli random variable*.

Individual terms and partial sums for various numerical values of p, n, and x for the binomial distribution are given in the reference by National Bureau of Standards (1950).

EXAMPLE 3.1 Given that the probability of drawing a tenant farm in a sample of farms is 1/3, if samples of 5 farms are drawn, then the respective probabilities of obtaining 0, 1, 2, 3, 4, 5 tenant farms in a single sample are $(2/3)^5$; $5(1/3)(2/3)^4$; $10(1/3)^2(2/3)^3$; $10(1/3)^3(2/3)^2$; $5(1/3)^4(2/3)$; $(1/3)^5$; or

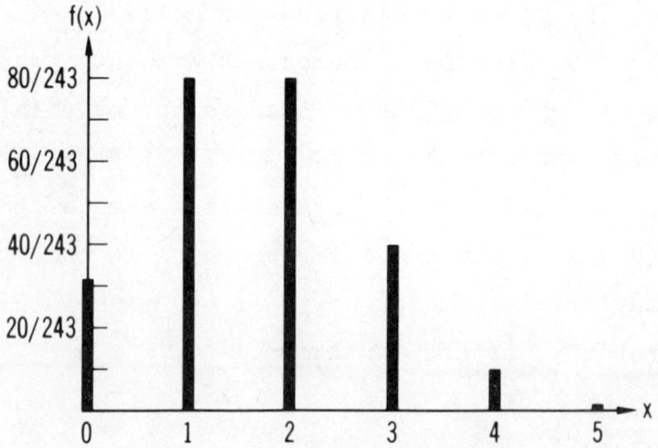

FIG. 3.1 Graph of probabilities of obtaining x tenant farms.

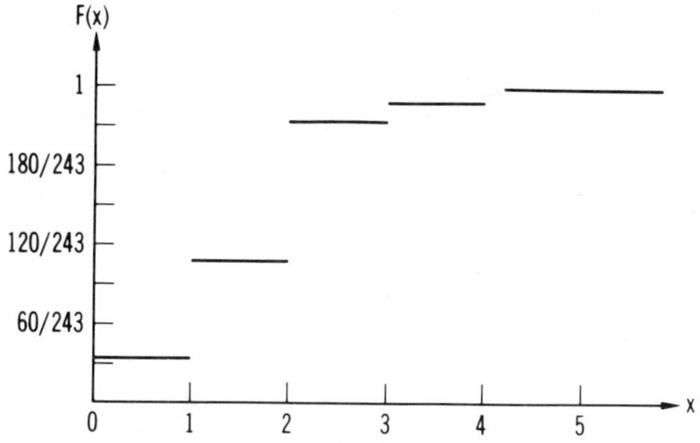

FIG. 3.2 Cumulative distribution of tenant farm probabilities.

$$\frac{1}{243}\ [32,\ 80,\ 80,\ 40,\ 10,\ 1]$$

The probabilities f(x) of obtaining x = 0, 1, 2, 3, 4, 5 tenant farms may be shown graphically as in Fig. 3.1.

If the probabilities are accumulated and graphed as in Fig. 3.2, then some F(a) value gives the probability of obtaining a value of X less than or equal to a, that is, $F(a) = P(X \leq a)$. This step function $F(x) = P(X \leq x)$ is called a *cumulative distribution function* or an ogive.

Note that points of discontinuity occur for each whole number on the x axis and that $F(5) = P(X \leq 5) = 1$.

3.4 POISSON DISTRIBUTION: Another discrete distribution of importance in applied statistics is the *Poisson distribution*. The Poisson distribution may be derived as a limiting form of the binomial distribution when p is very small but n is so large that np is a finite constant, equal to m, say. To see this, consider the binomial distribution

$$f(x) = \frac{n(n-1)(n-2)\cdots(n-x+1)}{x!}\ p^x q^{n-x}$$

Since p = m/n,

$$f(x) = \frac{n(n-1)(n-2)\cdots(n-x+1)}{x!}\left(\frac{m}{n}\right)^x\left(1-\frac{m}{n}\right)^{n-x}$$

$$= \frac{(1-1/n)(1-2/n)\cdots[1-(x-1)/n]m^x(1-m/n)^{n-x}}{x!}$$

Then

$$\lim_{n\to\infty} f(x) = \frac{m^x e^{-m}}{x!}$$

Obviously, this function is also a function of x; hence the Poisson distribution may be written as

$$f(x) = e^{-m}\frac{m^x}{x!} \qquad x = 0, 1, 2, \ldots \tag{3.3}$$

Since

$$\sum_{x=0}^{\infty} \frac{m^x}{x!} = e^m$$

then

$$\sum_{x=0}^{\infty} f(x) = 1$$

The distribution was named for Poisson, having been given first by him in 1837. Individual terms and partial sums for various numerical values of m and x for the Poisson distribution have been made available by Molina (1942) and Pearson and Hartley (1958).

It should be noted that the Poisson distribution is a one-parameter family, m being the parameter. Sometimes λ is used as the parameter instead of m.

The Poisson distribution usually arises in random phenomena that events occur at random in an interval of time or a given space. For example, the number of customers joining a waiting line per minute, the number of flaws per foot of wire, the number of telephone calls per hour in an office, etc. can be assumed to have the Poisson distribution. A comprehensive treatment of the Poisson distribution is given by Haight (1967).

EXAMPLE 3.2 A bag of clover seed is known to contain 1% weed seeds. A sample of 100 seeds is drawn. Since $m = np = 100(.01) = 1$ and $e^{-1} = .3679$, the probabilities of 0, 1, 2, 3, ... weed seeds being in the sample are

Number of weed seeds	0	1	2	3	4	5	6	7
Probability	.3679	.3679	.1839	.0613	.0153	.0031	.0005	.0001

3.5 HYPERGEOMETRIC DISTRIBUTION: In discussing the binomial distribution, it was stated that the n trials are independent. Consider now that the n trials are not independent and they are dependent in the following way. Suppose an urn contains N balls of which M are red and N - M are black. We draw n balls in succession without replacement, i.e., after the first ball is drawn, it is not returned to the urn before the second drawing; after the first 2 balls are drawn, they are not returned before the third drawing; etc. Let the random variable X be the number of red balls drawn; then the probability density function of X is

$$f(x) = \frac{\binom{M}{x}\binom{N-M}{n-x}}{\binom{N}{n}} \tag{3.4}$$

where x ranges from the larger of [0, n - (N - M)] and the smaller of [n, M]. This distribution is called the *hypergeometric distribution.* Since the balls are not returned at each drawing, the trials are dependent in that the probability of drawing a red ball in the next trial depends on whether the previous ball drawn was red or black. When the drawings are with replacement, i.e., balls are returned at each drawing, the trials are independent and X follows a binomial distribution. In sampling from finite populations without replacement, the investigator usually encounters the hypergeometric distribution. Sampling from finite populations will be discussed in Chapter 8.

We realize that when N and M are very large, the difference would be negligible whether or not the ball is returned at each drawing. Therefore, the hypergeometric distribution can be approximated by the binomial distribution with $p = M/N$ provided M and N are very large. This is shown as follows:

$$\frac{\binom{M}{x}\binom{N-M}{n-x}}{\binom{N}{n}} = \frac{n!}{x!(n-x)!}\,\frac{M\cdots(M-x+1)}{N\cdots(n-x+1)}$$

$$\times \frac{(N-M)(N-M-1)\cdots(N-M-n+x+1)}{(N-x)(N-x-1)\cdots(N-n+1)}$$

$$\doteq \binom{n}{x} p^x (1-p)^{n-x}$$

EXAMPLE 3.3 A committee of 5 people is to be selected from 4 females and 7 males. The probability that the committee has x female members is $\binom{4}{x}\binom{7}{5-x}\Big/\binom{11}{5}$, that is,

x	0	1	2	3	4
Probability	.0455	.3030	.4545	.1818	.0152

The hypergeometric distribution is tabulated by Lieberman and Owen (1961).

3.6 GEOMETRIC DISTRIBUTION: Suppose there is a sequence of independent Bernoulli trails in which the probability of success on any trial is p. Let X be the number of failures preceding the first success. Then

$$f(x) = p(1-p)^x \qquad x = 0, 1, 2, \ldots \tag{3.5}$$

This is called the *geometric distribution* because the probabilities form a geometric progression. The probability is

$$\sum_{x=0}^{\infty} f(x) = p[1 + (1-p) + (1-p)^2 + \cdots]$$

$$= \frac{p}{1-(1-p)} = 1$$

provided $p \neq 0$.

EXAMPLE 3.4 Suppose the proportion of retarded children in a community is .01. Children are sampled in sequence at random with replacement. The probability that 9 children are observed to be normal children before the first retarded child is observed is $(.01)(.99)^9$ or .0091.

3.7 CONTINUOUS DISTRIBUTIONS: If *measurements* instead of *counts* constitute the data under consideration, then the hypothetical parent population distribution is usually that of a *continuous* variate instead of a *discrete* variate. Snedecor and Cochran (1967) give the histogram in Fig. 3.3 for the gains in weight of 100 swine. Before powerful mathematical methods may be applied to derive a methodology providing techniques for statistical inferences, it is desirable to "idealize" the histogram into a curve which may be represented by a mathematical function. Such a process takes place in other branches of applied mathematics, for example, in surveying. Before the surveyor can be furnished with a powerful methodology for the solution of his practical problems in mensuration, it is necessary for the geometer to idealize the *physical points, lines,* and *planes*. A geometrical point is defined as having no dimensions but simply an indicator of position. Again, a geometrical line has no width, and a

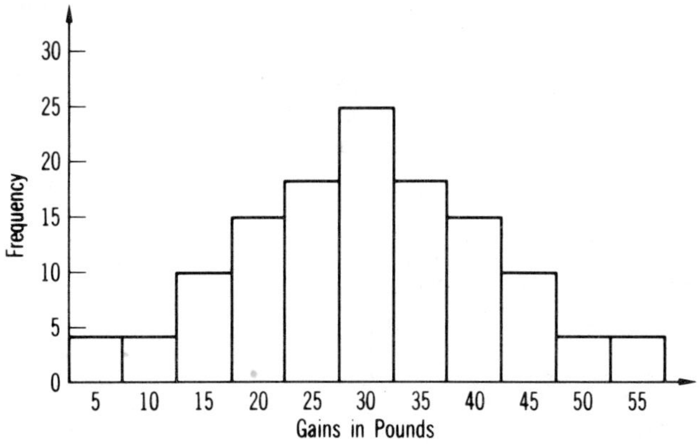

FIG. 3.3 Histogram showing frequency of gains in weight of 100 swine.

geometrical plane has no thickness. With these idealized definitions and certain assumptions called *axioms* the geometer is able to prove *theorems* concerning relationship and properties of geometrical configurations. These theorems in turn form the bases for a practical surveying methodology.

How shall we idealize a histogram of the type given in Fig. 3.3? First of all, instead of a finite population of possible values of the variate, we assume an infinite population of gains in weight. Next, instead of the class marks differing by 5 pounds, suppose that they are selected closer and closer together. It is not difficult to see that the histogram might reasonably be expected to approach some continuous smooth curve of the type shown in Fig. 3.4. The function representing the curve is denoted by f(x), which is called the *probability density function* of X, or simply the *density function* of X.

In Fig. 3.1 the probability of obtaining some particular x for the discrete distribution was represented by an ordinate. Here, x represents only whole numbers, but because of the limitations imposed by measuring devices, the best that can be said concerning the "true" value of some observed x value of a continuous variate X is that it lies in some interval (x, x + dx). If the area under the continuous curve be made equal to 1, corresponding to the similar requirement that the sum of the probabilities equal 1 for the discrete distributions, then the probability of X lying in the interval (x, x + dx)

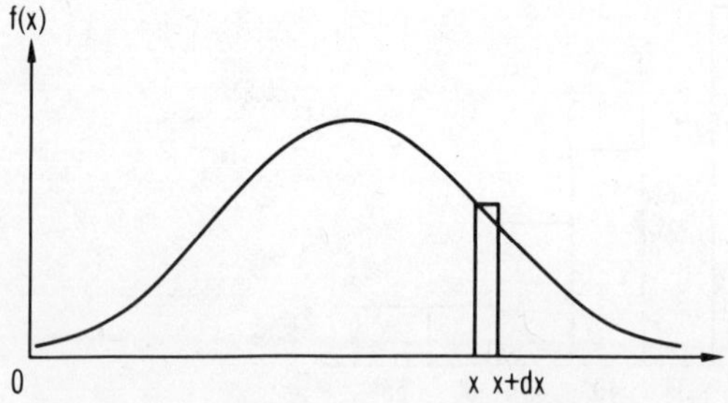

FIG. 3.4 "Idealized," or theoretical, probability density function.

will be f(x) dx. This would be the *theoretical* probability corresponding to the *empirical* probability, say, of a gain of weight lying between 37.5, and 42.5, that is, 13/100 = .13.

The range of x may be thought of as extending from $-\infty$ to $+\infty$, even through the curve may actually contact the x axis at some finite value, since the area under the curve in the contact interval would be zero. Hence, the density function f(x) is always nonnegative. As was pointed out, the total probability or area under the curve is 1; symbolically,

$$\int_{-\infty}^{\infty} f(x)\, dx = 1 \tag{3.6}$$

The probability of X being equal to or less than some constant a is expressed as

$$P(X \le a) = \int_{-\infty}^{a} f(x)\, dx \tag{3.7}$$

Again, the probability of X lying between a and b is given by

$$P(a \le X \le b) = \int_{a}^{b} f(x)\, dx \tag{3.8}$$

It is possible to omit the equal signs in the left sides of (3.7) and (3.8), since the probability of obtaining any *particular* value of X is equal to the area above the width of a geometrical line, which is zero.

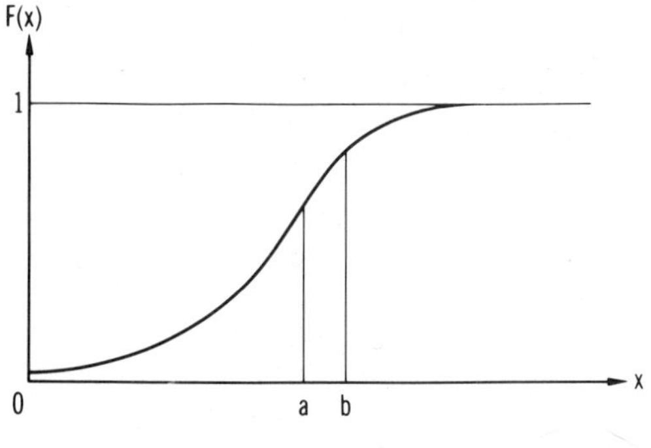

FIG. 3.5 Cumulative distribution function for a continuous variate.

The *cumulative distribution function* for the continuous variate corresponding to the probability density function of Fig. 3.4 would appear as in Fig. 3.5. For this curve, $F(a) = P(X \leq a)$, and hence $F(b) - F(a) = P(a < X \leq b)$. From (3.7) we have that the relationship between the probability density function and the cumulative distribution function is

$$f(x) = \frac{dF(x)}{dx}$$

3.8 UNIFORM DISTRIBUTION: The probability density function of the uniform distribution is a constant in the interval $a < x < b$, where a and b are real parameters. Hence

$$f(x) = \begin{cases} \dfrac{1}{b - a} & a < x < b \\ 0 & \text{otherwise} \end{cases} \tag{3.9}$$

The corresponding cumulative distribution function is

$$F(x) = \begin{cases} 0 & x \leq a \\ \dfrac{x - a}{b - a} & a < x < b \\ 1 & x \geq b \end{cases} \tag{3.10}$$

The uniform distribution is also known as the *rectangular distribution* since the density function together with the base forms a rectangle.

3.9 EXPONENTIAL DISTRIBUTION: The density function of the exponential distribution (or the negative exponential distribution) is

$$f(x) = \begin{cases} \lambda e^{-\lambda x} & x > 0 \\ 0 & \text{otherwise} \end{cases} \tag{3.11}$$

where $\lambda > 0$ is the parameter. This distribution is used extensively in reliability theory; for example, the waiting time to the failure of a component in a system may be assumed to have an exponential distribution. In queueing theory, the "input distribution" and the "service time distribution" may be of the form of the exponential distribution. In fact, if the number of customers joining a waiting line follows the Poisson distribution, then the time between arrivals follows the exponential distribution.

The cumulative distribution function is

$$F(x) = \begin{cases} 0 & x \leq 0 \\ 1 - e^{-\lambda x} & x > 0 \end{cases} \tag{3.12}$$

EXAMPLE 3.5 Customers are arriving at a service station at random. The waiting time between arrivals for the service station attendant is assumed to have an exponential distribution with λ = 1/2. Under this assumption, the probability that the service station attendant waits more than 1 minute between arrivals is

$$\int_1^{\infty} \frac{1}{2} e^{-x/2} \, dx = e^{-1/2} = .6065$$

3.10 NORMAL DISTRIBUTION: The most important continuous distribution in applied statistics is the *normal distribution*, also known as the *Gaussian distribution*. The histogram of Fig. 3.3 and the theoretical distribution of Fig. 3.4 are those for data specified as being normally distributed. Data arising from many different measurements taken on plants and animals are specified as following the normal distribution. There is empirical justification for this assumption. Similarly, distributions of certain data of the physical and social sciences are found to be satisfactorily represented by the normal distribution. It should not be assumed, however, that every continuous distribution representing actual data should be normal. For example, it is known that the distribution of sizes of cumulus clouds should be represented by a U-shaped curve.

The density function of the standard normal distribution is of the form

$$f(x) = \frac{1}{\sqrt{2\pi}} e^{-x^2/2} \qquad -\infty < x < \infty \tag{3.13}$$

This curve is bell-shaped and symmetrical about zero. To show that this function serves as a density function, let the total area under the curve be

$$A = \int_{-\infty}^{\infty} \frac{1}{\sqrt{2\pi}} e^{-x^2/2} dx$$

Then

$$A^2 = \int_{-\infty}^{\infty} \frac{1}{\sqrt{2\pi}} e^{-x^2/2} dx \int_{-\infty}^{\infty} \frac{1}{\sqrt{2\pi}} e^{-y^2/2} dy$$

$$= \int_{-\infty}^{\infty} \int_{-\infty}^{\infty} \frac{1}{\sqrt{2\pi}} e^{-(x^2+y^2)/2} dx\, dy$$

Let us transform to the polar coordinates, i.e.,

$$x = r \cos \theta$$

$$y = r \sin \theta$$

The Jacobian of the transformation is

$$J = \begin{vmatrix} \cos\theta & -r\sin\theta \\ \sin\theta & r\cos\theta \end{vmatrix} = r(\cos^2\theta + \sin^2\theta) = r$$

So

$$A^2 = \frac{1}{2\pi} \int_0^{2\pi} \int_0^{\infty} re^{-r^2/2} dr\, d\theta = \frac{2\pi}{2\pi} = 1$$

Hence the total area under the curve is unity.

Table C.1 (Appendix C) gives ordinates and areas for the standard normal distribution.

The density function of the general normal distribution has the form

$$f(x) = \frac{1}{\sigma\sqrt{2\pi}} e^{-(x-\mu)^2/2\sigma^2} \qquad -\infty < x < \infty \tag{3.14}$$

where $-\infty < \mu < \infty$ and $0 < \sigma < \infty$. This function may be obtained from the density function of the standard normal distribution. Let $Y = \sigma X + \mu$; then

$$F(y) = P(Y \le y) = P\left(X \le \frac{y - \mu}{\sigma}\right)$$

$$= \int_{-\infty}^{(y-\mu)/\sigma} \frac{1}{\sqrt{2\pi}} e^{-x^2/2}\, dx$$

Hence

$$f(y) = \frac{dF(y)}{dy} = \frac{1}{\sigma\sqrt{2\pi}} e^{-(y-\mu)^2/2\sigma^2}$$

which is of the same form as (3.14).

The curve for the normal density function is symmetrical about $x = \mu$ and bell-shaped as in Fig. 3.4. The inflection points are at $x = \mu \pm \sigma$, and the tails of the curve, although approaching the x axis quite rapidly, extend indefinitely far in both directions. The function (3.14) represents a two-parameter family, μ and σ, of continuous distributions, that is, as μ and σ vary in magnitude a family of distributions is generated. Usually the normal distribution is denoted by $N(\mu,\sigma^2)$, and the standard normal distribution is denoted by $N(0,1)$.

We have shown that (3.14) may be obtained from (3.13). Conversely, if X is distributed as $N(\mu,\sigma^2)$, then $(X - \mu)/\sigma$ is distributed as $N(0,1)$. So $P(X < x)$ can be obtained from the table of the standard normal distribution and no new table is needed.

EXAMPLE 3.6 Suppose the gain in weight, denoted by X, of a certain animal over a month after feeding a certain diet follows a normal distribution with mean 20 grams and standard deviation 4 grams. Then the probability that an animal will gain weight between 18 and 24 grams is

$$P(18 < X < 24) = P\left(\frac{18 - 20}{4} < \frac{X - 20}{4} < \frac{24 - 20}{4}\right) = P(-.05 < Z < 1)$$

$$= P(Z < 1) - P(Z < -.05)$$

where Z has the standard normal distribution. From Table C.1.b (Appendix C), the probability is 0.8413 - 0.3085 = 0.5328.

3.11 PROBABILITY DISTRIBUTIONS AS SPECIALIZED MATHEMATICAL FUNCTIONS: We have noticed that theoretical probability distributions are mathematical functions possessing certain requirements. In order to give a complete formal definition of the requirements necessary for a mathematical function to be a probability distribution of statistics, it is convenient and sufficient to consider the cumulative distribution function F(x). It is sufficient since, given the cumulative distribution, it is possible to find the density function by taking the differential, that is,

$$d[F(x)] = f(x)\ dx$$

A mathematical function F(x) may be used as a cumulative distribution function of a random variable provided that

1. $F(-\infty) = 0$, $F(+\infty) = 1$, and $0 \leq F(x) \leq 1$ for all x.

2. F(x) is a nondecreasing function, that is, if $x_1 > x_2$, $F(x_1) \geq F(x_2)$.

3. F(x) is defined at every point in a continuous range and is continuous, except possibly at a denumerable number of points. F(x) is always continuous from the right. I.e., $F(x) = F(x^+)$ at every point x.

The following notation should be kept clearly in mind:

f(x) is the density function.

F(x) is the cumulative distribution function

More specifically, one may denote the density function and the cumulative distribution function by $f_X(x)$ and $F_X(x)$, respectively, for the random variable X.

The general properties of the density function f(x) are

1. $f(x) \geq 0$ for all x.

2. $\sum_{\text{all } x} f(x) = 1$ or $\int_{-\infty}^{\infty} f(x)\ dx = 1$.

3. $P(a < X \leq b) = \sum_{a<x\leq b} f(x)$ or $\int_a^b f(x)\ dx$.

3.12 SOME MATHEMATICAL FUNCTIONS USEFUL AS PROBABILITY DISTRIBUTIONS: Karl Pearson has suggested the differential equation

$$\frac{dy}{dx} = \frac{(d - x)y}{a + bx + cx^2}$$

as a generator of possible parent population distributions useful in applied statistics. For example, if $d = \mu$, $b = c = 0$, and $a = \sigma^2$, then the differential equation becomes

$$\frac{dy}{y} = \frac{(\mu - x)\ dx}{\sigma^2}$$

Solving for y, we have

$$\log y = -\frac{(\mu - x)^2}{2\sigma^2} + \log k$$

In this book we use log for natural logarithm. Then

$$y = ke^{-(x-\mu)^2/2\sigma}$$

Upon setting the integral between the limits from $-\infty$ to ∞ equal to 1, we find

$$k = \frac{1}{\sigma\sqrt{2\pi}}$$

Hence

$$y = \frac{1}{\sigma\sqrt{2\pi}} e^{-(x-\mu)^2/2\sigma^2} \qquad -\infty < x < \infty$$

which is the normal density function. This is Type VII of the Pearson system of density functions.

Another method of obtaining a mathematical representation of a frequency function is furnished by the Gram-Charlier series. This latter method will not be discussed here, but the interested reader should consult Kendall and Stuart (1969), Chap. 6.

3.13 GAMMA AND BETA DISTRIBUTIONS: The gamma and beta functions are two useful functions in statistics of which extensive use will be made in subsequent chapters. The gamma function of the positive number n is defined by

$$\Gamma(n) = \int_0^\infty x^{n-1} e^{-x}\, dx \qquad n > 0 \tag{3.15}$$

The properties of the gamma function will be exhibited in Exercises 3.22 to 3.27. An important result is that $\Gamma(n) = (n - 1)!$ when n is a positive integer.

The incomplete gamma function, defined by

$$F(x) = I_x(n) = \frac{1}{\Gamma(n)} \int_0^x u^{n-1} e^{-u}\, du \qquad 0 < x < \infty, \qquad n > 0$$

$$= \frac{\Gamma_x(n)}{\Gamma(n)} \tag{3.16}$$

furnishes a useful cumulative distribution function. Such a distribution is called the *gamma distribution*. The density function of the gamma distribution is

$$f(x) = \begin{cases} \dfrac{1}{\Gamma(n)} x^{n-1} e^{-x} & 0 < x < \infty \quad n > 0 \\ 0 & \text{otherwise} \end{cases} \tag{3.17}$$

This is Type III of the Pearson system of density functions. Figure 3.6 depicts the density functions for several values of n.

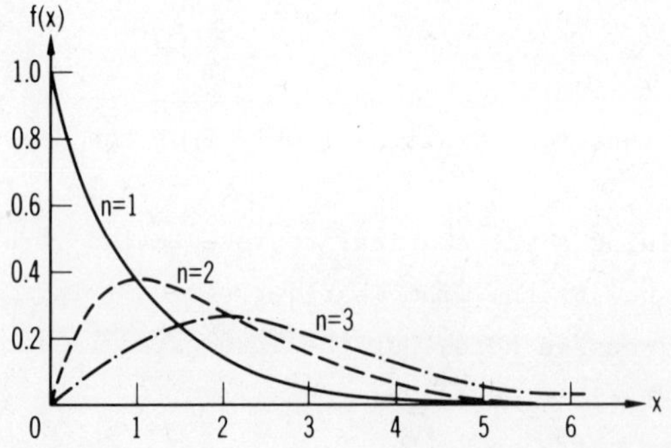

FIG. 3.6 Density function for the gamma distribution.

The density function of the general gamma distribution is

$$f(x) = \begin{cases} \frac{1}{\Gamma(n)} \theta^n x^{n-1} e^{-\theta x} & 0 < x < \infty \\ 0 & \text{otherwise} \end{cases} \tag{3.18}$$

where $n > 0$, $\theta > 0$; n is a shape parameter and θ a scale parameter. When $n = 1$ the gamma distribution reduces to the exponential distribution. So the gamma distribution may be viewed as an extension of the exponential distribution.

The beta function, defined by

$$B(m,n) = \int_0^1 x^{m-1}(1 - x)^{n-1}\, dx \qquad m,n > 0$$

is also of importance in theoretical and applied statistics. The properties of the beta function will be exhibited in Exercises 3.28 to 3.30. It is noted that the relationship between the gamma function and the beta function is

$$B(m,n) = \frac{\Gamma(m)\Gamma(n)}{\Gamma(m + n)}$$

The cumulative incomplete beta function is defined by

$$F(x) = I_x(m,n) = \frac{1}{B(m,n)} \int_0^x u^{m-1}(1 - u)^{n-1}\, du \qquad 0 < x < 1$$

$$= \frac{B_x(m,n)}{B(m,n)}$$

This is the cumulative distribution function of the *beta distribution*. The corresponding density function is

$$f(x) = \begin{cases} \frac{1}{B(m,n)} x^{m-1}(1 - x)^{n-1} & 0 < x < 1 \\ 0 & \text{otherwise} \end{cases}$$

This is Type I of the Pearson system of density functions. When $m = n = 1$, the beta distribution reduces to the uniform distribution. Figure 3.7 depicts the density functions for some selected values of m and n.

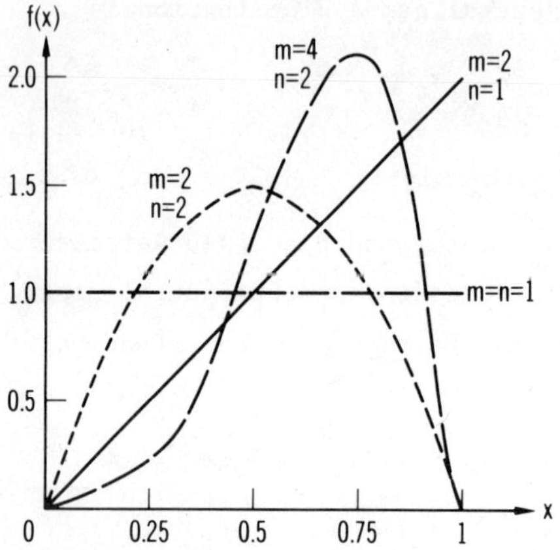

FIG. 3.7 Density function for the beta distribution.

Both $I_x(n)$ and $I_x(m,n)$ have been tabulated by Karl Pearson and his staff at the Biometric Laboratory, University College, London.

EXAMPLE 3.7 Suppose the life of a certain electronic machine, denoted by X, has a gamma distribution with n = 2 and $\theta = 1/100$. The probability that the machine will last more than 200 units of time is

$$P(X > 200) = \int_{200}^{\infty} \frac{x}{(100)^2} e^{-x/100}\, dx$$

$$= e^{-2} = 0.135$$

In this chapter we have studied some univariate, discrete, and continuous distributions. For other distributions, the reader may consult the books by Johnson and Kotz (1969, 1970).

EXERCISES

3.1 If you have a set of random numbers from 0 to 999, how would you set up a sampling scheme to select a random sample of 50 from 490 farms, the farms being numbered 0 to 489?

3.2 Construct a parent population probability distribution for the sum of numbers appearing when two dice are tossed. Give (a) a table of x and f(x); (b) a graph of (a); and (c) the cumulative distribution graph.

3.3 Follow the instructions in Exercise 3.2 for the binomial distribution with $p = 2/5$, $n = 5$.

3.4 Follow the instructions in Exercise 3.2 for the Poisson distribution with $m = 1.5$.

3.5 Given: the Poisson approximation to the binomial distribution with $n = 2048$ and $p = 1/1024$. Obtain the probabilities of there being 0, 1, 2, 3, ... times that 10 tails appear in 2048 tosses of 10 coins.

3.6 A lot consists of 100 machines of which 10 are defective. If 20 machines are selected without replacement for inspection, what is the probability of obtaining 1 defective machine? 2 defective machines? At least 3 defective machines?

3.7 An urn contains 6 white balls and 3 red balls. Suppose balls are drawn in succession with replacement and let X be the number of drawings before a white ball appears the first time. Find the distribution of X. This sampling scheme is called inverse sampling and the distribution of X is a geometric distribution.

3.8 Find the distribution of X in Exercise 3.7 if the balls are drawn without replacement.

3.9 Given the density function $f(x) = 2x$, $0 < x < 1$, $f(x) = 0$ for $x < 0$ or $x > 1$. Show that $\int_0^1 f(x)\,dx = 1$. What is the functional form of the cumulative distribution function?

3.10 Show that the area under the curve is 1 for the *triangular distribution* with density function

$$f(x) = \begin{cases} \dfrac{2(b - x)}{b^2} & 0 < x < b \\ 0 & \text{otherwise} \end{cases}$$

3.11 Repeat Exercise 3.10 for the distribution with

$$f(x) = \begin{cases} e^{-(x-\theta)} & \theta < x < \infty \\ 0 & \text{otherwise} \end{cases}$$

3.12 Repeat Exercise 3.10 for the *Cauchy distribution* with

$$f(x) = \frac{1}{\pi(1 + x^2)} \qquad -\infty < x < \infty$$

3.13 A random variable X, which lies between the limits 0 and 10, has the density function $f(x) = kx^3$. Determine the value of k so that the total probability is 1. What is the probability that X lies between 2 and 5? that X is less than 3?

3.14 A variate has the density function $f(x) = e^{-x}$, $0 < x < \infty$, and 0 otherwise. The probability is 1/2 that x will exceed what value?

3.15 If $P(X < x_1) = 1 - 1/(1 + x_1^2)$, X being a continuous variate with range $0 < x < \infty$, and 0 otherwise, find the density function f(x).

3.16 May $f(x) = -1/(x - 2)^2$, $0 < x < 4$, and 0 otherwise, serve as a density function? Why?

3.17 For the distribution with the density function $f(x) = 2x$, $0 < x < 1$, and 0 otherwise, find the number a such that the probability of $x > a$ is 3 times the probability of $x < a$.

3.18 If two values of x are drawn at random from the distribution with the density function $f(x) = e^{-x}$, $0 < x < \infty$, and 0 otherwise, what is the probability that both are greater than 1?

3.19 In Exercise 3.18 what is the probability that at least one value of x is greater than 1?

3.20 A random variable X follows the normal distribution with mean μ and variance σ^2. Determine the maximum of the density function and show that $\mu \pm \sigma$ are the inflection points.

3.21 Use the table of areas of the normal curve to determine, for the normal distribution given in Exercise 3.20, the probability of (a) $X > \mu$; (b) $\mu - \sigma < X < \mu + \sigma$; (c) $X > \mu + 2\sigma$.

3.22 Use integration by parts to show that $\Gamma(n + 1) = n\Gamma(n)$.

3.23 Show that $\Gamma(n + 1) = n(n - 1)\cdots(n - k)\Gamma(n - k)$, where k is a positive integer less than n.

3.24 If n is also a positive integer, show that $\Gamma(n + 1) = n!$

3.25 (a) Find $\Gamma(1)$, $\Gamma(2)$, $\Gamma(3)$, and $\Gamma(4)$. (b) Using $\Gamma(n + c) = (n + c - 1)\cdots(n + 1)n\Gamma(n)$, show that $\Gamma(n)$ becomes infinite when $n \to 0$.

3.26 Show that $\Gamma(n) = 2 \int_0^\infty y^{2n-1}e^{-y^2}\,dy$ by setting $x = y^2$ in the integral defining the gamma function.

3.27 Using the result of Exercise 3.26, show that:

(a) $\Gamma\left(\frac{1}{2}\right) = 2 \int_0^\infty e^{-u^2}\,du$

(b) $\left[\Gamma\left(\frac{1}{2}\right)\right]^2 = 4 \int_0^\infty \int_0^\infty e^{-(u^2+v^2)}\,du\,dv$

(c) $\left[\Gamma\left(\frac{1}{2}\right)\right]^2 = 4 \int_0^{\pi/2} \int_0^\infty e^{-r^2}\,r\,dr\,d\theta$ in polar coordinates

(d) $\Gamma\left(\frac{1}{2}\right) = \sqrt{\pi}$

3.28 By setting $x = \sin^2\theta$ in the integral defining $B(m,n)$, show that

$$B(m,n) = 2 \int_0^{\pi/2} \sin^{2m-1}\theta \cos^{2n-1}\theta\,d\theta$$

3.29 By setting $x = 1 - y$ in the integral defining $B(m,n)$, show that $B(m,n) = B(n,m)$.

3.30 Show that:

(a) $\Gamma(n)\Gamma(m) = 4 \int_0^\infty \int_0^\infty x^{2n-1}y^{2m-1}e^{-(x^2+y^2)}\,dx\,dy$

(b) $\Gamma(n)\Gamma(m) = 4 \int_0^{\pi/2} \sin^{2m-1}\theta \cos^{2n-1}\theta\,d\theta \int_0^\infty r^{2(m+n)-1}e^{-r^2}\,dr$

(c) $\Gamma(n)\Gamma(m) = B(m,n)\Gamma(m + n)$ or

$$B(m,n) = \frac{\Gamma(m)\Gamma(n)}{\Gamma(m + n)}$$

3.31 Let the density function of X be

$$f(x) = \begin{cases} kx^2 + 6x & 0 < x < 1 \\ 0 & \text{otherwise} \end{cases}$$

(a) Find the value of k. (b) Show that this is the density of a beta distribution and find the values of the parameters in this distribution.

3.32 Use integration by parts to show that the cumulative distribution function $I_x(n)$ for the gamma distribution can be written as

$$I_x(n) = 1 - \sum_{i=0}^{n-1} \frac{x^i}{i!} e^{-x}$$

where n is an integer. Hence the lower-tail probability for the gamma distribution may be expressed as the upper-tail probability of the Poisson distribution.

3.33 Use integration by parts to show that the cumulative distribution function $I_x(m,n)$ for the beta distribution can be written as

$$I_x(m,n) = 1 - \sum_{i=0}^{m-1} \binom{n + m - 1}{i} x^i(1 - x)^{n+m-1-i}$$

where m and n are integers. Hence the lower-tail probability of the beta distribution may be expressed as the upper-tail probability of the binomial distribution.

REFERENCES

Fisher, R. A. (1922). Mathematical foundations of theoretical statistics, Phil. Trans. Roy. Soc. A222:309-368.

Haight, F. A. (1967). *Handbook of the Poisson Distribution.* Wiley, New York.

Johnson, N. L., and Kotz, S. (1969). *Distribution in Statistics: Discrete Distributions.* Houghton Mifflin, Boston.

Johnson, N. L., and Iotz, S. (1970). *Distribution in Statistics: Continuous Univariate Distributions 1 and 2.* Houghton Mifflin, Boston.

Kendall, M. G., and Stuart, A. (1969). *The Advanced Theory of Statistics,* Vol. 1, 3rd ed. Charles Griffin, London.

Lieberman, G. J., and Owen, D. B. (1961). *Tables of the Hypergeometric Probability Distribution.* Stanford University Press, Stanford, Calif.

Molina, E. C. (1942). *Poisson's Exponential Binomial Limit.* Van Nostrand, New York.

National Bureau of Standards (1950). *Tables of the Binomial Probability Distribution.* U. S. Government Printing Office, Washington, D.C.

Pearson, E. S., and Hartley, H. O. (1958). *Biometrika Tables for Statisticians,* Vol. 1. Cambridge University Press, London.

Snedecor, G. W., and Cochran, W. G. (1967). *Statistical Methods,* 6th ed. Iowa State University Press, Ames.

4

PROPERTIES OF UNIVARIATE DISTRIBUTION FUNCTIONS

4.1 INTRODUCTION: In this chapter certain important properties of *parent population distributions* will be discussed. The discussion will also apply to similar properties of *derived sampling distributions*. These properties will be found useful in describing parent population distributions and derived sampling distributions.

4.2 MATHEMATICAL EXPECTATION: The mathematical expectation of any random variable X, which can assume values $x_1, x_2, \ldots, x_n$ with probabilities $p_1, p_2, \ldots, p_n$, respectively, where $\Sigma_1^n p_i = 1$, is defined to be

$$E(X) = \sum_{i=1}^{n} x_i p_i \tag{4.1}$$

For the discrete distribution this becomes

$$E(X) = \sum_{i=1}^{n} x_i f(x_i) \tag{4.2}$$

for distributions with $p_i = f(x_i)$. For the continuous distribution

$$E(X) = \int_{-\infty}^{\infty} xf(x)\, dx \tag{4.3}$$

where as noted before f(x) may vanish over part of the range $(-\infty, +\infty)$.

The term *mathematical expectation* may be shortened to *expected value* or *average value*.

The definition of the expected value of any random variable X may be generalized to include functions of X. The expected value of any function of X, say u(X), is

$$E(u(X)) = \Sigma u(x)f(x) \qquad \text{or} \qquad \int u(x)f(x)\,dx \tag{4.4}$$

over the range of x and depending upon whether X is a discrete or continuous variate.

It is possible, by introducing a generalized form of summation called a *Stieltje's integral* to replace the Σ or $\int$ by a single integral sign. Although the concept of the Stieltje's integral simplifies statements concerning probabilities and expected values, the mathematical concepts behind this refinement are beyond the scope of this text. A distinction in all statements will be made between the discrete and continuous distributions.

If the indicated integration, necessary to obtain the expected value, cannot be performed directly, various methods of numerical integration are available. The numerical integrations are usually evaluated on a computer.

EXAMPLE 4.1 The expected value of X, where X gives the various sums possible to be made by throwing two ordinary dice, may be found from the following table:

i	1	2	3	4	5	6	7	8	9	10	11
x_i	2	3	4	5	6	7	8	9	10	11	12
$36f(x_i)$	1	2	3	4	5	6	5	4	3	2	1

Then

$$E(X) = \sum_{i=1}^{11} x_i f(x_i) = \frac{2 + 6 + 12 + \cdots + 12}{36} = 7$$

EXAMPLE 4.2 If $f(x) = 3x^2$, $0 < x < 1$, then

$$E(X) = \int_0^1 x \cdot 3x^2\,dx = \frac{3}{4}$$

Also

$$E(X^2) = \int_0^1 x^2 \cdot 3x^2 \, dx = \frac{3}{5}$$

4.3 *OPERATIONS WITH EXPECTED VALUES*: The rules stated below will be found useful in operating with expected values:

1. The expected value of a constant is the constant itself: $E(c) = c$.

2. The expected value of a constant times a variate or a function of a variate is the constant times the expected value of the variate or the function of the variate: $E[cu(X)] = cE[u(X)]$.

3. The expected value of a sum (or difference) of two variates or functions is the sum (or difference) of the expected values of the separate parts: $E(u_1(X) \pm u_2(X)) = E(u_1(X)) \pm E(u_2(X))$.

The proof of these statements is left as an exercise for the student.

4.4 *MOMENTS:* The expected value of X^k is called the kth moment of X about the origin and is represented by the symbol μ'_k. Hence,

$$\mu'_k = E(X^k) = \Sigma x^k f(x) \quad \text{or} \quad \int x^k f(x) \, dx \tag{4.5}$$

over the range of x. The first moment of X about the origin is referred to as the *mean* of X, and is denoted by μ'_1. To simplify writing, let $\mu'_1 = \mu$.

The expected value of $(X - \mu)^k$ is called the kth moment of X about the mean and is designated by the symbol μ_k. Hence,

$$\mu_k = E(X - \mu)^k = \Sigma(x - \mu)^k f(x) \quad \text{or} \quad \int (x - \mu)^k f(x) \, dx \tag{4.6}$$

over the range of x. It is easily seen that

$$\mu_1 = E(X - \mu) = 0$$

The second moment about the mean, μ_2, is called the *variance* and is usually designated by the symbol σ^2. Hence,

$$\sigma^2 = \mu_2 = E(X - \mu)^2 = \Sigma(x - \mu)^2 f(x) \qquad \text{or} \qquad \int (x - \mu)^2 f(x)\, dx \tag{4.7}$$

over the range of x. The formula for σ^2 may be written as follows:

$$\sigma^2 = E(X - \mu)^2 = E(X^2) - 2\mu E(X) + \mu^2 = E(X^2) - [E(X)]^2 \tag{4.8}$$

The positive square root of σ^2, or σ, is referred to as the *standard deviation*.

The third moment about the mean, μ_3, furnishes a measure of *skewness,* or departure from symmetry about the mean of the distribution. One of the most generally accepted measures of skewness is the nondimensional quantity

$$\alpha_3 = \frac{\mu_3}{\sigma^3} \tag{4.9}$$

It is seen that α_3 will be zero for a symmetrical distribution.

A measure of the relative flatness or peakedness of the distribution, called the *kurtosis*, is given by the nondimensional quantity

$$\alpha_4 = \frac{\mu_4}{\sigma^4} \tag{4.10}$$

EXAMPLE 4.3 Using the table of values of Example 4.1, we see that $\mu = 7$. Also,

$$\mu_2' = \frac{4 + 18 + 48 + \cdots + 144}{36} = \frac{1974}{36}$$

Hence,

$$\sigma^2 = \frac{1974}{36} - 49 = \frac{35}{6}$$

EXAMPLE 4.4 Using the distribution function of Example 4.2, that is, $f(x) = 3x^2$, $0 < x < 1$,

$$\mu = \frac{3}{4}, \qquad \sigma^2 = \frac{3}{5} - \frac{9}{16} = \frac{3}{80}$$

and

$$\mu_3 = \int_0^1 \left(x - \frac{3}{4}\right)^3 \cdot 3x^2 \, dx = -\frac{1}{160}$$

EXAMPLE 4.5 In order that the function

$$f(x) = ke^{-c(x-m)^2} \qquad -\infty < x < \infty$$

represent a density function, it is necessary that

$$\int_{-\infty}^{\infty} f(x) \, dx = 1$$

Letting $z = \sqrt{2c}(x - m)$, we have

$$\int_{-\infty}^{\infty} \frac{k}{\sqrt{2c}} e^{-z^2/2} \, dz = \frac{k\sqrt{\pi}}{\sqrt{c}} \int_{-\infty}^{\infty} \frac{1}{\sqrt{2\pi}} e^{-z^2/2} \, dz = \frac{k\sqrt{\pi}}{\sqrt{c}} = 1$$

Hence

$$k = \sqrt{c/\pi}$$

Also,

$$\mu = \int_{-\infty}^{\infty} xf(x) \, dx = \int_{-\infty}^{\infty} \sqrt{\frac{c}{\pi}} \, xe^{-c(x-m)^2} \, dx$$

$$= \int_{-\infty}^{\infty} \frac{1}{\sqrt{2\pi}}(z - m)e^{-z^2/2} \, dz = m$$

and

$$\sigma^2 = \int_{-\infty}^{\infty} (x - \mu)^2 f(x) \, dx$$

$$= \int_{-\infty}^{\infty} \sqrt{\frac{c}{\pi}} (x - \mu)^2 e^{-c(x-\mu)^2} \, dx$$

$$= \frac{2}{c\sqrt{\pi}} \int_{-\infty}^{\infty} y^2 e^{-y^2} \, dy$$

$$= \frac{1}{c\sqrt{\pi}} \Gamma\left(\frac{3}{2}\right) = \frac{1}{2c}$$

by using Exercise 3.26 and $y = \sqrt{c}(x - \mu)$. Substituting these values in the original function, we find

$$f(x) = \frac{1}{\sigma\sqrt{2\pi}} e^{-(x-\mu)^2/2\sigma^2} \qquad -\infty < x < \infty$$

which is the normal distribution. Hence for the normal distribution $N(\mu,\sigma^2)$, the mean is μ and the variance is σ^2.

EXAMPLE 4.6 For the binomial distribution,

$$f(x) = \binom{n}{x} p^x q^{n-x}$$

$$\mu = \sum_{x=0}^{n} x \frac{n!}{(n-x)!x!} p^x q^{n-x}$$

$$= np \sum_{x=1}^{n} \frac{(n-1)!}{(n-x)!(x-1)!} p^{x-1} q^{n-x}$$

$$= np(q+p)^{n-1}$$

Therefore, $\mu = np$.

To obtain the variance, we first compute

$$E(X(X-1)) = \sum_{x=0}^{n} x(x-1)\frac{n!}{(n-x)!x!} p^x q^{n-x}$$

$$= n(n-1)p^2 \sum_{x=2}^{n} \frac{(n-2)!}{(n-x)!(x-2)!} p^{x-2} q^{n-x}$$

$$= n(n-1)p^2(p+q)^{n-2}$$

$$= n(n-1)p^2$$

Now

$$\sigma^2 = E(X(X-1)) + E(X) - [E(X)]^2$$

$$= n(n-1)p^2 + np - (np)^2 = npq$$

4.5 MOMENT-GENERATING FUNCTIONS: The expected value of e^{tX} often provides a convenient shortcut in evaluating the moments of X. Since

$$e^{tx} = 1 + \frac{tx}{1!} + \frac{(tx)^2}{2!} + \frac{(tx)^3}{3!} + \cdots = \sum_{i=0}^{\infty} \frac{t^i}{i!} x^i$$

then

$$E(e^{tX}) = \sum_{i=0}^{\infty} \mu_i' \frac{t^i}{i!}$$

where $\mu_0' = 1$. We set, say,

$$m(t) = E(e^{tX}) \tag{4.11}$$

and designate *m(t) the moment-generating function of X* or, more specifically, $m_X(t)$. If m(t) be differentiated k times with respect to t and then evaluated at t = 0, we note that

$$\left.\frac{\partial^k m(t)}{\partial t^k}\right|_{t=0} = \mu_k' \tag{4.12}$$

Also we have m(0) = 1.

To obtain the moment-generating function of X - μ, we consider

$$e^{t(x-\mu)} = 1 + \frac{t(x - \mu)}{1!} + \frac{t^2(x - \mu)^2}{2!} + \frac{t^3(x - \mu)^3}{3!} + \cdots$$

Then,

$$E\left(e^{t(X-\mu)}\right) = 1 + \sum_{i=1}^{\infty} \mu_i \frac{t^i}{i!} \tag{4.13}$$

If we set $M(t) = E\left(e^{t(X-\mu)}\right)$ [= $M_X(t)$], then

$$\mu_k = \left.\frac{\partial^k M(t)}{\partial t^k}\right|_{t=0} \tag{4.14}$$

In general, the moment-generating function of any function u(X) may be defined as

$$m_{u(X)}(t) = E\left(e^{tu(X)}\right) \tag{4.15}$$

In evaluating the variance of a discrete random variable, it is often simpler by first obtaining the factorial moment E(X(X - 1)) as seen in Example 4.6. Then the variance may be written as

$$\sigma^2 = E(X(X - 1)) + E(X) - [E(X)]^2 \tag{4.16}$$

In general the *kth factorial moment* is defined as

$$E(X(X - 1)(X - 2)\cdots(X - k + 1)) \tag{4.17}$$

The corresponding *factorial moment-generating function* is

$$E(t^X) = \Sigma t^x f(x) \quad \text{or} \quad \int t^x f(x)\, dx \tag{4.18}$$

over the range of x. It is easily seen that differentiating the factorial moment-generating function k times and evaluating the result at t = 1, we obtain Eq. (4.17).

EXAMPLE 4.7 For the distribution,

$$f(x) = e^{-x} \qquad 0 < x < \infty$$

$$m(t) = \int_0^\infty e^{-(1-t)x}\, dx = (1 - t)^{-1} = \sum_{i=0}^{\infty} t^i \qquad t < 1$$

Hence,

$$\mu_k' = k!$$

EXAMPLE 4.8 For the binomial distribution,

$$m(t) = \sum_{x=0}^{n} e^{tx}\binom{n}{x}p^x q^{n-x} = \sum_{x=0}^{n}\binom{n}{x}(pe^t)^x q^{n-x} = (q + pe^t)^n \tag{4.19}$$

Then,

$$\mu_1' = \mu = npe^t(q + pe^t)^{n-1}\Big|_{t=0} = np$$

$$\mu_2' = np[e^t(n - 1)(q + pe^t)^{n-2}pe^t + (q + pe^t)^{n-1}e^t]\Big|_{t=0}$$

$$= np[(n - 1)p + 1] \tag{4.20}$$

Hence,

$$\mu_2 = \sigma^2 = \mu_2' - (\mu)^2 = np(1 - p) = npq \tag{4.21}$$

EXAMPLE 4.9 Consider the distribution in Example 4.7. The moment-generating function of $u(X) = X/\lambda$, where λ is a positive constant, is

$$m_{X/\lambda}(t) = E(e^{X/\lambda}) = \int_0^\infty e^{-(1-t/\lambda)x}\,dx = \left(1 - \frac{t}{\lambda}\right)^{-1}$$

$$= \sum_{i=0}^{\infty} \left(\frac{t}{\lambda}\right)^i \qquad t < \lambda \tag{4.22}$$

This is the moment-generating function of the exponential distribution. The moments are

$$\mu_k' = \frac{k!}{\lambda^k} \qquad k = 1, 2, \ldots$$

Hence $\mu = 1/\lambda$ and $\sigma^2 = 2/\lambda^2 - (1/\lambda)^2 = 1/\lambda^2$.

The moment-generating function of $X - \mu$ is denoted by $M(t)$, which is defined as $M(t) = E[e^{t(X-\mu)}]$. So we may obtain the following relation between $M(t)$ and $m(t)$:

$$M(t) = E\left(e^{t(X-\mu)}\right) = e^{-t\mu}[E(e^{tX})]$$

Therefore,

$$M(t) = e^{-t\mu}m(t) \tag{4.23}$$

Besides furnishing a simple method of obtaining the moments in certain cases, the moment-generating function is of use in deriving distribution functions and in comparing distributions. These latter two uses follow from Theorems 4.1 and 4.2 of Section 4.7.

The expected value of e^{tX} may not exist for real values of t for many discrete and continuous distributions, for example, for the Cauchy distribution:

$$f(x) = \frac{1}{\pi(1 + x^2)} \qquad -\infty < x < \infty$$

A more general function, which can be proved always to exist, is the characteristic function defined as $E(e^{itX})$, where t is real. The characteristic function for the Cauchy distribution is $e^{-|t|}$.

However, the evaluation of the integral, necessary to obtain the characteristic function, makes use of more advanced mathematical methods than are assumed for this course. Our uses of Theorems 4.1 and 4.2 will be confined to the moment-generating function, which will be assumed to exist in such cases.

4.6 CUMULANTS: Suppose that we define

$$\log m(t) = K(t) = \kappa_1 t + \kappa_2 \frac{t^2}{2!} + \cdots + \kappa_r \frac{t^r}{r!} \tag{4.24}$$

But

$$\log m(t) = t\mu + \log M(t)$$

and

$$\log M(t) = \log\left(1 + \sum_{i=1}^{\infty} \mu_i \frac{t^i}{i!}\right)$$

Hence

$$\log m(t) = t\mu + \log\left[1 + \left(\mu_2 \frac{t^2}{2!} + \mu_3 \frac{t^3}{3!} + \cdots\right)\right]$$

Since

$$\log(1 + x) = x - \frac{1}{2}x^2 + \frac{1}{3}x^3 - \cdots$$

then

$$\begin{aligned}\log m(t) &= t\mu + \left(\mu_2 \frac{t^2}{2!} + \mu_3 \frac{t^3}{3!} + \mu_4 \frac{t^4}{4!} + \cdots\right) \\ &\qquad - \frac{1}{2}\left(\mu_2 \frac{t^2}{2!} + \mu_3 \frac{t^3}{3!} + \cdots\right)^2 + \cdots \\ &= \mu t + \mu_2 \frac{t^2}{2!} + \mu_3 \frac{t^3}{3!} + (\mu_4 - 3\mu_2^2) \frac{t^4}{4!} \cdots \end{aligned} \tag{4.25}$$

Hence, equating coefficients of like powers of t in (4.24) and (4.25) we find $\kappa_1 = \mu$, $\kappa_2 = \mu_2$, $\kappa_3 = \mu_3$, $\kappa_4 = \mu_4 - 3\mu_2^2$, etc. The function

K(t) or $K_X(t)$ is called the *cumulant generating function.* The rth *cumulant* is obtained by differentiating K(t) r times with respect to t and then evaluating at t = 0:

$$K_r = \left. \frac{\partial^r K(t)}{\partial t^r} \right|_{t=0}$$

EXAMPLE 4.10 For the normal distribution,

$$f(x) = \frac{1}{\sigma\sqrt{2\pi}} e^{-(x-\mu)^2/2\sigma^2} \qquad -\infty < x < \infty$$

$$m(t) = \frac{1}{\sigma\sqrt{2\pi}} \int_{-\infty}^{\infty} e^{tx-(x-\mu)^2/2\sigma^2} \, dx$$

Let x = μ + y; then

$$m(t) = \frac{e^{t\mu}}{\sigma\sqrt{2\pi}} \int_{-\infty}^{\infty} e^{ty-(y^2/2\sigma^2)} \, dx$$

$$= \frac{e^{t\mu}}{\sigma\sqrt{2\pi}} \int_{-\infty}^{\infty} e^{-(y-t\sigma^2)^2/2\sigma^2} e^{t^2\sigma^2/2} \, dy$$

Therefore,

$$m(t) = e^{t\mu+(t^2\sigma^2/2)} \tag{4.26}$$

Now in order to read off the moments, we need the expansion of m(t) in series, but this is not very simple. However, the cumulants may be found quite simply, since for this case

$$K(t) = \log m(t) = t\mu + \frac{t^2\sigma^2}{2} \tag{4.27}$$

Hence, by (4.24) for the normal distribution,

$$\kappa_1 = \mu \quad \kappa_2 = \sigma^2 \quad \kappa_i = 0 \qquad \text{for } i = 3, 4, \ldots$$

It is important to remember that all cumulants after and including κ_3 for the normal distribution are zero.

Hence, for the normal distribution, it is simpler to read the cumulants κ_i for the cumulant-generating function K(t). If the moments are desired, they may be obtained easily from the cumulants. Hence the use of either the moment-generating function or the cumulant-generating function depends on the form of the distribution function.

4.7 AN INVERSE PROBLEM: It was seen that if we are given the theoretical distribution, then we may obtain a set of moments (μ, μ_2', μ_3', ...). In applied work we may have a large sample of observations and wish to determine from the data some evidence regarding an appropriate theoretical function to represent the assumed parent population distribution.

From the table of values giving the empirical frequency distribution it would be possible to obtain sample moments for the large sample. If it is assumed that these *sample* moments are "good" estimates of the corresponding moments of some *theoretical* distribution, then we have the inverse problem of determining uniquely a theoretical distribution having given the moments. This problem is discussed in texts on advanced mathematical statistics and is beyond the scope of this course. The "Pearson system of curves" is an assumed set of continuous functions whose parameters may be expressed in terms of the moments. Hence, estimates of these parameters may be obtained from a large sample and some one of the theoretical curves "fitted" to the empirical frequency distribution. A test of "goodness of fit" may then be accomplished by the use of chi-square (see Chapter 11).

Closely related to the moment problem mentioned above is the inverse relation of the characteristic function or the moment-generating function to a possible corresponding distribution function. Two theorems from advanced theoretical statistics will be stated without proof and made use of in subsequent derivations.

THEOREM 4.1 A distribution function is uniquely determined by its characteristic function or, where it exists, the moment-generating function.

THEOREM 4.2 If a distribution function has a characteristic function (moment-generating function) which approaches the characteristic function (moment-generating function) of another distribution, the two distributions approach each other.

EXERCISES

4.1 Find the mathematical expectation for the following distribution:

y	10	20	30	40
p	.1	.3	.5	.1

4.2 Given the following density functions, find μ, σ^2 for each: (a) $f(x) = 10x^9$, $0 < x < 1$, and 0 otherwise; (b) $f(x) = x/50$, $0 < x < 10$, and 0 otherwise.

4.3 A random variable can assume only two values 1 and 2. Its mathematical expectation is 3/2. Find p_1 and p_2.

4.4 A random variable has the density function $f(x) = a + bx$, $0 < x < 1$, and 0 otherwise. The mathematical expectation is 1/2. Find the constants a and b.

4.5 Express μ_3 and μ_4 in terms of moments about zero.

4.6 Use the cumulants for the normal distribution to determine the first four moments about the mean.

4.7 Two other measures of skewness and kurtosis, or departure from normality, are $\gamma_1 = \kappa_3/(\kappa_2)^{3/2}$ and $\gamma_2 = \kappa_4/\kappa_2^2$. Show that $\gamma_1 = \alpha_3$ $\gamma_2 = \alpha_4 - 3$. Find γ_1 and γ_2 for the normal distribution.

4.8 Determine the first four cumulants for the binomial distribution. Verify that $\kappa_{r+1} = pq(d\kappa_r/dp)$, $r > 1$, for the cumulants obtained.

4.9 Show that $E(X - a)^2$ is a minimum if $a = E(X)$.

4.10 If X has the density function $f(x) = 1/2$ on the interval (0,2), find the moment-generating function of X, and determine the mean and variance of X.

4.11 An unbiased penny is tossed 64 times. Find (a) the expected number of heads; (b) the theoretical standard deviation.

4.12 A pair of dice is thrown 60 times. Find (a) the expected number of times that the sum 10 appears; (b) the expected value of the square of the standard deviation.

4.13 There are 6 urns containing, respectively, 1 white, 9 black; 2 white, 8 black; 3 white, 7 black; 4 white, 6 black; 5 white, 5 black; 4 white, 6 black balls. One ball is to be drawn from each urn. What is the expected number of white balls taken? *Hint:* Let X_i be a variate which assumes the values 1 or 0 according as to whether the trial results in success or failure. Then, $Y = X_1 + X_2 + \cdots + X_n$ is the number of successes in n trials. But $E(X_i) = p_i \cdot 1 + (1 - p_i) \cdot 0 = p_i$ for $i = 1, 2, \ldots, n$. Hence $E(Y) = p_1 + p_2 + \cdots + p_n$.

4.14 An urn contains a red balls and b black balls, and c balls are drawn simultaneously. What is the expected number of red balls drawn?

4.15 An urn contains r tickets numbered from 1 to r, and s tickets are drawn at a time. What is the expected sum of the numbers on the tickets drawn? *Hint:* Let X_i be the variable attached to the ith ticket which may assume any of the values 1, 2, ..., r. Then

$$E(X_i) = \frac{1}{r}(1 + 2 + \cdots + r)$$

Set $Y = X_1 + X_2 + \cdots + X_s$, and complete the solution by finding $E(Y)$.

4.16 Find the moment-generating function m(t) for the triangular distribution with density function

$$f(x) = \begin{cases} x & 0 < x \leq 1 \\ 2 - x & 1 < x < 2 \\ 0 & \text{otherwise} \end{cases}$$

4.17 Find the moment-generating function m(t) for the rectangular distribution $f(x) = 1/\theta$, $0 < x < \theta$, and 0 otherwise. Verify that when $\theta = 1$, the square of the moment-generating function of the rectangular distribution equals the moment-generating function of the triangular distribution in Exercise 4.16.

4.18 Find μ_2', μ_3', μ_4' for the binomial distribution using the formulas for the definitions of these moments.

$$x^2 = x(x - 1) + x$$

$$x^3 = x(x - 1)(x - 2) + 3x^2 - 2x$$

$$x^4 = x(x - 1)(x - 2)(x - 3) + 6x^3 - 11x^2 + 6x$$

4.19 Find α_3 and α_4 for the binomial distribution.

4.20 Find μ, σ^2, α_3, and α_4 for the following binomial distributions: (a) $n = 7$, $p = 1/6$; (b) $n = 18$, $p = 1/3$.

4.21 Show that the moment-generating function for the Poisson distribution is $m(t) = e^{m(e^t-1)}$.

4.22 Show that $\kappa_i = m$ ($i = 1, 2, 3, \ldots$) for the Poisson distribution.

4.23 Find μ, μ_2', μ_3', μ_4' for the Poisson distribution using (a) the formulas for the definitions of the moments; (b) the moment-generating function.

4.24 Find μ_2, α_3, and α_4 for the Poisson distribution.

4.25 Prove the general formula connecting the moments about μ with the moments about the origin:

$$\mu_k = \mu_k' - k\mu\mu_{k-1}' + \frac{k(k-1)}{2!}\mu^2\mu_{k-2}' - \cdots$$

Use the formula to obtain

$$\mu_1 = 0$$
$$\mu_2 = \mu_2' - (\mu)^2$$
$$\mu_3 = \mu_3' - 3\mu\mu_2' + 2(\mu)^3$$
$$\mu_4 = \mu_4' - 4\mu\mu_3' + 6(\mu)^2\mu_2' - 3(\mu)^4$$

4.26 Let the random variable X have the density function

$$f(x) = \begin{cases} 1/2 & -1 < x \leq 0 \\ x & 0 < x \leq 1 \\ 0 & \text{otherwise} \end{cases}$$

Find μ_1', μ_2', μ_3', and μ_4'.

4.27 Let the random variable X have the density function

$$f(x) = \begin{cases} \dfrac{x}{\theta^2} e^{-x/\theta} & x > 0, \quad \theta > 0 \\ 0 & \text{otherwise} \end{cases}$$

(a) Find the moment-generating function of X. (b) Find the cumulant-generating function and hence determine the cumulants of X.

5

BIVARIATE AND MULTIVARIATE DISTRIBUTIONS AND THEIR PROPERTIES

5.1 INTRODUCTION: In the previous chapters single-variate, or univariate, distributions and their properties have been discussed. It is proposed now to extend the discussion to cases of two or more variates, that is, *bivariate and multivariate distributions*. The discussion will apply alike to bivariate and multivariate *parent population* distributions and bivariate and multivariate *derived sampling* distributions. The later distributions will be discussed in Chapter 6.

5.2 DISCRETE BIVARIATE DISTRIBUTIONS: Suppose that for every value of a given variate X we also know the values which a second variate Y can take. Then it will be possible to construct a joint probability distribution, from which can be obtained the probability that any combination of X and Y will occur in random draws. The *bivariate density function* will be represented symbolically by f(x,y). The *joint cumulative distribution function* is defined as $F(x,y) = P(X \leq x, Y \leq y)$. The conditions required for a mathematical function F(x,y) to be used as a joint cumulative distribution function are analogous to those in Section 3.11 for the univariate case. These conditions imply the following conditions for f(x,y):

1. f(x,y) is nonnegative over the (x,y) plane.
2. $\Sigma_W f(x,y) = 1$, where W is the entire (x,y) region.
3. $\Sigma_w f(x,y)$ can be computed for any subregion w of W.

The probability that X and Y lie, at the same time, in ranges $a \leq x \leq b$, $c \leq y \leq d$ is given by

$$P(w) = \Sigma_w f(x,y) \tag{5.1}$$

where w is the subregion defined by $\{a \leq x \leq b,\ c \leq y \leq d\}$. By the definition a cumulative distribution function for the bivariate case is given in a manner similar to that for P(w) in (5.1) above. If F(b,d) represents the probability that $X \leq b$, $Y \leq d$, then

$$F(b,d) = \Sigma_w f(x,y) \tag{5.2}$$

where w is the subregion $\{x \leq b,\ y \leq d\}$. If the subregion w is allowed to assume all possible values, then X and Y will assume all possible pairs of values and (5.2) in three dimensions becomes analogous to the two-dimensional step function of Section 3.3 and may be written as

$$F(a,b) = \Sigma_w f(x,y) \tag{5.3}$$

over the range of values of the subregion $w = \{x \leq a,\ y \leq b\}$.

EXAMPLE 5.1 Consider the two-dice problem. Let x represent the number of spots showing on die 1 and y the number on die 2 in any one toss of the two dice. The joint probability distribution is given by Table 5.1 (p = 1/36).

TABLE 5.1 Joint Probability Distribution in Two-dice Problem

y \ x	1	2	3	4	5	6	Total
1	p	p	.	.	.	p	6p
2	p	p	.	.	.	p	6p
3	.	.	.	.	.	.	.
4	.	.	.	.	.	.	.
5	.	.	.	.	.	.	.
6	p	p	.	.	.	p	6p
Total	6p	6p	.	.	.	6p	36p

The probability density function is $f(x,y) = 1/36$ for any pair of values (x,y) for $x = 1, 2, \ldots, 6$ and $y = 1, 2, \ldots, 6$. Note that $\Sigma f(x,y) = 1$ over the entire ranges of both variates. Also we can compute the probability of any subregion. For example,

$$P(1 < X < 3,\ 2 \le Y \le 4) = \frac{9}{36} = \frac{1}{4}$$

and

$$F(3,2) = P(X \le 3,\ Y \le 2) = \frac{6}{36} = \frac{1}{6}$$

EXAMPLE 5.2 In a social study for a community, let X denote the random variable for sex and Y the level of education. We shall let $X = 0$ for male, $X = 1$ for female; $Y = 1$ if a person's education is less than high school graduate, $Y = 2$ for high school graduate but not college graduate, and $Y = 3$ for college graduate. The probability distribution in this study is given in Table 5.2.

The first entry $f(0,1) = .1$ is the proportion of males with education less than high school graduate. Other entries can be similarly interpreted. The total probability is

$$\sum_{x=0}^{1} \sum_{y=1}^{3} f(x,y) = 1$$

Also the probability of a person (male or female) whose education is less than college graduate is

$$F(1,2) = P(X \le 1,\ Y \le 2) = .7$$

TABLE 5.2

x \ y	1	2	3	Total
0	.1	.3	.2	.6
1	.1	.2	.1	.4
Total	.2	.5	.3	1

Suppose we extend the region w in (5.3). Add all of the probabilities for a given value of X, say x_1; then the range of values of the subregion w is simply the linear range of Y, and hence we may write

$$\Sigma_y f(x_1,y) = g(x_1)$$

which is the probability that $X = x_1$. If this is thought of as being done for all values of x, then we may write the probability density function of X as

$$g(x) = \Sigma_y f(x,y) \qquad (5.4)$$

which is called the *marginal distribution of X*. Similarly the *marginal distribution of Y* is

$$h(y) = \Sigma_x f(x,y) \qquad (5.5)$$

EXAMPLE 5.3 Consider the two-dice problem in Example 5.1. As can be seen from the border totals in Table 5.1, the distributions (5.4) and (5.5) may be exhibited as in Table 5.3.

Finally, the marginal distribution and the joint distribution may be used to define the *conditional distribution*, corresponding to the *conditional probability* discussed in Section 2.6. Using the notation $f(y|x)$ to mean the probability of Y, given X = x, we know that

$$f(y|x) = \frac{f(x,y)}{g(x)} \qquad g(x) \neq 0 \qquad (5.6)$$

since

$$f(x,y) = g(x)f(y|x)$$

TABLE 5.3 Marginal Distributions of X and Y in Two-dice Problem

x = y	1	2	3	4	5	6
g(x) = h(y)	$\frac{1}{6}$	$\frac{1}{6}$	$\frac{1}{6}$	$\frac{1}{6}$	$\frac{1}{6}$	$\frac{1}{6}$

by Law 3 of Section 2.6. The distribution (5.6) is called the *conditional distribution* of Y given X = x. Similarly, the *conditional distribution* of X, given Y = y, is

$$f(x|y) = \frac{f(x,y)}{h(y)} \qquad h(y) \neq 0 \tag{5.7}$$

Now, if $f(y|x)$ does not depend on X, then Y and X are said to be independent variates, since

$$f(x,y) = g(x) \cdot h(y) \tag{5.8}$$

The mean value for the conditional distribution of Y given X = x is called the *conditional expectation* or the *conditional mean* of Y given X = x, which is denoted by $\mu_{Y|x}$. For the discrete case,

$$\mu_{Y|x} = \Sigma y f(y|x) \tag{5.9}$$

where the summation is taken over all possible values of y in the conditional distribution.

In general, the conditional expectation of a function u(Y) is

$$E(u(Y)|X = x) = \Sigma u(y) f(y|x)$$

When $u(Y) = (Y - \mu_{Y|x})^2$, we obtain the conditional variance, which is denoted by $\sigma^2_{Y|x}$:

$$\sigma^2_{Y|x} = \Sigma(y - \mu_{Y|x})^2 f(y|x)$$

$$= \Sigma y^2 f(y|x) - \mu^2_{Y|x} \tag{5.11}$$

Similarly, the conditional mean and variance of X given Y = y are denoted by $\mu_{X|y}$ and $\sigma^2_{X|y}$, respectively. The conditional expectation of h(X) given Y = y is $\Sigma h(x) f(x|y)$.

EXAMPLE 5.4 Again consider the two-dice problem. For any $x = x_1$, $f(y|x_1) = 1/6$; hence by (5.8),

$$f(x,y) = \frac{1}{6} \cdot \frac{1}{6} = \frac{1}{36}$$

for x = 1, 2, ..., 6 and y = 1, 2, ..., 6. So X and Y are independently distributed.

EXAMPLE 5.5 For the distribution given in Table 5.2, the marginal distribution of x is

x	0	1
g(x)	.6	.4

The marginal distribution of Y is

y	1	2	3
h(y)	.2	.5	.3

The conditional distribution of Y (education level) given X = 0 (male) is

y	1	2	3
$f(y \mid x = 0)$	$\frac{1}{6}$	$\frac{1}{2}$	$\frac{1}{3}$

Since $f(y|x) \neq h(y)$, X and Y are dependent random variables. The conditional expectation of Y given X = 0 is $\mu_{Y|x} = 1 \times 1/6 + 2 \times 1/2 + 3 \times 1/3 = 13/6$. The conditional variance is computed as follows:

$$\sigma^2_{Y|x} = \Sigma y^2 f(y|x) - \mu^2_{Y|x}$$

$$= 1 \times \frac{1}{6} + 4 \times \frac{1}{2} + 9 \times \frac{1}{3} - \left(\frac{13}{6}\right)^2 = \frac{17}{36}$$

EXAMPLE 5.6 *The trinomial distribution.* Consider that a taxi driver approaches an intersection where he must make either a left turn, or make a right turn, or drive straight ahead. Let p_1 be the probability that he makes a left turn, p_2 the probability of making a right turn, and p_3 the probability of driving straight ahead. Since he must select one of the three choices (left turn, right turn, straight ahead), the three probabilities must add up to unity, i.e., $\Sigma_{i=1}^{3} p_i = 1$. Supposing that the taxi driver will pass this intersection n times from the same direction and the n passes are mutually independent, the probability that among the n trials he will make x left turns, y right turns, and drive straight ahead (n - x - y) times is

$$f(x,y) = \frac{n!}{x!y!(n-x-y)!} p_1^x p_2^y (1 - p_1 - p_2)^{n-x-y} \quad (5.12)$$

where x = 0, 1, ..., n, y = 0, 1, ..., n, and $0 \leq x + y \leq n$. This distribution is called the *trinomial distribution* and is an extension of the binomial distribution. Even though the trinomial distribution describes an experiment with three outcomes, there are only two random variables. If we let z = n - x - y, because of the linear relationship x + y + z = n, the value of z is completely determined by x and y.

The marginal distribution of X is obtained by summing over y from 0 to n - x. Therefore,

$$\begin{aligned} g(x) &= \sum_{y=0}^{n-x} f(x,y) \\ &= \frac{n!}{x!(n-x)!} p_1^x \left\{ \sum_{y=0}^{n-x} \frac{(n-x)!}{y!(n-x-y)!} p_2^y [(1-p_1) - p_2]^{n-x-y} \right\} \\ &= \frac{n!}{x!(n-x)!} p_1^x (1-p_1)^{n-x} \end{aligned}$$

for x = 0, 1, ..., n. This shows that the marginal distribution of X is a binomial distribution. Similarly, the marginal distribution of Y is a binomial distribution:

$$h(y) = \frac{n!}{y!(n-y)!} p_2^y (1-p_2)^{n-y} \qquad y = 0, 1, \ldots, n$$

The conditional distribution of X is

$$\begin{aligned} f(x|y) &= \frac{f(x,y)}{h(y)} \\ &= \frac{(n-y)!}{x!(n-y-x)!} \left(\frac{p_1}{1-p_2}\right)^x \left(1 - \frac{p_1}{1-p_2}\right)^{n-y-x} \end{aligned}$$

for x = 0, 1, ..., n - y. Therefore, the conditional distribution of X is also a binomial distribution. The probability of the occurrence of X in the conditional distribution has been "revised" to $p_1/(1 - p_2)$ and the total number of trials is reduced to n - y In terms of the taxi driver example, given the driver made y right turns, he has two choices, namely, left turn or straight ahead, among the remaining n - y trials. The probability that he will make a left turn for the remaining n - y trials is $p_1/(1 - p_2)$. The divisor $(1 - p_2)$ is to make the total probability for the two outcomes be unity as a binomial distribution requires.

Since the conditional distribution of X is not the same as the marginal distribution of X for all Y, the variates X and Y are dependent.

5.3 CONTINUOUS BIVARIATE DISTRIBUTIONS: In the case of two continuous variates X and Y, the probability that X is in the interval (x, x + dx) and Y in the interval (y, y + dy) is

$$f(x,y)\, dx\, dy \tag{5.13}$$

The conditions for the bivariate density function f(x,y), similar to those for the discrete distributions, are

1. f(x,y) is nonnegative.
2. $\int_{-\infty}^{\infty} \int_{-\infty}^{\infty} f(x,y)\, dx\, dy = 1.$ (5.14)
3. The probability that (X,Y) will fall in some subregion w of the (x,y) plane W is given by

$$P(w) = \int_W \int f(x,y)\, dx\, dy \tag{5.15}$$

The cumulative distribution function is given by

$$F(x,y) = \int_{-\infty}^{x} \int_{-\infty}^{y} f(u,v)\, dv\, du \tag{5.16}$$

The marginal distribution of X is given by

$$g(x) = \int_{-\infty}^{\infty} f(x,y)\, dy$$

and similarly for h(y).

Finally, the conditional probability that Y lies in the interval (y, y + dy), given that X is the interval (x, x + dx), is

$$f(y|x)\ dy = \frac{f(x,y)\ dy\ dx}{g(x)\ dx} = \frac{f(x,y)\ dy}{g(x)}$$

so

$$f(y|x) = \frac{f(x,y)}{g(x)} \tag{5.17}$$

Again, if $f(y|x)$ is independent of x, that is, if the right side of (5.17) does not contain x after algebraic simplification, then X and Y are said to be independent variates and

$$f(x,y) = g(x)\cdot h(y) \tag{5.18}$$

The conditional expectation of Y given X = x is given by

$$\mu_{Y|x} = \int_{-\infty}^{\infty} yf(y|x)\ dy \tag{5.19}$$

In general, the conditional expectation of a function u(Y) is

$$E[u(Y)|X = x] = \int_{-\infty}^{\infty} u(y)f(y|x)\ dy \tag{5.20}$$

EXAMPLE 5.7 Given $f(x,y) = e^{-x-y}(x > 0,\ y > 0)$:
(a) $F(x,y) = \int_0^x \int_0^y e^{-u-v}\ dv\ du$. For x = 1, y = 1, $F(1,1) = P(X \le 1,\ Y \le 1) = (1 - e^{-1})^2 = \left(\frac{e - 1}{e}\right)^2 = .3996$. (b) $g(x) = \int_0^{\infty} e^{-x-y}\ dy = e^{-x}$. (c) $f(y|x) = \frac{e^{-x-y}}{e^{-x}} = e^{-y}$ (independent of x).

EXAMPLE 5.8 The bivariate normal density function is

$$f(x,y) = \frac{1}{2\pi\sigma_X\sigma_Y\sqrt{1 - \rho^2}} \exp\left\{-\frac{1}{2(1 - \rho^2)}\left[\frac{(x - \mu_X)^2}{\sigma_X^2} + \frac{(y - \mu_Y)^2}{\sigma_Y^2} - \frac{2\rho(x - \mu_X)(y - \mu_Y)}{\sigma_X\sigma_Y}\right]\right\} \tag{5.21}$$

where $-\infty < x < \infty$, $-\infty < y < \infty$.

(a) $$g(x) = \int_{-\infty}^{\infty} f(x,y)\, dy$$

$$= \int_{-\infty}^{\infty} \frac{1}{\sigma_X\sqrt{2\pi}} \exp\left[-\frac{(x - \mu_X)^2}{2\sigma_X^2}\right]$$

$$\times \frac{1}{\sigma_y\sqrt{2\pi(1 - \rho^2)}} \exp\left\{-\frac{1}{2(1 - \rho^2)}\left[\frac{y - \mu_Y}{\sigma_Y}\right.\right.$$

$$\left.\left. - \frac{\rho(x - \mu_X)}{\sigma_X}\right]^2\right\} dy$$

$$= \frac{1}{\sigma_X\sqrt{2\pi}} \exp\left[-\frac{(x - \mu_X)^2}{2\sigma_X^2}\right]$$

(b) $$f(y|x) = \frac{f(x,y)}{g(x)} = \frac{1}{\sigma_Y\sqrt{2\pi(1 - \rho^2)}} \exp\left\{-\frac{1}{2(1 - \rho^2)}\left[\frac{y - \mu_Y}{\sigma_Y}\right.\right.$$

$$\left.\left. - \frac{\rho(x - \mu_X)}{\sigma_X}\right]^2\right\}$$

If $\rho = 0$,

$$f(y|x) = \frac{1}{\sigma_Y\sqrt{2\pi}} \exp\left[-\frac{(y - \mu_Y)^2}{2\sigma_Y^2}\right] = h(y)$$

hence *in this case*, X and Y are independent. Hence ρ may be used as a measure of relationship between X and Y. It is, in fact, the population *correlation coefficient* between X and Y.

5.4 DISTRIBUTIONS OF FUNCTIONS OF DISCRETE VARIATES: In order to obtain certain properties of bivariate and multivariate discrete or continuous distributions, such as expected values, moments, and moment-generating functions, it is sometimes necessary to make transformations of the variates so that the summations or integrals may be evaluated. The discrete case will be considered first.

The distribution of a function of X, say $Z = z(X)$, given the distribution of X, is simple if there is a one-to-one correspondence between x and z, that is, if for every value of x there is only one value of z, and vice versa. In this case, the same probabilities hold for Z as for X. For example, consider a single die with $f(x) = 1/6$, for $x = 1, 2, \ldots, 6$. Suppose $z = x^2$. In general, there is not a one-to-one correspondence between x and z because $x = \pm\sqrt{z}$, resulting in two values of x for each z. However, in our die problem, x must be positive; hence $x = +\sqrt{z}$ only. Therefore, $f(z) = 1/6$, for $z = 1, 4, \ldots, 36$.

On the other hand, suppose $z = (x - 1)(x - 2)$, or $x = (3 \pm \sqrt{1 + 4z})/2$. Then, even for x always positive, there is not a one-to-one correspondence between x and z, since, say for $z = 0$, $x = 1$ or 2. Hence, in this case

$$f(z) = \begin{cases} \frac{1}{3} & \text{for } z = 0\text{, for two integral values} \\ & \text{of } x \text{ in the range } x = 1, 2, \ldots, 6 \\ \frac{1}{6} & \text{for } z = 2, 6, 12, 20\text{, for one integral value} \\ & \text{of } x \text{ in the range } x = 1, 2, \ldots, 6 \\ 0 & \text{elsewhere, no integral value} \\ & \text{of } x \text{ in the range } x = 1, 2, \ldots, 6 \end{cases}$$

Again, if $z = (x - 1)(x - 2)\ldots(x - 6)$,

$$f(z) = \begin{cases} 1 & \text{for } z = 0\text{, for six values of } x \\ & \text{in the range } x = 1, 2, \ldots, 6 \\ 0 & \text{elsewhere, no values of } x \\ & \text{in the range } x = 1, 2, \ldots, 6 \end{cases}$$

If we consider a bivariate distribution, such as that of the two-dice problem, the distribution of a function of the two variates is still simple. For example, consider the distribution $f(w)$, where $w = xy$; then

$$f(1) = f(1, 1) = \frac{1}{36}$$

$$f(2) = f(2, 1) + f(1, 2) = \frac{2}{36}$$

$$f(3) = f(3, 1) + f(1, 3) = \frac{2}{36}$$

$$f(4) = f(4, 1) + f(2, 2) + f(1, 4) = \frac{3}{36}$$

$$\vdots$$

$$f(7) = 0$$

$$\vdots$$

$$f(36) = f(6, 6) = \frac{1}{36}$$

Here again $\Sigma f(w) = 1$ over the possible range.

5.5 DISTRIBUTIONS OF FUNCTIONS OF CONTINUOUS VARIATES: We now consider the transformation of continuous variates. Let the density function of X be f(x), which is positive in the range x_1 to x_2. We seek the distribution of $Z = z(X)$. If there is a one-to-one correspondence between x and z and x can be solved uniquely in terms of z, then $x = \Psi(z)$, $dx = \Psi'(z)\, dz$, $f(x) = f(\Psi(z))$, and the limits are $z_1 = z(x_1)$, $z_2 = z(x_2)$. The density function of Z, with these conditions, will be $f[\Psi(z)]|\Psi'(z)|$ over the range z_1 to z_2 where the absolute value of the function $\Psi'(z)$ is used.

EXAMPLE 5.9 If $f(x) = 2(1 - x)$ for $0 < x < 1$ and $z = x^2$, we find the distribution of Z as follows:

Since z is always positive, there is a one-to-one correspondence between z and x, that is, $x = +\sqrt{z}$; hence

$$f(z) = 2(1 - \sqrt{z})\, \frac{1}{2\sqrt{z}}$$

or

$$f(z) = (z^{-1/2} - 1) \qquad 0 < z < 1$$

EXAMPLE 5.10 If $f(x) = (1 - x)/2$, $-1 < x < 1$, then $x = \pm\sqrt{z}$ for $z = x^2$. For positive values of x, $x = +\sqrt{z}$, and for negative values of x, $x = -\sqrt{z}$. In this case

$$f(z) = \begin{cases} \frac{1}{4}(z^{-1/2} - 1) & x > 0 \\ \frac{1}{4}(z^{-1/2} + 1) & x < 0 \end{cases} \qquad 0 < z < 1$$

If we wish, we may add the two functions to obtain a single function

$$f(z) = \frac{1}{2} z^{-1/2} \qquad 0 < z < 1$$

but this is not possible in all cases.

The distribution of a function of *two* continuous variates X and Y is more complex mathematically. We wish to derive the simultaneous distribution of U and V, $f(u,v)$, where the transformations are $u = u(x,y)$ and $v = v(x,y)$. In case we wish only the distribution of U, then V is integrated out between its limits of integration, leaving some $f(u)$. In such cases it is necessary in general to assume a one-to-one correspondence between (x,y) and (u,v). It will then be possible to make the inverse solution:

$$x = x(u,v) \quad \text{and} \quad y = y(u,v)$$

The density function of U and V is

$$f(x(u,v),\ y(u,v))\,|J| \tag{5.22}$$

where J is the Jacobian of the transformation, that is,

$$J = \begin{vmatrix} \frac{\partial x}{\partial u} & \frac{\partial y}{\partial u} \\ \frac{\partial x}{\partial v} & \frac{\partial y}{\partial v} \end{vmatrix} \quad \text{or} \quad \frac{1}{J} = \begin{vmatrix} \frac{\partial u}{\partial x} & \frac{\partial u}{\partial y} \\ \frac{\partial v}{\partial x} & \frac{\partial v}{\partial y} \end{vmatrix} \tag{5.23}$$

This implies, of course, that these partial derivatives exist. If the second form is used, then J must be evaluated at $x = x(u,v)$, $y = y(u,v)$.

The limits for u and v must be determined individually for each problem. The limits for the first variable in the integral may be functions of the second variable.

EXAMPLE 5.11 Let us find f(u,v), where $f(x,y) = e^{-x-y}$, $0 < x < \infty$, $0 < y < \infty$, $u = x + y$, $v = x/y$.

Then

$$x = \frac{uv}{1 + v} \qquad y = \frac{u}{1 + v}$$

and

$$J = \begin{vmatrix} \dfrac{v}{1 + v} & \dfrac{1}{1 + v} \\ \dfrac{u}{(1 + v)^2} & \dfrac{-u}{(1 + v)^2} \end{vmatrix} = \frac{-u}{(1 + v)^2}$$

or

$$\frac{1}{J} = \begin{vmatrix} 1 & 1 \\ \dfrac{1}{y} & -\dfrac{x}{y^2} \end{vmatrix} = -\frac{(x + y)}{y^2} = -\frac{(1 + v)^2}{u}$$

Therefore

$$f(u,v) = e^{-u} \frac{u}{(1 + v)^2}$$

Since, if u is fixed, v can assume any positive value, then the limits of integration for u and v are $0 < u < \infty$, $0 < v < \infty$.

To find the distribution of U, we have

$$f(u) = ue^{-u} \int_0^\infty \frac{1}{(1 + v)^2} dv = ue^{-u} \qquad 0 < u < \infty$$

Similarly,

$$f(v) = \frac{1}{(1 + v)^2} \qquad 0 < v < \infty$$

Note that U and V are independent in this case.

EXAMPLE 5.12 Find f(u,v), where

$$f(x,y) = \frac{2e^2}{(1 - e)^2} e^{-x-y}$$

$0 < x < y$, $0 < y < 1$, $u = x + y$, and $v = x - y$.

We find easily $J = 1/2$, and hence

$$f(u,v) = \frac{e^2}{(1 - e)^2} e^{-u}$$

However, the assignment of limits for u and v in this case is more involved. First we plot as in Fig. 5.1 the region on the (x.y) plane with the following boundaries: $x = y$, $y = 1$, and $x = 0$.

Since $x = (u + v)/2$, $y = (u - v)/2$, then the boundary $x = y$ becomes $v = 0$; $y = 1$ becomes $u - v = 2$; and $x = 0$ becomes $u = -v$. We now plot the new region determined by these new boundaries on the (u,v) plane as in Fig. 5.1. In the new region, u varies from $-v$ to $v + 2$, and v varies from 0 to -1. Hence the new limits are $-v < u < v + 2$, $-1 < v < 0$.

Another way to find the distribution of a function of random variables is to use the moment-generating function. It is stated in Theorem 4.1 that the moment-generating function determines the distribution uniquely. If we are able to find the moment-generating function of the variate, then its distribution can be obtained.

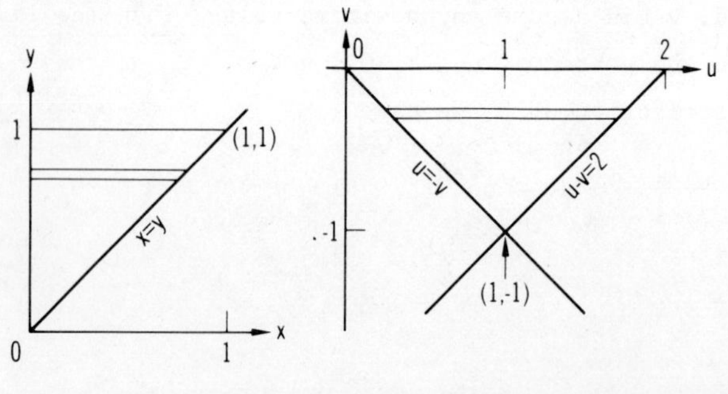

FIG. 5.1 Regions of integration for Example 5.8.

EXAMPLE 5.13 In Example 5.11, if we want to find the distribution of U = X + Y only, the moment-generating function of U is

$$m_U(t) = E(e^{tU}) = E(e^{tX+tY}) = E(e^{tX})E(e^{tY})$$

The last equality holds because X and Y are independently distributed in the example. From Example 4.7, both $E(e^{tX})$ and $E(e^{tY})$ are $(1 - t)^{-1}$, hence

$$m_U(t) = (1 - t)^{-2}$$

It can be shown that this is the moment-generating function of the gamma distribution with n = 2. Therefore, the density function of U is

$$f(u) = ue^{-u} \qquad 0 < u < \infty$$

The distribution of a function of random variables can also be found by directly evaluating the cumulative distribution function if the integration is not difficult to evaluate. The procedure is first to determine $F_U(u) = P\{U \leq u\}$. If U is a function of the random variables X and Y, say, then $P\{U \leq u\}$ can be evaluated by using the joint distribution of X and Y. The density function of U is then obtained by taking the derivative of the cumulative distribution function.

EXAMPLE 5.14 Consider the problem in Example 5.13. The cumulative distribution function of U is

$$\begin{aligned} F(u) &= P\{U \leq u\} = P\{X + Y \leq u\} \\ &= \int_0^u \int_0^{u-x} e^{-x-y}\, dy\, dx \\ &= \int_0^u (e^{-x} - e^{-u})\, dx \\ &= 1 - e^{-u} - ue^{-u} \end{aligned}$$

Hence

$$f(u) = ue^{-u} \qquad 0 < u < \infty$$

5.6 EXPECTED VALUES FOR BIVARIATE DISTRIBUTIONS: For any bivariate distribution with density function $f(x,y)$, the expected value of any function of X and Y, say $\Psi(X,Y)$, is

$$E(\Psi(x,y)) = \Sigma_W\Sigma\ \Psi(x,y)f(x,y) \quad \text{or} \quad \int_W\int \Psi(x,y)f(x,y)\ dx\ dy \tag{5.24}$$

where W is the entire region of (x,y). The following specializations of $\Psi(X,Y)$ will enable us to derive some simple rules for operating with expected values. Continuous variates will be used in the derivations, but similar rules will hold for discrete variates.

1. $E(c) = c$, where c is a constant.

2. $E(cX) = c\int_{-\infty}^{\infty} x[\int_{-\infty}^{\infty} f(x,y)\ dy]\ dx = c\int_{-\infty}^{\infty} xg(x)\ dx = cE(X)$

3. $E(X + Y) = \int_{-\infty}^{\infty} x[\int_{-\infty}^{\infty} f(x,y)\ dy]\ dx + \int_{-\infty}^{\infty} y[\int_{-\infty}^{\infty} f(x,y)\ dx]\ dy$

$$= \int_{-\infty}^{\infty} xg(x)\ dx + \int_{-\infty}^{\infty} yh(y)\ dy = E(X) + E(Y)$$

4. $E(XY) = \int_{-\infty}^{\infty} x[\int_{-\infty}^{\infty} yf(x,y)\ dy]\ dx = \int_{-\infty}^{\infty} xh_1(x)\ dx$

where

$$h_1(x) = \int_{-\infty}^{\infty} yf(x,y)\ dy$$

Note that $E(XY)$ can be evaluated, if the integrations can be performed, even though X and Y are not independent.

5. If X and Y are independent, then $f(x,y) = g(x)\cdot h(y)$ and hence

$$E(XY) = \int_{-\infty}^{\infty} xg(x)\ dx \int_{-\infty}^{\infty} yh(y)\ dy = E(X)E(Y)$$

5.7 MOMENTS: The product moment of X^rY^s about the origin is given by

$$\mu'_{rs} = E(X^rY^s) = \int_{-\infty}^{\infty}\int_{-\infty}^{\infty} x^ry^sf(x,y)\ dx\ dy \tag{5.25}$$

Let the mean of X be μ'_{10} and the mean of Y be μ'_{01}; then the product moment about the mean is

$$\mu_{rs} = E((X - \mu'_{10})^r (Y - \mu'_{01})^s)$$

$$= \int_{-\infty}^{\infty} \int_{-\infty}^{\infty} (x - \mu'_{10})^r (y - \mu'_{01})^s \, f(x,y) \, dx \, dy \qquad (5.26)$$

EXAMPLE 5.15 To find σ_X^2, defined as equal to μ_{20}, we have

$$\mu_{20} = E(X - \mu'_{10})^2 = \mu'_{20} - \mu'^{\,2}_{10}$$

Therefore,

$$\sigma_X^2 = \mu'_{20} - \mu'^{\,2}_{10}$$

Similarly,

$$\sigma_Y^2 = \mu_{02} = \mu'_{02} - \mu'^{\,2}_{01}$$

EXAMPLE 5.16 To find σ_{XY}, defined as equal to μ_{11} and called the *covariance* of X and Y, we have

$$\mu_{11} = E((X - \mu'_{10})(Y - \mu'_{01})) = E(XY) - \mu'_{10}\mu'_{01}$$

Therefore

$$\sigma_{XY} = \mu'_{11} - \mu'_{10}\mu'_{01} \qquad (5.27)$$

If X and Y are independent $E(XY) = \mu'_{11} = \mu'_{10}\mu'_{01}$, hence $\sigma_{XY} = 0$ for this condition. The correlation coefficient ρ_{XY} between X and Y is defined as the nondimensional quantity

$$\rho_{XY} = \frac{\sigma_{XY}}{\sigma_X \sigma_Y} \qquad (5.28)$$

If $\sigma_{XY} = 0$, then $\rho_{XY} = 0$. The correlation coefficient ρ_{XY} is invariant under the transformation $u = ax + b$, $v = cy + d$, where a, b, c, d are constants and a, c are nonzero and have the same sign. That is,

$$\sigma_{UV} = ac\sigma_{XY} \qquad \sigma_U^2 = a^2\sigma_X^2 \qquad \sigma_V^2 = c^2\sigma_Y^2$$

So

$$\rho_{UV} = \frac{\sigma_{UV}}{\sigma_U \sigma_V} = \frac{ac\sigma_{XY}}{\sqrt{a^2\sigma_X^2 c^2 \sigma_Y^2}} = \frac{\sigma_{XY}}{\sqrt{\sigma_X^2 \sigma_Y^2}} = \rho_{XY}$$

For simplicity, we let $\rho_{XY} = \rho$. It can be shown that $-1 \leq \rho \leq 1$. Let us consider the nonnegative quantity $E[(X - kY)^2]$. After squaring and taking expectation,

$$k^2E(Y^2) - 2kE(XY) + E(X^2) \geq 0$$

In view of the invariance property of ρ stated above, we may let $E(X) = E(Y) = 0$. [If $E(X) \neq 0$ and $E(Y) \neq 0$, we may make the transformation $U = X - E(X)$, $U = Y - E(Y)$, so $E(U) = E(V) = 0$.] Then $\sigma_{XY} = E(XY)$, $\sigma_X^2 = E(X^2)$, and $\sigma_Y^2 = E(Y^2)$. The above equation becomes

$$k^2\sigma_Y^2 - 2k\sigma_{XY} + \sigma_X^2 \geq 0$$

Since this holds for all real values of k, the determinant must be nonpositive. So we have

$$\sigma_{XY}^2 - \sigma_X^2\sigma_Y^2 \leq 0$$

or

$$\rho^2 = \frac{\sigma_{XY}^2}{\sigma_X^2\sigma_Y^2} \leq 1$$

Note that we have essentially proved the Cauchy-Schwarz inequality. The case $\rho^2 = 1$ means that $E((X - kY)^2) = 0$ or $X = kY$ with probability 1. Hence X and Y are linearly related.

EXAMPLE 5.17 To find the variance of $(X + Y)$, we have

$$\begin{aligned}\sigma_{X+Y}^2 = E(X - \mu'_{10} + Y - \mu'_{01})^2 &= \sigma_X^2 + \sigma_Y^2 + 2\sigma_{XY} \\ &= \sigma_X^2 + \sigma_Y^2 + 2\rho\sigma_X\sigma_Y\end{aligned}$$

If X and Y are independent, that is, $\sigma_{XY} = 0$, then $\sigma_{X+Y}^2 = \sigma_X^2 + \sigma_Y^2$.

EXAMPLE 5.18 To find E(XY) for the bivariate normal distribution, with $\mu'_{10} = 0$, $\mu'_{01} = 0$, we have

$$E(XY) = k \int_{-\infty}^{\infty} \int_{-\infty}^{\infty} xye^{-\theta}\, dx\, dy$$

where

$$k = \frac{1}{2\pi\sigma_X\sigma_Y\sqrt{1-\rho^2}} \qquad \text{and} \qquad \theta = \frac{1}{2(1-\rho^2)}\left(\frac{x^2}{\sigma_X^2} + \frac{y^2}{\sigma_Y^2} - \frac{2\rho xy}{\sigma_X\sigma_Y}\right)$$

Note that the correlation ρ_{XY} has been abbreviated to ρ.

The function θ may be written

$$\theta = \frac{1}{2(1-\rho^2)}\left(\frac{x^2}{\sigma_X^2} - \frac{2\rho xy}{\sigma_X\sigma_Y} + \frac{\rho^2 y^2}{\sigma_Y^2} + \frac{y^2}{2\sigma_Y^2}\right)$$

$$= \frac{1}{2}\left(\frac{t^2}{1-\rho^2} + \frac{y^2}{\sigma_Y^2}\right)$$

where

$$t = \frac{x}{\sigma_X} - \frac{\rho y}{\sigma_Y}$$

Then, using the methods of Section 5.5,

$$E(XY) = k\int_{-\infty}^{\infty} y\left\{\int_{-\infty}^{\infty} \frac{\sigma_X^2(t\sigma_Y + \rho y)}{\sigma_Y} \exp\left[-\frac{t^2}{2(1-\rho^2)}\right] dt\right\} \exp\left(\frac{-y^2}{2\sigma_Y^2}\right) dy$$

$$= \frac{\rho\sigma_X}{\sqrt{2\pi}}\int_{-\infty}^{\infty} \frac{y^2}{\sigma_Y^2} \exp\left(\frac{-y^2}{2\sigma_Y^2}\right) dy = \rho\sigma_X\sigma_Y$$

EXAMPLE 5.19 Consider

$$f(x,y) = e^{-x-y} \qquad 0 < x < \infty,\ 0 < y < \infty$$

Then,

$$E(X) = \mu'_{10} = \int_0^\infty e^{-y}[\int_0^\infty xe^{-x}\, dx]\, dy = 1$$

$$E(Y) = \mu'_{01} = 1$$

$$\mu'_{20} = \mu'_{02} = 2$$

$$\sigma_X^2 = \sigma_Y^2 = 1$$

$$\mu'_{11} = \int_0^\infty xe^{-x}\, dx \int_0^\infty ye^{-y}\, dy = 1$$

$$\sigma_{XY} = \mu'_{11} - \mu'_{10}\mu'_{10} = 0$$

and

$$\rho_{XY} = 0$$

EXAMPLE 5.20 Consider the following discrete bivariate distribution with p = 1/12:

x \ y	0	1	2	Total
0	2p	2p	2p	6p
1	p	4p	p	6p
Total	3p	6p	3p	12p

Then

$$\mu'_{10} = \sum_0^2 \sum_0^1 xf(x,y) = .5 \qquad \mu'_{01} = \sum_0^2 \sum_0^1 yf(x,y) = 1.0$$

$$\mu'_{20} = \sum_0^2 \sum_0^1 x^2 f(x,y) = .5 \qquad \mu'_{02} = \sum_0^2 \sum_0^1 y^2 f(x,y) = 1.5$$

$$\sigma_X^2 = .5 - .25 = .25 \qquad \sigma_Y^2 = 1.5 - .10 = .5$$

$$\mu'_{11} = \sum_0^2 \sum_0^1 xyf(x,y) = .5 \qquad \sigma_{XY} = .5 - .5 = 0$$

and

$$\rho = 0$$

Note that we cannot state that X and Y are independent in this case even though $\rho = 0$, since $f(x|y)$ is not the same for all values of y. It can be seen that

x	$f(x\|y = 0)$	$f(x\|y = 1)$	$f(x\|y = 2)$
0	$\frac{2}{3}$	$\frac{1}{3}$	$\frac{2}{3}$
1	$\frac{1}{3}$	$\frac{2}{3}$	$\frac{1}{3}$

It should be emphasized that if X and Y are independent then $\rho = 0$; but if $\rho = 0$, X and Y may or may not be independent.

5.8 MOMENT- AND CUMULANT-GENERATING FUNCTIONS: For the bivariate case, the moment-generating function about the origin is defined to be $m(t_x,t_y) = E(e^{Xt_x+Yt_y})$. The moment-generating function about μ'_{10} and μ'_{01} is defined as

$$M(t_x,t_y) = \int\int e^u f(x,y)\ dx\ dy \tag{5.29}$$

where $u = (x - \mu'_{10})t_x + (y - \mu'_{01})t_y$ and the double integration is performed over the ranges of x and y.

Then

$$M(t_x,t_y) = \sum_{r=0}^{\infty}\sum_{s=0}^{\infty} \mu_{rs} \frac{t_x^r t_y^s}{r!s!}$$

and

$$\mu_{rs} = \left.\frac{\partial^{r+s}M(t_x,t_y)}{\partial t_x^r \partial t_y^s}\right|_{t_x=t_y=0} \tag{5.30}$$

Note that $\mu_{00} = 1$, $\mu_{01} = \mu_{10} = 0$.

The cumulant-generating function in this case is given by

$$K(t_x,t_y) = \log m(t_x,t_y) = \log M(t_x,t_y) + \mu'_{10}t_x + \mu'_{01}t_y$$

$$= \sum_{r=0}^{\infty}\sum_{s=0}^{\infty} {}_{rs}\frac{t_x^r t_y^s}{r!s!} \tag{5.31}$$

and

$$\kappa_{rs} = \left.\frac{\partial^{r+s}K(t_x,t_y)}{\partial t_x^r \partial t_y^s}\right|_{t_x=t_y=0} \tag{5.32}$$

If $f(x,y) = g(x)\cdot h(y)$, the moments of X and Y may be computed separately. In this case,

$$M(t_x,t_y) = \int e^{u_x}g(x)\ dx \cdot \int e^{u_y}h(y)\ dy = M_X(t_x)\cdot M_Y(t_y)$$

where

$$u_x = (x - \mu'_{10})t_x \qquad u_y = (y - \mu'_{01})t_y$$

and the two integrations are taken over the respective ranges of x and y.

For the bivariate distribution and X,Y independent,

$$K(t_x,t_y) = \log m(t_x) + \log m(t_y) = K(t_x) + K(t_y)$$

THEOREM 5.1 The moment-generating function about the origin of the sum of two independent variates is the product of the moment-generating functions of each.

Proof:

$$m_{X+Y}(t) = E(e^{t(X+Y)}) = E(e^{tX})E(e^{tY}) = m_X(t)\cdot m_Y(t)$$

The theorem is also true for the moment-generating function about the mean of the sum of two independent variates.

5.9 EXTENSION TO k VARIATES: Once we understand the properties of the bivariate distributions, it is easy to generalize them to multivariate distributions. Let $\underline{X} = (X_1,X_2,\ldots,X_k)'$ be a random vector in which each component is a random variable and $f(x_1, x_2, \ldots, x_k) = f(\underline{x})$ represent the joint density function of $\underline{X}$. The following properties are those for the continuous case, but they can be applied equally well to discrete variates. (In order to discuss distributions in multidimensions, it is necessary to use matrix notation. A brief description of matrix algebra is given in Appendix A.)

$$\text{(a)} \quad P(w) = \int \underset{w}{\cdots} \int f(\underline{x}) \prod_{i=1}^{k} dx_i \tag{5.33}$$

where w is a subregion in the k-dimensional space W, and $\Pi_{i=1}^{k} dx_i = dx_1\ dx_2 \cdots dx_k$.

$$\text{(b)} \quad \underline{\mu} = E(\underline{X}) = (E(X_1), \ldots, E(X_k))' \tag{5.34}$$

is the *mean vector* of $\underline{X}$. Note that the expectation $E(\underline{X})$ indicates the expected value for each element of the vector (or matrix). The expected value of $\underline{Y} = \underline{CX}$, where $\underline{C}$ is a matrix of constants, is $E(\underline{Y}) = \underline{C}E(\underline{X}) = \underline{C\mu}$.

$$\text{(c)} \quad \underline{\Sigma} = E(\underline{X} - \underline{\mu})(\underline{X} - \underline{\mu})' \tag{5.35}$$

is called the *covariance matrix* of $\underline{X}$. The (i,j)th element of $\underline{\Sigma}$ is $\sigma_{ij} = E(\underline{X}_i - \underline{\mu}_i)(\underline{X}_j - \underline{\mu}_j)$, which is called the covariance of X_i and X_j. Note that the covariance matrix is a symmetrical matrix. The covariance matrix of $\underline{Y} = \underline{CX}$ is $E(\underline{CX} - \underline{C\mu})(\underline{CX} - \underline{C\mu})' = \underline{C\Sigma C}'$.

$$\text{(d)} \quad g(x_1) = \int \underset{W}{\cdots} \int f(\underline{x}) \sum_{i=2}^{k} dx_i$$

is the marginal density function of X_1.

$$\text{(e)} \quad f(x_1 | x_2, \ldots, x_k) = \frac{f(\underline{x})}{f(x_2, \ldots, x_k)}$$

is the conditional density function of X_1 given $X_2 = x_2, \ldots, X_k = x_k$, where

$$f(x_2, \ldots, x_k) = \int_{-\infty}^{\infty} f(\underline{x})\ dx_1$$

(f) If $X_1, X_2, \ldots, X_k$ are mutually independent, then

$$f(x_1,x_2,\ldots,x_k) = g(x_1)g(x_2), \ldots, g(x_k)$$

and

$$f(x_i|x_1,\ldots,x_{i-1},x_{i+1},\ldots,x_k) = g(x_i) \qquad i = 1, \ldots, k$$

where $g(x_i)$ represents the marginal density function of X_i

(g) For a transformation

$$u_i = u_i(\underline{x}) \qquad i = 1, 2, \ldots, k$$

$$\frac{1}{J} = \begin{vmatrix} \frac{\partial u_1}{\partial x_1} & \frac{\partial u_1}{\partial x_2} & \cdots & \frac{\partial u_1}{\partial x_k} \\ \vdots & & & \vdots \\ \frac{\partial u_k}{\partial x_1} & \frac{\partial u_k}{\partial x_2} & \cdots & \frac{\partial u_k}{\partial x_k} \end{vmatrix}$$

If the transformation is one-to-one and the inverse transformation is

$$x_i = x_i(\underline{u}) \qquad i = 1, \ldots, k$$

where $\underline{u} = (u_1,u_2,\ldots,u_k)$, then the joint density function of $U_1, U_2, \ldots, U_k$ is

$$f(x_1(\underline{u}),\ldots,x_k(\underline{u}))|J|$$

(h) $$E(\Psi(X_1,\ldots,X_k)) = \int \underset{W}{\cdots} \int \Psi(x_1,\ldots,x_k) f(x_1,\ldots,x_k) \prod_{i=1}^{k} dx_i$$

(i) $$E(e^{\Psi(X_1,\ldots,X_k)}) = \int \underset{W}{\cdots} \int e^{\Psi(x_1,\ldots,x_k)} f(x_1,\ldots,x_k) \prod_{i=1}^{k} dx_i$$

(j) The moment-generating function about the origin is $E(e^{\underline{t}'\underline{x}})$ where $\underline{t} = (t_1, \ldots, t_k)'$.

5.10 MULTINOMIAL DISTRIBUTION: The multinomial distribution is a generalization of the binomial and trinomial distributions. Any one of the events $A_1, A_2, \ldots, A_k$ can occur with respective probabilities

$p_1, p_2, \ldots, p_k$ on a single trial, $\Sigma_{i=1}^{k} p_i = 1$. If n independent trials are made, the probability that A_1 occurs x_1 times, A_2 occurs x_2 times, etc., $\Sigma_{i=1}^{k} x_i = n$, is given by

$$f(\underline{x}) = \frac{n!}{\Pi_{i=1}^{k} x_i!} \prod_{i=1}^{k} p_i^{x_i} \tag{5.36}$$

This is the general term of the expansion of

$$(p_1 + p_2 + \cdots + p_k)^n$$

Note that since $\Sigma x_i = n$, there are only k - 1 random variables. For example, a single die can show the number 1, 2, ..., 6 on the upper face with equal probabilities, $p_i = 1/6$.

The moment-generating function about the origin for the multinomial distribution is

$$m(\underline{t}) = E(e^{\underline{t}'\underline{X}}) = \sum_{x} \frac{n!}{\Pi_1^k x_i!} e^{\Sigma_1^k t_i x_i} \prod_1^k p_i^{x_i}$$

where $\underline{t} = (t_1, t_2, \ldots, t_k)'$. The symbol Σ_x is to be interpreted as meaning the summation over all values of x such that $\Sigma x_i = n$; therefore,

$$m(\underline{t}) = \sum_{x} \frac{n!}{\Pi_1^k x_i!} \prod_{i=1}^{k} (p_i e^{t_i})^{x_i} = (p_1 e^{t_1} + p_2 e^{t_2} + p_3 e^{t_3} + \cdots$$
$$+ p_k e^{t_k})^n \tag{5.37}$$

Then

$$E(X_i) = \left. \frac{\partial m(\underline{t})}{\partial t_i} \right|_{\underline{t}=0} = np_i \tag{5.38}$$

$$E(X_i^2) = \left. \frac{\partial^2 m(\underline{t})}{\partial t_i^2} \right|_{\underline{t}=0} = np_i + n(n - 1)p_i^2$$

and

$$\sigma_i^2 = E(X_i^2) - [E(X_i)]^2 = np_i - np_i^2 = np_i(1 - p_i) \tag{5.39}$$

Also,

$$E(X_iX_j) = \left.\frac{\partial^2 m(\underline{t})}{\partial t_i \partial t_j}\right|_{\underline{t}=0} = n(n - 1)p_ip_j$$

Hence,

$$\sigma_{ij} = E(X_iX_j) - E(X_i)E(X_j) = n(n - 1)p_ip_j - n^2p_ip_j = -np_ip_j \tag{5.40}$$

EXAMPLE 5.21 Suppose patients receiving a certain treatment are classified into one of the following 4 groups: (1) no improvement, (2) mild improvement, (3) great improvement, and (4) cured, with probabilities $p_1 = .1$, $p_2 = .2$, $p_3 = .2$, and $p_4 = .5$, respectively. A random sample of 20 patients is taken. Let X_i be the number of patients in the ith group; then X_1, X_2, X_3, and X_4 have a multinomial distribution. The probability that $X_1 = 2$, $X_2 = 1$, $X_3 = 8$, and $X_4 = 9$ is

$$\frac{20!}{2!1!8!9!}(.1)^2(.2)^1(.2)^8(.5)^9$$

The moment-generating function is

$$m(\underline{t}) = (.1e^{t_1} + .2e^{t_2} + .2e^{t_3} + .5e^{t_4})$$

So

$$E(X_1) = 20 \times .1 = 2$$

$$E(X_2) = 20 \times .2 = 4$$

$$\sigma_1^2 = 20 \times .1 \times .9 = 1.8$$

$$\sigma_2^2 = 20 \times .2 \times .8 = 3.2$$

$$\sigma_{12} = -20 \times .1 \times .2 = -0.4$$

The means, variances, and covariances of other variables can be obtained similarly.

5.11 MULTIVARIATE NORMAL DISTRIBUTION: As the univariate normal distribution plays an important role in univariate analysis, the multivariate normal distribution assumes an equally important role in multivariate analysis. Letting $Z_1, Z_2, \ldots, Z_k$ be distributed independently as N(0,1), the joint density function of $\underline{Z} = (Z_1, \ldots, Z_k)'$ is

$$f(\underline{z}) = \frac{1}{(2\pi)^{k/2}} \exp\left(-\frac{1}{2}\sum_{i=1}^{k} z_i^2\right)$$

$$= \frac{1}{(2\pi)^{k/2}} \exp\left(-\frac{1}{2}\underline{z}'\underline{z}\right)$$

where $\underline{z} = (z_1, \ldots, z_k)'$. Let us make a transformation, with $k \times k$ nonsingular matrix $\underline{B}$,

$$\underline{x} = \underline{\mu} + \underline{Bz}$$

or

$$\underline{z} = \underline{B}^{-1}(\underline{x} - \underline{\mu})$$

The Jacobian of the transformation is $J = 1/|\underline{B}|$. Therefore, the density function of $\underline{X} = (\underline{X}_1, \ldots, \underline{X}_k)'$ is

$$f(\underline{x}) = \frac{1}{(2\pi)^{k/2}|\underline{B}|} \exp\left[-\frac{1}{2}(\underline{x} - \underline{\mu})'(\underline{BB}')^{-1}(\underline{x} - \underline{\mu})\right]$$

Letting $\underline{\Sigma} = \underline{BB}'$, then

$$f(\underline{x}) = \frac{1}{(2\pi)^{k/2}|\underline{\Sigma}|^{1/2}} \exp\left[-\frac{1}{2}(\underline{x} - \underline{\mu})'\underline{\Sigma}^{-1}(\underline{x} - \underline{\mu})\right] \tag{5.41}$$

which is the density function of a multivariate normal distribution.

It can easily be shown that the mean vector of $\underline{X}$ is $\underline{\mu}$ and the covariance matrix of $\underline{X}$ is $\underline{\Sigma}$, since

$$\begin{aligned} E(\underline{X}) &= \underline{\mu} + \underline{B}E(\underline{Z}) = \underline{\mu} \\ E(\underline{X} - \underline{\mu})(\underline{X} - \underline{\mu})' &= \underline{B}E(\underline{ZZ}')\underline{B}' \\ &= \underline{BB}' = \underline{\Sigma} \end{aligned} \tag{5.42}$$

We denote the multivariate normal distribution with mean $\underline{\mu}$ and covariance matrix $\underline{\Sigma}$ by $N(\underline{\mu},\underline{\Sigma})$.

Let us partition the vectors $\underline{X},\underline{\mu}$ and the covariance matrix $\underline{\Sigma}$:

$$\underline{X} = \begin{pmatrix} \underline{X}^{(1)} \\ \\ \underline{X}^{(2)} \end{pmatrix} \qquad \underline{\mu} = \begin{pmatrix} \underline{\mu}^{(1)} \\ \\ \underline{\mu}^{(2)} \end{pmatrix} \qquad \underline{\Sigma} = \begin{pmatrix} \underline{\Sigma}_{11} & \underline{\Sigma}_{12} \\ \\ \underline{\Sigma}_{21} & \underline{\Sigma}_{22} \end{pmatrix}$$

where $\underline{X}^{(1)}$, $\underline{\mu}^{(1)}$ are $m \times 1$ vectors, $\underline{X}^{(2)}$, $\underline{\mu}^{(2)}$ are $(k - m) \times 1$ vectors, and $\underline{\Sigma}_{11}$ and $\underline{\Sigma}_{22}$ are $m \times m$ and $(k - m) \times (k - m)$ matrices, respectively. The joint density of $\underline{X}$ can be written as

$$f(\underline{x}) = f(\underline{x}^{(1)})f(\underline{x}^{(2)}|\underline{x}^{(1)}) \tag{5.43}$$

where $f(\underline{x}^{(1)})$ is the marginal density function of $\underline{X}^{(1)}$ and $f(\underline{x}^{(2)}|\underline{x}^{(1)})$ is the conditional density function of $\underline{X}^{(2)}$ given $\underline{X}^{(1)} = \underline{x}^{(1)}$. These density functions are of the following forms [a derivation may be found in Anderson (1958)]:

$$f(\underline{x}^{(1)}) = \frac{1}{(2\pi)^{m/2}|\underline{\Sigma}_{11}|^{1/2}} \exp\left[-\frac{1}{2}(\underline{x}^{(1)} - \underline{\mu}^{(1)})'\underline{\Sigma}_{11}^{-1}(\underline{x}^{(1)} - \underline{\mu}^{(1)})\right]$$

$$f(\underline{x}^{(2)}|\underline{x}^{(1)}) = \frac{1}{(2\pi)^{(k-m)/2}|\underline{\Sigma}_{22\cdot 1}|^{1/2}} \exp\left[-\frac{1}{2}(\underline{x}^{(2)} - \underline{\mu}_{2\cdot 1})'\underline{\Sigma}_{22\cdot 1}^{-1}(\underline{x}^{(2)} - \underline{\mu}_{2\cdot 1})\right]$$

where

$$\underline{\mu}_{2\cdot 1} = \underline{\mu}^{(2)} + \underline{\Sigma}_{21}\underline{\Sigma}_{11}^{-1}[\underline{x}^{(1)} - \underline{\mu}^{(1)}]$$

$$\underline{\Sigma}_{22\cdot 1} = \underline{\Sigma}_{22} - \underline{\Sigma}_{21}\underline{\Sigma}_{11}^{-1}\underline{\Sigma}_{12}$$

Therefore, the marginal and conditional distributions are also normal. For the marginal distribution, the mean vector and covariance matrix are determined by taking the corresponding components of $\underline{\mu}$ and $\underline{\Sigma}$. The conditional mean $\underline{\mu}_{2\cdot 1}$ is often called the *regression* of $\underline{X}^{(2)}$ on $\underline{x}^{(1)}$. Regression analysis will be discussed in detail in Chapter 13. Note that the conditional covariance matrix $\underline{\Sigma}_{22\cdot 1}$ does not depend on the value of $\underline{x}^{(1)}$.

The distribution of any linear function of $\underline{X}$ is also normal. Let $\underline{Y} = \underline{CX} + \underline{d}$ where $\underline{C}$ is a $k \times k$ nonsingular matrix and $\underline{d}$ is a $k \times 1$ vector, the Jacobian of the transformation $\underline{y} = \underline{Cx} + \underline{d}$ is $J = 1/|C|$. The inverse transformation is $\underline{x} = \underline{C}^{-1}(\underline{y} - \underline{d})$. Therefore, the density function of $\underline{Y}$ is

$$f(\underline{y}) = \frac{1}{(2\pi)^{k/2}(|\Sigma||C|^2)^{1/2}} \exp\left[-\frac{1}{2}(\underline{y} - \underline{C\mu} - \underline{d})'\underline{C}'^{-1}\underline{\Sigma}^{-1}\underline{C}^{-1}(\underline{y} - \underline{C\mu} - \underline{d})\right]$$

$$= \frac{1}{(2\pi)^{k/2}|\underline{C\Sigma C}'|^{1/2}} \exp\left[-\frac{1}{2}(\underline{y} - \underline{C\mu} - \underline{d})'(\underline{C\Sigma C}')^{-1}(\underline{y} - \underline{C\mu} - \underline{d})\right]$$

So $\underline{Y}$ is normally distributed with mean vector $\underline{C\mu} + \underline{d}$ and covariance matrix $\underline{C\Sigma C}'$. This result also holds when $\underline{C}$ is an $\ell \times k$ matrix ($\ell < k$) and $\underline{d}$ is an $\ell \times 1$ vector.

EXAMPLE 5.22 Let $\underline{X} = (X_1, X_2, X_3)'$ have a trivariate normal distribution with

$$\underline{\mu} = \begin{pmatrix} 1 \\ 4 \\ 2 \end{pmatrix} \quad \text{and} \quad \underline{\Sigma} = \begin{pmatrix} 10 & 4 & 5 \\ 4 & 30 & 11 \\ 5 & 11 & 16 \end{pmatrix}$$

Suppose $\underline{X}$ is partitioned as $X^{(1)} = X_1$ and $\underline{X}^{(2)} = (X_1, X_2)'$, then the corresponding partitions of $\underline{\mu}$ and $\underline{\Sigma}$ are

$$\underline{\mu}^{(1)} = \mu_1 = 1 \qquad \underline{\mu}^{(2)} = \begin{pmatrix} \mu_2 \\ \mu_3 \end{pmatrix} = \begin{pmatrix} 4 \\ 2 \end{pmatrix}$$

$$\Sigma_{11} = \sigma_1^2 = 10 \qquad \underline{\Sigma}_{22} = \begin{pmatrix} \sigma_2^2 & \sigma_{23} \\ \sigma_{32} & \sigma_3^2 \end{pmatrix} = \begin{pmatrix} 30 & 11 \\ 11 & 16 \end{pmatrix}$$

$$\underline{\Sigma}_{12} = \underline{\Sigma}'_{21} = [\sigma_{12} \quad \sigma_{13}] = [4 \quad 5]$$

So the marginal distribution of X_1 is N(1,10); the marginal distribution of $\underline{X}^{(2)}$ is the bivariate normal distribution with mean vector $\underline{\mu}^{(2)}$ and covariance matrix $\underline{\Sigma}_{22}$ given above.

The conditional distribution of $\underline{X}^{(2)}$ given that $X_1 = x_1$ is also a bivariate normal distribution. Its mean vector and covariance matrix are computed as

$$\underline{\mu}_{2\cdot 1} = \begin{pmatrix} 4 \\ 2 \end{pmatrix} + \begin{pmatrix} 4 \\ 5 \end{pmatrix} \frac{1}{10}(x_1 - 1) = \begin{pmatrix} 4 + 0.4(x_1 - 1) \\ 2 + 0.5(x_1 - 1) \end{pmatrix}$$

$$\underline{\Sigma}_{22\cdot 1} = \begin{pmatrix} 30 & 11 \\ 11 & 16 \end{pmatrix} - \begin{pmatrix} 4 \\ 5 \end{pmatrix} \frac{1}{10} (4 \quad 5)$$

$$= \begin{pmatrix} 28.4 & 9 \\ 9 & 14.4 \end{pmatrix}$$

EXERCISES

5.1 As an example of the distribution of a function of two variables for a discrete bivariate distribution, consider the distribution of the sum of two independent Poisson variates X and Y. The density function of X and Y is

$$f(x,y) = e^{-m_1} \frac{m_1^x}{x!} e^{-m_2} \frac{m_2^y}{y!}$$

Let z = x + y, or x = z - y; then

$$f(z,y) = e^{-(m_1+m_2)} \frac{m_1^{(z-y)} m_2^y}{(z - y)!y!}$$

Sum out the variable y over the range y = 0, 1, 2, ..., z, and show that

$$f(z) = e^{-(m_1+m_2)} \frac{(m_1 + m_2)^2}{z!}$$

Hence, show that the sum of two independent Poisson variates is distributed as a single Poisson variate with a mean equal to the sum of the two single means ($m = m_1 + m_2$).

5.2 If X is a discrete variate having the Poisson distribution

$$g(x) = \frac{m^x e^{-m}}{x!} \quad x = 0, 1, 2, \ldots$$

and Y is another discrete variate having the binomial distribution, given X = x,

$$f(y|x) = \binom{x}{y} p^y q^{x-y} \quad y = 0, 1, \ldots, x$$

show that the marginal distribution of Y is

$$\frac{(mp)^y e^{-mp}}{y!} \quad y = 0, 1, 2, \ldots$$

5.3 If X and Y represent the number of dots appearing on dice A and B, respectively, what is the probability that in throwing the two dice $X + 2Y \leq 6$?

5.4 For the trinomial distribution in Example 5.6, what is the distribution of $Z = X + Y$?

5.5 Given $f(x|y) = y^x e^{-y}/x!$ and $h(y) = e^{-y}$, where X is discrete ($x = 0, 1, \ldots$) and Y continuous ($y \geq 0$), show that $g(x) = (1/2)^{x+1}$.

5.6 In Example 5.7, find (a) $P(0 < X < 2)$; (b) $P(X < Y)$; (c) $E(X + Y)$; (d) σ^2_{X+Y}; (e) σ_{XY}.

5.7 Suppose X and Y are independently distributed as gamma distributions with parameters n and m, respectively. Find the joint distribution of $U = X + Y$ and $V = X/(X + Y)$. Show that the marginal distribution of U is gamma and that of V is beta. Further show that U and V are independently distributed.

5.8 Given $f(x,y) = 6(1 - x - y)$ for (x,y) contained within the triangle bounded by $x = 0$, $y = 0$, $x + y = 1$, (a) find the means and variances of X and Y and the covariance of X and Y; (b) find the equation of the regression line of Y on x and $\sigma^2_{Y \cdot x}$.

5.9 Given $f(x,y) = kxy(1 - x - y)$ over the same triangle as in Exercise 5.8, (a) find the value of k which makes $f(x,y)$ a density function; (b) find the marginal distribution $g(x)$; (c) find $\mu_{Y|x}$.

5.10 Given two continuous variates X and Y with the joint density function $f(x,y) = 2(1 + x + y)^{-3}$, $x > 0$, $y > 0$, and 0 otherwise, find $g(u)$ and $P(U \leq 1)$, where $U = X + Y$.

5.11 Given $f(x,y) = 1$ over the square $(0 < x < 1, 0 < y < 1)$, show that $P(XY > u) = 1 - u + u \log u$.

5.12 Given the bivariate density function $f(x,y) = 2/a^2$, $0 < x < y$, $0 < y < a$, show that (a) $\int\int f(x,y)\, dx\, dy = 1$ over the respective ranges of x and y; (b) $g(x) = 2(a - x)/a^2$; (c) $h(y) = (2/a^2)y$; (d) $f(x|y) = 1/y$; (e) $\mu'_{10} = a/3$, $\mu'_{01} = 2a/3$; (f) $\rho = 1/2$; (g) $\mu_{Y|x} = (a + x)/2$; (h) $\mu_{X|y} = y/2$.

5.13 The marginal density function of X for a bivariate distribution is $g(x) = \int f(x,y)\, dy$, where $f(x,y)$ is the joint density function of X and Y. The limits for y must be determined by the region W within which $f(x,y)$ is defined. If W includes the entire (x,y) plane, the limits are $(-\infty,\infty)$. However, consider a problem of this nature: $f(x,y) = kxy$ for $0 < x < y$, $0 < y < 1$; then

$$g(x) = \int_x^1 kxy\, dy \qquad 0 < x < 1$$

and

$$h(y) = \int_0^y kxy\, dx \qquad 0 < y < 1$$

Find the explicit values of $g(x)$ and $h(y)$.

5.14 Given a continuous bivariate density function $f(x,y)$ and the corresponding marginal densities $g(x)$ and $h(y)$, set up the integrals for the following: (a) mean and variance for Y for a given x, $(\mu_{Y|x}, \sigma^2_{Y|x})$; (b) the mean and variance of X for a given y.

5.15 In Exercise 5.14(a) the mean value of Y for a given x, $\mu_{Y|x}$, is a function of x and hence defines a curve in the (x,y) plane called the *curve of regression of Y on x*. If the regression of Y on x is linear ($\mu_{Y|x} = \alpha + \beta x$), then

$$\mu_{Y|x} = \int_{-\infty}^{\infty} y\frac{f(x,y)}{g(x)}\,dy = \alpha + \beta x$$

or

$$\int_{-\infty}^{\infty} yf(x,y)\,dy = \alpha g(x) + \beta x g(x)$$

By integrating each side of the last equation with respect to x, show that

$$\mu'_{01} = \alpha + \beta\mu'_{10}$$

Before integrating, as above, multiply both sides by x, and show that

$$\mu'_{11} = \alpha\mu'_{10} + \beta\mu'_{20}$$

Hence, find the values of α and β in terms of the moments of the original distribution, and show that

$$\mu_{Y|x} = \mu'_{01} + \rho\frac{\sigma_Y}{\sigma_X}(x - \mu'_{10})$$

5.16 In Exercise 5.14(a) if the variance of Y for a given x is averaged over all values of X, we have

$$\sigma^2_{Y\cdot x} = \int_{-\infty}^{\infty}\int_{-\infty}^{\infty} g(x)\left[(y - \mu_{Y|x})^2\,\frac{f(x,y)}{g(x)}\,dy\right]dx$$

$$= \int_{-\infty}^{\infty}\int_{-\infty}^{\infty} (y - \mu_{Y|x})^2\,f(x,y)\,dy\,dx$$

If $\mu_{Y|x} = \alpha + \beta x$, as given in Exercise 5.15, show that

$$\sigma^2_{Y\cdot x} = \sigma^2_Y(1 - \rho^2)$$

5.17 Work Exercises 5.14(a), 5.15, and 5.16 for a bivariate normal distribution.

5.18 Let X and Y have a bivariate normal distribution. Determine the values of the parameters for the following density functions:

(a) $f(x,y) = \frac{1}{12\pi} \exp\left(-\frac{1}{8}x^2 - \frac{1}{18}y^2 + x - 2\right)$

(b) $f(x,y) = \frac{10}{12\pi} \exp\left[-\frac{1}{72}(100x^2 + 100y^2 + 160xy)\right]$

(c) $f(x,y) = \frac{1}{3.2\pi} \exp\left\{-\frac{1}{1.28}\left[(x - 3)^2 + \frac{1}{4}(y - 1)^2 - 0.6(x - 3)(y - 1)\right]\right\}$

5.19 Given that X and Y are bivariately normally distributed with zero means, variances σ_X^2 and σ_Y^2, correlation coefficient ρ, find the distribution of $Z = X + Y$.

5.20 Show that the joint moment-generating function of X_1 and X_2 from the bivariate normal distribution with means μ_1 and μ_2, variances σ_1^2 and σ_2^2, and correlation ρ is given by

$$m(t_1,t_2) = \exp\left[t_1\mu_1 + t_2\mu_2 + \frac{1}{2}(t_1^2\sigma_1^2 + 2\rho t_1 t_2 \sigma_1\sigma_2 + t_2^2\sigma_2^2)\right]$$

5.21 Use the results of Exercise 5.20 to find the variances and covariance of X_1 and X_2.

5.22 Let X_1 and X_2 have a bivariate normal distribution with means μ_1, μ_2, the variances of X_1 and X_2 being the same and equal to σ^2, and the correlation coefficient ρ. Define $Y_1 = X_1 + X_2$ and $Y_2 = X_1 - X_2$. Find the joint distribution of Y_1 and Y_2. Show that Y_1 and Y_2 are independently distributed and each has a normal distribution.

5.23 Let $X_1, \ldots, X_n$ be independently and identically distributed as a Poisson distribution with parameter m. Use the moment-generating function to find the distribution of $U = \Sigma_{i=1}^n X_i$.

5.24 Let $X_1, \ldots, X_n$ be independently and identically distributed as $N(\mu,\sigma^2)$. Use the moment-generating function to find the distribution of $\bar{X} = \Sigma_{i=1}^n X_i/n$.

5.25 Suppose X_1 and X_2 are independently distributed as a uniform distribution on (0,1). Use the cumulative distribution to find the distribution of $U = X + Y$.

5.26 Let X_1, X_2, X_3, and X_4 be independently identically distributed as $N(\mu,\sigma^2)$. Define $U_1 = X_1 - X_4$, $U_2 = X_2 - X_4$, $U_3 = X_3 - X_4$, and $U_4 = X_1 + X_2 + X_3 + X_4$. Determine the distribution of $\underline{U} = (U_1, U_2, U_3, U_4)'$. Show that U_4 is independent of (U_1,U_2,U_3). *Hint:* Write $\underline{U} = \underline{CX}$.

5.27 Let $(X_1,X_2,\ldots X_k)'$ be distributed as $N(\underline{\mu},\underline{\Sigma})$ where $\underline{\mu},\underline{\Sigma}$ have the form $\mu_i = \mu$, $\sigma_i^2 = \sigma^2$, and $\sigma_{ij} = \rho\sigma^2$, i.e., all the means are the same, all the variances are the same, and all the covariances are the same. (This distribution is called a symmetric normal distribution. This covariance matrix is said to have an intraclass correlation structure.) Suppose we make the Helmert transformation

$$y_1 = \frac{x_1 + x_2 + \cdots + x_k}{\sqrt{k}}$$

$$y_2 = \frac{x_1 - x_2}{\sqrt{2}}$$

$$y_3 = \frac{x_1 + x_2 - 2x_3}{\sqrt{6}}$$

$$\vdots$$

$$y_k = \frac{x_1 + x_2 + \cdots + x_{k-1} - (k - 1)x_k}{\sqrt{k(k - 1)}}$$

Show that $Y_1, Y_2, \ldots, Y_k$ are independently distributed and each has a univariate normal distribution. *Hint:* Write $\underline{\Sigma} = \sigma^2[(1 - \rho)\underline{I} + \rho\underline{1}\underline{1}']$ where $\underline{I}$ is the identity matrix and $\underline{1}$ is a $k \times 1$ vector of ones.

REFERENCE

Anderson, T. W. (1958). *An Introduction to Multivariate Statistical Analysis.* Wiley, New York.

6

DERIVED SAMPLING DISTRIBUTIONS

6.1 INTRODUCTION: For the sequence of the statistical method—*specification, distribution, estimation,* and *tests of hypotheses*—we have considered certain parent populations and their properties which are usually specified in applied statistical investigations. After problems of specification it would seem logical to discuss problems of estimation next in order. Such problems involve determining what functions of the sample observations should be used to "best" estimate the parameters of the specified population distribution, where "best" must be defined in some exact manner. For example, let $X_1, X_2, \ldots, X_n$ be a sample of n observations, specified as having been drawn in some manner from a normal distribution with mean μ and variance σ^2, what two functions of the observations should be used to "best" estimate μ and σ^2? A function of the observations used to estimate a population parameter is called an *estimator*. The numerical value obtained by using the estimator is called an *estimate*. It seemed appropriate to earlier workers in statistics to use the sample mean $\bar{X} = \Sigma X_i/n$ and $\hat{\sigma}^2 = \Sigma(X_i - \bar{X})^2/n$ as estimators of the two population parameters μ and σ^2. However, as will be shown later, in cases where the sample is a *random sample*, the "best" sample estimate of σ^2 is $S^2 = \Sigma(X_i - \bar{X})^2/(n - 1)$. A *random sample* is a sample consisting of independently and identically distributed random variables; therefore, it is a sample drawn in such a manner that the probability of obtaining any member is independent of the probability of obtaining any other member.

A less restrictive method of estimating the value of a parameter θ is a method which derives limits C_1 and C_2 which are functions of the sample $X_1, X_2, \ldots, X_n$. The interval (C_1, C_2) will contain the parameter θ a certain percentage of the time. The limits are thus functions of the sample and this percentage is called the *confidence probability* or confidence coefficient. The concepts of confidence limits were introduced by Neyman (1935).

Various methods of estimating the confidence interval will be discussed in later chapters. Fisher (1974) uses the term *fiducial limits* to indicate essentially the same concept.

In the chronological development of modern statistical methodology, however, the distribution of estimators and the distributions of certain functions of these estimators used in making tests of hypotheses and setting confidence limits did not wait upon the development of a sound theory of estimation. In this chapter, then, we shall assume, for the time being, that certain estimators are the "best" estimates of the corresponding population parameters and that certain functions of these estimators used in making tests of hypotheses and setting confidence limits are the "best" such functions. In later chapters on estimation and tests of hypotheses, a discussion will be given of what properties a "best" estimator or an "appropriate" test criterion should have.

6.2 DERIVED SAMPLING DISTRIBUTION PROBLEMS: In a typical applied statistical investigation, for example, a *sample survey* or an *experiment*, we specify that the observations obtained by some sampling process are drawn from some particular parent population distribution. We then calculate certain functions of the observations as estimates of the parameters of the specified population. These functions of the random variables from a sample are called statistics. A statistic does not involve any unknown parameters of the population. Now, in order to set confidence limits for the population parameters or to make tests of hypotheses concerning the population parameters, it is necessary to know the probability distributions of the estimators or statistics, for example, of $\bar{X}$ and S^2 for a normal parent population, and of certain functions of the statistics, for example, of χ^2, t, and F. (See Sections 6.9 to 6.12 for their definitions.) Mathematically these probability distributions are *derived* from the specified parent population distributions and hence are called *derived sampling distributions*. Some uses of these sampling distributions in problems of estimation and tests of hypotheses will be indicated in subsequent sections. However, the justifications of the estimator and the test criterion will be discussed in later chapters.

6.3 LINEAR FUNCTIONS: In the derivations of sampling distributions it will frequently be necessary to know the distribution of some linear functions of the members of a sample. Let such a linear function be given by

$$L = a_1X_1 + a_2X_2 + \cdots + a_nX_n = \sum_{i=1}^{n} a_iX_i \tag{6.1}$$

where a_i is some fixed constant and the sample, here not necessarily random, is represented by $X_1, X_2, \ldots, X_n$ or $\{X_i\}$. In order to obtain some general results, in the discussion following, each X_i is assumed to be drawn from a population with mean μ_i and variance σ_i^2. Also the covariance between X_i and X_j is denoted by σ_{ij}.

It follows that

$$E(L) = \sum_{i=1}^{n} a_iE(X_i) = \sum_{i=1}^{n} a_i\mu_i \tag{6.2}$$

Also,

$$\sigma_L^2 = E(L - E(L))^2 = E\left(\sum_{i=1}^{n} a_i(X_i - \mu_i)\right)^2$$

$$= \sum_{i=1}^{n} a_i^2\sigma_i^2 + 2\sum_{i<j}\sum a_ia_j\sigma_{ij} \tag{6.3}$$

If $X_1, X_2, \ldots, X_n$ is a random sample, then $\sigma_{ij} = 0$ and

$$\sigma_i^2 = \sum_{i=1}^{n} a_i^2\sigma_i^2$$

Finally, if also all $\mu_i = \mu$ and all $\sigma_i^2 = \sigma^2$, then

$$E(L) = \mu \sum_{i=1}^{n} a_i$$

and

$$\sigma_L^2 = \sigma^2 \sum_{i=1}^{n} a_i^2$$

EXAMPLE 6.1 If each $a_i = 1/n$, then $L = (1/n) \Sigma_{i=1}^n X_i$, which is the sample mean $\bar{X}$. Further, for any sample,

$$E(\bar{X}) = \frac{1}{n}(n\mu) = \mu$$

Also, if the sample is random,

$$\sigma_{\bar{X}}^2 = E(\bar{X} - \mu)^2 = n\sigma^2 \frac{1}{n^2} = \frac{\sigma^2}{n}$$

In section 6.1 mention was made of a "best" estimator of a population parameter. One property usually desired in a "best" estimator is that of being *unbiased*. An estimator of a population parameter is said to be unbiased if its expected value is equal to the population parameter. It follows from Example 6.1 that $\bar{X}$ is an unbiased estimator of μ.

EXAMPLE 6.2 Let $L = \Sigma_{i=1}^n a_i(X_i - \mu)$, for a random sample from a population with $\mu_i = \mu$, $\sigma_i^2 = \sigma^2$; then

$$E(L) = \sum_{i=1}^{n} a_i E(X_i - \mu) = 0$$

Also,

$$\sigma_L^2 = E(L - E(L))^2 = E\left[\sum_{i=1}^{n} a_i^2 (X_i - \mu)^2 \right.$$

$$\left. + 2 \sum_{i<j} \sum a_i a_j (X_i - \mu)(X_j - \mu)\right] = \sigma^2 \sum_{i=1}^{n} a_i^2$$

EXAMPLE 6.3 To find E(V) where $V = \Sigma_{i=1}^{n} (X_i - \bar{X})^2$, under conditions given in Example 6.2, we find

$$V = \sum_{i=1}^{n} [(X_i - \mu) - (\bar{X} - \mu)]^2 = \sum_{i=1}^{n} (X_i - \mu)^2 - n(\bar{X} - \mu)^2$$

But

$$E\left(\sum_{i=1}^{n} (X_i - \mu)^2\right) = n\sigma^2$$

Also,

$$E\left((\bar{X} - \mu)^2\right) = \frac{\sigma^2}{n}$$

by Example 6.1. Hence,

$$E(V) = n\sigma^2 - \sigma^2 = (n - 1)\sigma^2$$

Now, if we set $S^2 = V/(n - 1)$, then $E(S^2) = \sigma^2$ and therefore S^2 is an unbiased estimator of σ^2 for a random sample. We shall refer to this unbiased estimator as the sample variance.

6.4 ORTHOGONAL LINEAR FORMS: When samples are taken from different populations, it is often desirable to make comparisons about the populations. These comparisons are usually linear functions of the means. In order to make the comparisons uncorrelated, we consider orthogonal linear forms.

Consider the two linear forms

$$L_1 = \sum_{i=1}^{n} a_i X_i \quad \text{and} \quad L_2 = \sum_{i=1}^{n} b_i X_i$$

Since

$$E(L_1) = \sum_{i=1}^{n} a_i \mu_i \quad \text{and} \quad E(L_2) = \sum_{i=1}^{n} b_i \mu_i$$

then

$$E(L_1 \pm L_2) = \sum_{i=1}^{n} (a_i \pm b_i)\mu_i$$

Now, let $X_1, X_2, \ldots, X_n$ denote a random sample with $\mu_i = \mu$, $\sigma_i^2 = \sigma^2$; then

$$\sigma_1^2 = \sigma^2 \sum_{i=1}^{n} a_i^2 \qquad \sigma_2^2 = \sigma^2 \sum_{i=1}^{n} b_i^2 \qquad \sigma_{12} = \sigma^2 \sum_{i=1}^{n} a_i b_i$$

where σ_i^2 is the variance of L_i and σ_{12} is the covariance of L_1 and L_2. Hence the condition that would make L_1 and L_2 uncorrelated is that $\Sigma_{i=1}^{n} a_i b_i = 0$. Two uncorrelated linear forms are said to be *orthogonal*.

EXAMPLE 6.4 For a random sample of 5 drawn from a normal population with mean μ and variance σ^2, the mean value and variance of

$$L_1 = X_1 + X_2 + X_3 + X_4 + X_5 = 5\bar{X}$$

are, respectively,

$$E(L_1) = \mu + \mu + \mu + \mu + \mu = 5\mu \qquad \text{and} \qquad \sigma_1^2 = 5\sigma^2$$

For a second linear form

$$L_2 = -2X_1 - X_2 + X_4 + 2X_5$$

we obtain

$$E(L_2) = -2\mu - \mu + \mu + 2\mu = 0$$

and also

$$\sigma_{12} = (1)(-2) + (1)(-1) + (1)(1) + (1)(2) = 0$$

For a third linear form

$$L_3 = 2X_1 - X_2 - 2X_3 - X_4 + 2X_5$$

we obtain

$$E(L_3) = 2\mu - \mu - 2\mu - \mu + 2\mu = 0$$

Also

$$\sigma_{13} = \sigma^2[(1)(2) + (1)(-1) + (1)(-2) + (1)(-1) + (1)(2)] = 0$$

and

$$\sigma_{23} = \sigma^2[(-2)(2) + (-1)(-1) + (0)(-2) + (1)(-1) + (2)(2)] = 0$$

Notice that the sum of the coefficients of L_2 and L_3 respectively is zero and that L_2 and L_3 are uncorrelated with L_1, illustrating the theory in the paragraph above.

EXAMPLE 6.5 To determine a fourth orthogonal linear form

$$L_4 = b_1X_1 + b_2X_2 + b_3X_3 + b_4X_4 + b_5X_5$$

to L_1, L_2, and L_3 of Example 6.4, we find

$$L_1L_4: \quad b_1 + b_2 + b_3 + b_4 + b_5 = 0$$

$$L_2L_4: \quad -2b_1 - b_2 + b_4 + 2b_5 = 0$$

$$L_3L_4: \quad 2b_1 - b_2 - 2b_3 - b_4 + 2b_5 = 0$$

Eliminating b_3 from the first and third equations and then b_2 from the remaining two equations, we find

$$b_1 + b_4 + 3b_5 = 0$$

Similarly,

$$b_2 - 3b_4 - 8b_5 = 0$$

$$b_3 + 3b_4 + 6b_5 = 0$$

The first condition will be satisfied if we let $b_5 = 1$, $b_4 = -2$, and $b_1 = -1$. Then, we find $b_2 = 2$ and $b_3 = 0$. Hence a desired fourth orthogonal form is

$$L_4 = -X_1 + 2X_2 - 2X_4 + X_5$$

There are an infinite number of such functions.

6.5 LINEAR FORMS WITH NORMALLY DISTRIBUTED VARIATES: Suppose that the X_i values in the linear form $L = \Sigma_{i=1}^{n} a_i X_i$ each follows a normal parent population distribution with mean μ_i and variance σ_i^2. Then the probability of getting the particular x_i values in a random sample of size n will be

$$\frac{1}{(2\pi)^{n/2}} \frac{1}{\Pi_{i=1}^{n} \sigma_i} \exp\left[- \sum_{i=1}^{n} \frac{(x_i - \mu_i)^2}{2\sigma_i^2}\right] \prod_{i=1}^{n} dx_i$$

It is desired to find the distribution of $[L - E(L)]$.

The moment-generating function of $[L - E(L)]$ is

$$M(t) = E\left(e^{t\{L-E(L)\}}\right)$$

$$= \frac{1}{(2\pi)^{n/2}} \frac{1}{\Pi_{i=1}^{n} \sigma_i} \int_{-\infty}^{\infty} \cdots \int_{-\infty}^{\infty} \exp\left(- \left\{ \sum_{i=1}^{n} \frac{(x_i - \mu_i)^2}{2\sigma_i^2} - t\left[\sum_{i=1}^{n} a_i(x_i - \mu_i)\right]\right\}\right) \prod_{i=1}^{n} dx_i$$

$$= \prod_{i=1}^{n} \frac{1}{\sigma_i \sqrt{2\pi}} \int_{\infty}^{\infty} \exp\left[- \frac{(x_i - \mu_i)^2}{2\sigma_i^2} + ta_i(x_i - \mu_i)\right] dx_i$$

By completing the square of the exponent we have

$$- \frac{1}{2\sigma_i^2} [(x_i - \mu_i)^2 - 2\sigma_i^2 ta_i(x_i - \mu_i) + \sigma_i^4 t^2 a_i^2] + \frac{\sigma_i^2 t^2 a_i^2}{2} = - \frac{1}{2\sigma_i^2} [(x_i - \mu_i) - \sigma_i^2 ta_i]^2 + \frac{\sigma_i^2 t^2 a_i^2}{2}$$

Hence,

$$M(t) = \prod_{i=1}^{n} e^{\sigma_i^2 t^2 a_i^2/2} = e^{(t^2/2) \Sigma_{i=1}^{n} a_i^2 \sigma_i^2}$$

Now, the moment-generating function for the normal distribution

$$\frac{1}{\sigma\sqrt{2\pi}}\, e^{-y^2/(2\sigma^2)} \qquad -\infty < y < \infty$$

is

$$M(t) = \frac{1}{\sigma\sqrt{2\pi}} \int_{-\infty}^{\infty} e^{-y^2/(2\sigma^2)+ty}\, dy = e^{t^2\sigma^2/2}$$

Hence, we may say that L is normally distributed with mean $E(L) = \Sigma_{i=1}^{n} a_i\mu_i$ and variance $\sigma_L^2 = \Sigma_{i=1}^{n} a_i^2\sigma_i^2$. This result is analogous to that found in Section 6.4 for the mean and variance for nonspecified parent population distributions.

6.6 DISTRIBUTION OF THE SAMPLE MEAN IN NORMAL POPULATIONS: Using the result in Section 6.5 we let $a_i = 1/n$, $\mu_i = \mu$, and $\sigma_i^2 = \sigma^2$; then $\Sigma_{i=1}^{n} a_i X_i = \bar{X}$, $\Sigma_{i=1}^{n} a_i\mu_i = \mu$, and $\Sigma_{i=1}^{n} a_i^2\sigma_i^2 = \sigma^2/n$. Therefore, the distribution of $\bar{X}$ is $N(\mu,\sigma^2/n)$. Let

$$Z = \frac{\bar{X} - \mu}{\sigma/\sqrt{n}} \tag{6.4}$$

The sampling distribution of Z is N(0,1), which is symmetrical about zero. Then, if σ be known, Z fulfills the requirements for a test criterion which may be used to test a hypotheses specifying a numerical value for μ, say $\mu = \mu_0$ with σ known. This will be studied in Chapter 10.

EXAMPLE 6.6 Suppose the test scores follow a normal distribution with mean 60 and variance 100. A random sample of size 16 is taken; then the probability that the sample mean of these 16 students will exceed 62 is

$$P(\bar{X} > 62) = P\left(Z > \frac{62 - 60}{\sqrt{100/16}}\right)$$
$$= P(Z > 0.8) = 0.212$$

6.7 LAW OF LARGE NUMBERS: In the preceding section it was shown that for a random sample of n from an arbitrary parent population, $E[\bar{X}] = \mu$ and $\sigma^2_{\bar{X}} = \sigma^2/n$. It follows that whatever the form of the parent population distribution (provided the variance is finite), the distribution of the sample mean becomes more and more concentrated about the population mean as the size of the sample increases. It is evident, then, that as the size of the sample is increased, the more confident we can be that the sample mean provides a "good" estimate of the population mean. Essentially this is the meaning of the *law of large numbers*. A more precise statement of this property is provided by *Tchebysheff's inequality*, which will be derived in Chapter 7.

6.8 THE CENTRAL LIMIT THEOREM: The central limit theorem states:

If an arbitrary population distribution has a mean μ and finite variance σ^2, then the distribution of the sample mean for a random sample approaches the normal distribution with mean μ and variance σ^2/n as the sample size n increases.

The proof of this important theorem is beyond the scope of this text except for a distribution possessing a moment-generating function. Now, if Y is distributed as $N(\mu,\sigma^2)$ and $U = (Y - \mu)/\sigma$, then

$$m_U(t) = E\left(e^{t(Y-\mu)/\sigma}\right)$$

$$= e^{-t\mu/\sigma}E[e^{(t/\sigma)Y}]$$

$$= e^{-t\mu/\sigma}\, e^{\mu(t/\sigma)+t^2/2}$$

$$= e^{t^2/2}$$

Therefore,

$$K_U(t) = \frac{1}{2}\, t^2$$

If X has an arbitrary distribution and $V = (X - \mu)/\sigma$, then

$$m_V(t) = E\left(e^{t(X-\mu)/\sigma}\right) = e^{-\mu t/\sigma}E\left(e^{X(t/\sigma)}\right)$$

Also, the moment-generating function of $W = (\bar{X} - \mu)(\sigma/\sqrt{n})$ is

$$m_W(t) = E\left(e^{t[(\bar{X}-\mu)/(\sigma/\sqrt{n})]}\right) = e^{-\mu(t\sqrt{n}/\sigma)}E\left(e^{(t/\sigma\sqrt{n})(X_1+X_2+\cdots X_n)}\right)$$

$$= \{e^{-\mu(t/\sigma\sqrt{n})}E[e^{(tX/\sigma\sqrt{n})}]\}^n$$

$$= \left[m_V\left(\frac{t}{\sqrt{n}}\right)\right]^n$$

It follows that

$$\log m_W(t) = n \log\left[m_V\left(\frac{t}{\sqrt{n}}\right)\right]$$

and

$$K_W(t) = n \log\left[1 + \frac{1}{n}\left(\frac{t^2}{2} + \frac{1}{3!\sqrt{n}}\frac{\mu_3}{\sigma^3}t^3 + \cdots\right)\right]$$

Since

$$\log(1 + x) = x - \frac{1}{2}x^2 + \frac{1}{3}x^3 - \cdots$$

we have

$$K_W(t) = n\left[\frac{1}{n}\left(\frac{t^2}{2} + \frac{1}{3!\sqrt{n}}\frac{\mu_3}{\sigma^3}t^3 + \cdots\right)\right.$$
$$\left. - \frac{1}{2}\frac{1}{n^2}\left(\frac{t^2}{2} + \frac{1}{3!\sqrt{n}}\frac{\mu_3}{\sigma^3}t^3 + \cdots\right)^2 + \cdots\right]$$

and

$$\lim_{n\to\infty} K_W(t) = \frac{t^2}{2}$$

Hence, $K_W(t)$ approaches $K_U(t)$, and it follows that the distribution of $\bar{X}$ approaches the normal distribution with mean μ and variance σ^2/n as the sample size n increases.

EXAMPLE 6.7 The central limit theorem may be used to obtain a normal approximation to the binomial distribution. The binomial distribution is the distribution of successes in n repeated independent trials. Let $X_i = 1$ if the trial is a success and $X_i = 0$ if it is a failure, $i = 1, \ldots, n$; $P(X_i = 1) = p$. Then $Y = \Sigma_{i=1}^{n} X_i$ has the binomial distribution with parameters n and p. By the central limit theorem, the distribution of Y/n approaches $N(p, p(1 - p)/n)$ as n increases. Therefore,

$$P(Y \leq y) = P\left(\frac{Y/n - p}{\sqrt{p(1 - p)/n}} \leq \frac{y/n - p}{\sqrt{p(1 - p)/n}}\right)$$
$$\doteq P\left(Z \leq \frac{y - np}{\sqrt{np(1 - p)}}\right)$$

where Z has the standard normal distribution.

A closer approximation is obtained by the use of "continuity correction." Since Y takes only integer values, then $P(Y = y) = P(y - 1/2 < Y \leq y + 1/2)$. The half-unit is used to correct the discreteness; therefore, a better approximation is

$$P(Y \leq y) \doteq P\left(Z \leq \frac{y + 1/2 - np}{\sqrt{np(1 - p)}}\right) \tag{6.5}$$

For the endpoints y = 0 or n, we use

$$P(Y = 0) = P\left(-\infty < Y \leq \frac{1}{2}\right)$$

and

$$P(Y = n) = P\left(n - \frac{1}{2} < Y < \infty\right)$$

6.9 CHI-SQUARE DISTRIBUTION: An important distribution that enters into the theory of derived sampling distributions is the chi-square distribution, defined as

$$f(x) = \frac{1}{2^{\nu/2}\Gamma(\nu/2)} x^{(\nu/2)-1} e^{-x/2} \qquad 0 < x < \infty \tag{6.6}$$

where ν is called the *degrees of freedom*. The chi-square distribution is denoted by $\chi^2(\nu)$ or χ^2. It is easy to see that

$$P(X \leq \chi_\alpha^2) = \int_0^{\chi_\alpha^2} f(x)\, dx = \frac{1}{\Gamma(\nu/2)} \int_0^{y_\alpha} y^{(\nu/2)-1} e^{(-y)}\, dy$$

$$= \frac{\Gamma_{y_\alpha}(\nu/2)}{\Gamma(\nu/2)}$$

which is the incomplete gamma function, $I_{y_\alpha}(\nu/2)$, $y_\alpha = \chi_\alpha^2/2$. This function has been tabulated in the form $I(u,p)$ by Karl Pearson (1922) for various values of $u = y_\alpha/\sqrt{p+1} = \chi_\alpha^2/\,2\sqrt{p+1}$ and $p = (\nu/2) - 1$. In this form, $P(X > \chi_\alpha^2) = 1 - I(u,p)$. These tables are described and their uses in statistics explained in detail by Bancroft (1950).

Catherine M. Thompson (1941) gives values of χ_α^2 for $\alpha = .995$, .990, .975, .950, .900, .750, .500, .250, .100, .050, .025, .010, and .005 and $\nu = 1\ (1)\ 30$, where $\alpha = P(X > \chi_\alpha^2)$. These values are presented in Table C.2 (Appendix C).

The moment-generating function of χ^2 is

$$m(t) = \frac{1}{\Gamma(\nu/2)} \int_0^{\infty} \left(\frac{x}{2}\right)^{(\nu/2)-1} e^{-(1-2t)(x/2)}\, d\left(\frac{x}{2}\right)$$

$$= (1 - 2t)^{-\nu/2} \qquad t < \frac{1}{2} \tag{6.7}$$

It follows that the cumulant-generating function is

$$K(t) = -\frac{\nu}{2}\log(1 - 2t) = \frac{\nu}{2}\sum_{i=1}^{\infty} \frac{(2t)^i}{i}$$

Hence

$$\kappa_i = (i - 1)!\,2^{i-1}\nu$$

So the χ^2 distribution has mean ν and variance 2ν.

Consider a new variable $(X - \nu)/\sqrt{2\nu}$, for which $\mu = 0$, $\sigma^2 = 1$. The cumulant-generating function for this new variable is

$$K(t) = \frac{\nu}{2} \sum_{i=2}^{\infty} \frac{(2t)^i}{i(2\nu)^{i/2}}$$

and

$$\kappa_i = \left(\frac{2}{\nu}\right)^{(i-2)/2} \cdot (i - 1)! \qquad i = 2, 3, \ldots$$

Then

$$\kappa_3 = \frac{2\sqrt{2}}{\sqrt{\nu}} \qquad \kappa_4 = \frac{12}{\nu} \qquad \text{etc.}$$

Also, for $i = 3, 4, \ldots,$

$$\lim_{\nu\to\infty} \kappa_i = 0$$

But this is a property of the normal distribution, indicating that the χ^2 distribution approaches the normal distribution for large ν.

Let us consider the distribution of a sum of squares of deviations. Let $X_1, X_2, \ldots, X_n$ be a random sample from $N(\mu,\sigma^2)$. Set

$$U_i^2 = \frac{(X_i - \mu)^2}{\sigma^2}$$

We wish to find the distribution of

$$V = \frac{\Sigma_{i=1}^n (X_i - \mu)^2}{\sigma^2} = \sum_{i=1}^{n} U_i^2$$

Now, since the X's are independent,

$$m_V(t) = m_1(t) \cdot m_2(t) \cdots m_n(t)$$

where $m_i(t)$ denotes the moment-generating function of U_i^2. Hence, since all the X's have the same distribution function,

$$M_V(t) = [m_i(t)]^n$$

Now

$$m_i(t) = \frac{1}{\sqrt{2\pi}} \int_{-\infty}^{\infty} e^{-u^2(1-2t)/2} \, du = (1 - 2t)^{-1/2} \qquad t < \frac{1}{2}$$

Hence,

$$m_V(t) = (1 - 2t)^{-n/2}$$

But this is the moment-generating function of a χ^2 with $\nu = n$ degrees of freedom; hence

$$\frac{\Sigma_{i=1}^{n} (X_i - \mu)^2}{\sigma^2}$$

is distributed as $\chi^2(n)$. In this case each X furnishes a single degree of freedom and each U_i^2 is distributed as $\chi^2(1)$; hence, the number of degrees of freedom is the number of independent values which make up χ^2. The number of degrees of freedom in general is the total number of observations less the number of independent restraints imposed on the observations in forming the distribution.

This χ^2 distribution can be used to test hypotheses and to set up confidence intervals concerning the population variance σ^2 for the normal population with μ known. When μ is unknown, one would use the sample variance S^2 and $(n - 1)S^2/\sigma^2$ is distributed as $\chi^2(n - 1)$. These will be discussed in Chapters 9 and 10. The χ^2 distribution is also used in goodness-of-fit tests and contingency tables which will be given in Chapter 11. A comprehensive treatment of the chi-square distribution is given by Lancaster (1969).

EXAMPLE 6.8 Suppose in Example 6.6 that the true mean of the population was $\mu = 60$. So the estimator of σ^2 is $(1/n)\ \Sigma_{i=1}^{16} (X_i - \mu)^2$. Now $\Sigma_{i=1}^{16} (X_i - \mu)^2/\sigma^2$ is distributed as $\chi^2(16)$. The probability that this estimator is greater than 50 when $\sigma^2 = 100$ is

$$P\left(\frac{1}{n}\ \Sigma(X_i - \mu)^2 > 50\right) = P\left(\frac{\Sigma(X_i - \mu)^2}{\sigma^2} > \frac{16 \times 50}{100}\right)$$

$$= P(\chi^2 > 8) = 0.949$$

Using the relationship

$$V = \frac{\Sigma_{i=1}^{n} (X_i - \mu)^2}{\sigma^2}$$

we may prove that the sum of n variates, each independently distributed as χ^2 with ν_i degrees of freedom, is itself distributed as χ^2 with $\nu = \Sigma_{i=1}^{n} \nu_i$ degrees of freedom. Suppose

$$V_1 = \frac{\Sigma_{i=1}^{\nu_1} (X_{1i} - \mu_1)^2}{\sigma_1^2}$$

and

$$V_2 = \frac{\Sigma_{i=1}^{\nu_1} (X_{2i} - \mu_2)^2}{\sigma_2^2}$$

The distribution of V_i is $\chi^2(\nu_i)$ and V_1 and V_2 are independent. Then $V = V_1 + V_2$ has the moment-generating function

$$m_V(t) = m_{V_1}(t)m_{V_2}(t) = (1 - 2t)^{-\nu_1/2}(1 - 2t)^{-\nu_2/2}$$

$$= (1 - 2t)^{-(\nu_1+\nu_2)/2}$$

But this is the moment-generating function for a chi-square with $\nu_1 + \nu_2$ degrees of freedom; hence V is distributed as $\chi^2(\nu_1 + \nu_2)$. This result may be extended to k variates.

6.10 SIMULTANEOUS DISTRIBUTION OF THE SAMPLE MEAN $\bar{X}$ AND THE SAMPLE VARIANCE S^2: It was shown in Section 6.9 that

$$\frac{\Sigma_{i=1}^{n} (X_i - \mu)^2}{\sigma^2}$$

is distributed as chi-square with $\nu = n$ degrees of freedom. We shall now obtain the simultaneous distribution of two parts of this sum when expressed as

$$\sum_{i=1}^{n} \frac{(X_i - \mu)^2}{\sigma^2} = \sum_{i=1}^{n} \frac{(X_i - \bar{X})^2}{\sigma^2} + \frac{n(\bar{X} - \mu)^2}{\sigma^2}$$

$$= \frac{(n - 1)S^2}{\sigma^2} + \frac{n(\bar{X} - \mu)^2}{\sigma^2}$$

where it will be recalled that $\bar{X}$ is an unbiased estimator of μ and S^2 is an unbiased estimator of σ^2.

In order to simplify the notation in the following derivation, let $U_i = (X_i - \mu)/\sigma$, so that U_i is N(0,1) and $\bar{U} = (\bar{X} - \mu)/\sigma$. Let $W = (n - 1)S^2/\sigma^2$.

The derivation will be accomplished by making use of orthogonal linear forms as discussed in Section 6.4. To this end we set up the following orthogonal linear transformation, known as the Helmert transformation:

$$y_1 = \frac{u_1 + u_2 + \cdots + u_n}{\sqrt{n}}$$

$$y_2 = \frac{u_1 - u_2}{\sqrt{2}}$$

$$y_3 = \frac{u_1 + u_2 - 2u_3}{\sqrt{6}}$$

$$\vdots$$

$$y_n = \frac{u_1 + u_2 + \cdots + u_{n-1} - (n - 1)u_n}{\sqrt{n(n - 1)}}$$

It is easily seen that each y_i is orthogonal to each y_j $(i \neq j)$. Also the denominators have been chosen to make the variances of the y's equal unity. This type of transformation is called *completely orthogonal*. From Section 6.5 we know that the random variables Y's are NID(0,1) where NID indicates normally independently and identically distributed; hence the joint distribution of $Y_1, Y_2, \ldots, Y_n$ is identical with that of $U_1, U_2, \ldots, U_n$ and $\Sigma_{i=1}^{n} Y_i^2 = \Sigma_{i=1}^{n} U_i^2$. Also, since $Y_1 = \sqrt{n}\bar{U}$, then

$$\sum_{i=2}^{n} Y_i^2 = W$$

Since Y_1 is independent of the other Y's, $\bar{U}$ is independent of W. Hence

$$f(\bar{u},w) = f_1(\bar{u})f_2(w)$$

We know that

$$f_1(\bar{u}) = \sqrt{\frac{n}{2\pi}}\, e^{-n\bar{u}^2/2} \tag{6.8}$$

It is also known that

$$\sum_{i=2}^{n} Y_i^2$$

is distributed as chi-square with n - 1 degrees of freedom since there are n - 1 independent normal variates. Hence

$$f_2(w) = \left[2^{(n-1)/2}\Gamma\left(\frac{n-1}{2}\right)\right]^{-1} w^{(n-3)/2}e^{-w/2} \tag{6.9}$$

Using the original X's, which were $N(\mu,\sigma^2)$, the joint density of $\bar{X}$ and S^2 is

$$f(\bar{x},s^2) = \frac{\sqrt{n/2\pi}\;[(n-1)/2]^{(n-1)/2}}{\sigma^3\Gamma[(n-1)/2]} \times \left(\frac{s^2}{\sigma^2}\right)^{(n-3)/2} \exp\left[-\frac{n(\bar{x}-\mu)^2 + (n-1)s^2}{2\sigma^2}\right]$$

$$-\infty < \bar{x} < \infty \qquad 0 < s^2 < \infty \tag{6.10}$$

6.11 DISTRIBUTION OF t: It is now proposed to obtain the distribution of a test criterion, independent of σ^2, to be used in testing the hypothesis that a sample $X_1, X_2, \ldots, X_n$ was drawn from a normal population with a specified mean μ. The distribution must be independent of σ^2 if it is to be of practical use, since we seldom know the value of σ^2.

One proposed test criterion of this hypothesis is

$$T = \frac{\bar{X} - \mu}{S/\sqrt{n}} \tag{6.11}$$

where $\bar{X}$ and S^2 are the sample mean and variance from a random sample of n from the parent population $N(\mu,\sigma^2)$. Two properties of T should be noted: (1) T is the ratio between a normal deviate and the square root of an unbiased estimate of its variance; (2) T is the ratio between a variate, $Z = \sqrt{n}(\bar{X} - \mu)/\sigma$, which is N(0,1), and the square root of a variate S^2/σ^2, which is independently distributed as χ^2/ν with ν degrees of freedom. Note that σ^2 cancels in the derivation of T. We shall take (2) as the definition of T.

The joint density of $\bar{X}$ and S^2 is given in (6.10). Making the transformation $t = \sqrt{n}(\bar{x} - \mu)/s$ and $w = (n - 1)s^2/\sigma^2$ and using the method in Section 5.5, we obtain the Jacobian

$$J = \frac{\sigma^2}{n - 1}\sqrt{\frac{w}{n(n - 1)}}$$

and

$$f(t,w) = \frac{1}{\sqrt{\pi(n - 1)}\,2^{n/2}\Gamma[(n - 1)/2]}\, w^{(n/2) - 1} e^{-(w/2)[1+t^2/(n-1)]}$$

$$-\infty < t < \infty \qquad 0 < w < \infty$$

Letting $u = (w/2)[1 + t^2/(n - 1)]$, we have

$$f(t,u) = \frac{1}{\sqrt{\pi(n - 1)}\;\Gamma[(n - 1)/2]}\left(1 + \frac{t^2}{n - 1}\right)^{-n/2} u^{(n/2)-1} e^{-u}$$

$$-\infty < t < \infty \qquad 0 < u < \infty$$

Integrating out u, we obtain the density function of T:

$$f_{n-1}(t) = \frac{\Gamma(n/2)}{\sqrt{\pi(n-1)}\ \Gamma[(n-1)/2]}\left(1 + \frac{t^2}{n-1}\right)^{-n/2} \qquad -\infty < t < \infty$$

Since S^2 has n - 1 degrees of freedom, $f_{n-1}(t)$ is designated as the distribution of t for n - 1 degrees of freedom. The density function for n degrees of freedom is

$$f_n(t) = \frac{\Gamma[(n+1)/2]}{\sqrt{\pi n}\ \Gamma(n/2)}\left(1 + \frac{t^2}{n}\right)^{-(n+1)/2} \qquad -\infty < t < \infty \qquad (6.12)$$

The t distribution with n degrees of freedom is denoted by t(n). The t distribution is symmetric about zero. When n goes to infinity, the t distribution tends to the standard normal distribution. It should be noted that

$$P\left(|t| \geq t_{\alpha/2}\right) = 2\int_{t_{\alpha/2}}^{\infty} f_n(t)\ dt = \alpha$$

Values of $t_{\alpha/2}$ are given for various values of α and various degrees of freedom in Table C.3 (Appendix C). For the location of other tables and methods of obtaining exact values, see Bancroft (1950). Note that the tabular values are for a two-tailed test. If it is desired to use $P(T > t_0)$, the tabular probabilities must be divided by 2.

6.12 DISTRIBUTION OF F: Another very useful statistic in making tests of hypotheses involves the ratio of two chi-squares. Let Y_1 and Y_2 be independently distributed as $\chi^2(n_1)$ and $\chi^2(n_2)$, respectively. We wish to find the distribution of $F = n_2Y_1/n_1Y_2$. The joint density function of Y_1 and Y_2 is

$$f(y_1,y_2) = \frac{1}{4\Gamma(n_1/2)\Gamma(n_2/2)}\left(\frac{y_1}{2}\right)^{(n_1/2)-1}\left(\frac{y_2}{2}\right)^{(n_2/2)-1} e^{-(y_1+y_2)/2}$$

Let

$$f = \frac{n_2 y_1}{n_1 y_2} \quad \text{or} \quad y_1 = \frac{n_1 f y_2}{n_2}$$

Then

$$g(f, y_2) = \frac{(n_1/n_2)^{n_1/2}}{2\Gamma(n_1/2)\Gamma(n_2/2)} f^{(n_1/2) - 1} \left(\frac{y_2}{2}\right)^{(n_1+n_2-2)/2}$$
$$\times e^{-y_2[1+(n_1/n_2)f]/2}$$

and

$$g(f) = \int_0^\infty g(f, y_2)\, dy_2$$
$$= \frac{\Gamma[(n_1+n_2)/2]}{\Gamma(n_1/2)(n_2/2)} \left(\frac{n_1}{n_2}\right)^{n_1/2}$$
$$\times f^{(n_1/2)-1} \left(1 + \frac{n_1}{n_2} f\right)^{-(n_1+n_2)/2} \qquad 0 < f < \infty \tag{6.13}$$

The F distribution with n_1 degrees of freedom in the numerator and n_2 degrees of freedom in the denominator is denoted by $F(n_1, n_2)$. This distribution is also known as the Snedecor's F distribution, as Snedecor (1937) has tabulated the distribution. Let us consider that

$$Y_1 = \frac{n_1 S_1^2}{\sigma_1^2} \quad \text{and} \quad Y_2 = \frac{n_2 S_2^2}{\sigma_2^2}$$

where S_1^2 and S_2^2 are two independent sample estimators of σ_1^2 and σ_2^2, based on n_1 and n_2 degrees of freedom respectively; then

$$F = \frac{S_1^2}{S_2^2} \cdot \frac{\sigma_2^2}{\sigma_1^2}$$

This function provides a test of the hypothesis $\sigma_1^2 = \sigma_2^2 = \sigma^2$. Then

$$F = \frac{s_1^2}{s_2^2} \tag{6.14}$$

which is the ratio of the independent sample estimators of two assumed equal variances.

Tabular values of $F_{.05}$ and $F_{.01}$ are represented in Table C.4 for various values of n_1 and n_2. Norton has computed values of $F_{.20}$ and Colcord and Deming of $F_{.001}$. All of these tables may be found in Fisher and Yates (1963). See Bancroft (1950) for a complete description of where such tables can be found. Further, Dixon and Massey (1969) compiled a rather extensive table for the F distribution.

The computations of many distribution functions studied in this chapter, such as the standard normal distribution, t distribution, chi-square distribution, and F distribution, are available on computer. Subroutines are written for researcher's use, for example, those given in the International Mathematical and Statistical Library.

6.13 DISTRIBUTIONS OF ORDER STATISTICS: In previous sections sampling distributions are mostly related to random samples from normal parent populations. This section will discuss the sampling distributions of order statistics which do not depend on any particular form of the parent population. Our treatment will be very brief. The readers are referred to David (1981) for detailed discussion.

Let $X_1, X_2, \ldots, X_n$ be a random sample of size n from a distribution with a cumulative distribution function F(x). If the sample is ordered in magnitude from the smallest to the largest, then $X_{(1)} \leq X_{(2)} \leq \cdots \leq X_{(n)}$ are called the order statistics of the sample. The kth order statistic is $X_{(k)}$, $k = 1, \ldots, n$. Note that the subscripts for order statistics are in parentheses and $X_{(k)}$ is not necessarily equal to X_k. Let us consider that the parent population of X is of the continuous type with density function f(x). The joint density function of $X_{(1)}, X_{(2)}, \ldots, X_{(n)}$ is

$$g(x_{(1)}, x_{(2)}, \ldots, x_{(n)}) = n! \prod_{i=1}^{n} f(x_{(i)})$$
$$-\infty < x_{(1)} < x_{(2)} < \cdots < x_{(n)} < \infty \tag{6.15}$$

since the transformation from $X_1, X_2, \ldots, X_n$ to $X_{(1)}, X_{(2)}, \ldots, X_{(n)}$ is 1 to n! The distribution of the kth order statistic may be obtained by integrating out the other order statistics from the joint density function, but a simpler way to find the distribution is by the following argument. The cumulative distribution function of $X_{(k)}$, $k = 1, 2, \ldots, n$, is

$$\begin{aligned} G(x_{(k)}) &= P(X_{(k)} \leq x_{(k)}) \\ &= P(\text{at least } k \text{ of the X values} \leq x_{(k)}) \\ &= \sum_{j=k}^{n} \binom{n}{j} [F(x_{(k)})]^j [1 - F(x_{(k)})]^{n-j} \end{aligned} \tag{6.16}$$

because the number of observations whose values are less than or equal to $x_{(k)}$ follows a binomial distribution with $p = F(x_{(k)})$. Differentiating $G(x_{(k)})$ with respect to $x_{(k)}$ we obtain the density function

$$\begin{aligned} g(x_{(k)}) &= \sum_{j=k}^{n} \binom{n}{j} j\, [F(x_{(k)})]^{j-1} [1 - F(x_{(k)})]^{n-j} f(x_{(k)}) \\ &\quad - \sum_{j=k}^{n} \binom{n}{j} (n - j)\, [F(x_{(k)})]^{j} [1 - F(x_{(k)})]^{n-j-1} f(x_{(k)}) \\ &= n \binom{n-1}{k-1} [F(x_{(k)})]^{k-1} [1 - F(x_{(k)})]^{n-k} f(x_{(k)}) \end{aligned} \tag{6.17}$$

after canceling terms.

Equations (6.16) and (6.17) provide the distribution of an order statistics of a random sample from any continuous distribution. In particular, when $k = 1$, we obtain the distribution of the smallest observation; when $k = n$, the distribution of the largest observation; when n is odd and $k = (n + 1)/2$, we obtain the distribution of the sample *median* $M = X_{((n+1)/2)}$ [the sample median for even sample size is defined to be the average of $X_{(n/2)}$ and $X_{((n/2)+1)}$].

EXAMPLE 6.9 Let $X_1, X_2, \ldots, X_n$ be a random sample from the exponential distribution with density function

$$f(x) = \begin{cases} e^{-x} & x > 0 \\ 0 & x \leq 0 \end{cases}$$

The corresponding cumulative distribution function is

$$F(x) = \begin{cases} 1 - e^{-x} & x > 0 \\ 0 & x \leq 0 \end{cases}$$

Hence the joint density function of the order statistics $X_{(1)}, X_{(2)}, \ldots, X_{(n)}$ is

$$g(x_{(1)}, x_{(2)}, \ldots, x_{(n)}) = n!e^{-\Sigma_{i=1}^{n} (x_{(i)})} \qquad 0 < x_{(1)} < x_{(2)} < \cdots < x_{(n)}$$

The density function of the kth order statistic is

$$g(x_{(k)}) = n\binom{n-1}{k-1}\left[1 - e^{-x_{(k)}}\right]^{k-1}\left[e^{-x_{(k)}}\right]^{n-k+1} \qquad 0 < x_{(k)} < \infty$$

EXERCISES

6.1 Determine a fifth linear form orthogonal to the forms in Examples 6.4 and 6.5. How many orthogonal linear forms make a complete set for a sample of n?

6.2 Given a random sample of size N, all members being drawn from the same population, determine the relationship between the cumulants for the total of the sample of the cumulants for a single member of the sample. *Hint:* $m_{\Sigma X}(t) = [m(t)]^N$.

6.3 Use the results of Exercise 6.2 to determine the first three cumulants for the total of N from a binomial distribution. (N represents the number of samples and n the number of independent trials per sample.)

6.4 What happens to Exercise 6.2 if each member of the random sample is drawn from different populations? *Hint:* Show that

$$K_{\Sigma X}(t) = \sum_{i=1}^{n} K_i(t)$$

and the rth cumulant for the sum is hence

$$\kappa_r = \sum_{i=1}^{N} \kappa_{ri}$$

Consider this problem for Poisson distributions with unlike means and binomial distributions with unlike values of n and p.

6.5 Given two independent estimators of θ, X_1 and X_2, with variances σ_1^2 and σ_2^2, respectively. If we desire to estimate θ as an unbiased linear function of the X's, find the coefficients of the X's which will minimize the variance of the estimator. *Hint:* $E(X_i) = \theta$. Let $\hat{\theta} = aX_1 + bX_2$.

6.6 If the X_i in the linear form $L = \Sigma a_i X_i$ are normally and independently distributed with means μ_i and variances σ_i^2 and all $a_i = 1/n$, $\mu_i = \mu$, and $\sigma_i = \sigma$, then $L = \bar{X}$. Show that the mean of a random sample from a normal distribution is normally distributed with mean μ and variance σ^2/n.

6.7 Suppose that $\bar{X}_1$ represents the mean of a random sample of size n_1 taken from a normal population with mean μ_1 and variance σ_1^2 and $\bar{X}_2$ the mean of a random sample of size n_2 taken from another normal population with mean μ_2 and variance σ_2^2. Let $L = \bar{X}_1 - \bar{X}_2$. Show that L is distributed as $N(\mu_1 - \mu_2, (\sigma_1^2/n_1) + (\sigma_2^2/n_2))$.

6.8 Consider the following applied problem: The effect of a nitrogen top-dressing on a crop is to be determined. In addition it is desired to discover at what time during the growing season the top-dressing, if beneficial, should be applied. The experiment was designed as follows: Use four plots, one with no nitrogen, one with nitrogen applied early, one applied in the middle of the growing season, and one applied later. Naturally, this entire experiment would be repeated several times, say r, in order to discover how consistent the differences were from replication to replication. The totals of the r plots are indicated as follows:

	Nitrogen Applied		
No N	Early	Middle	Late
T_0	T_1	T_2	T_3

$$T_0 = X_{01} + X_{02} + \cdots + X_{0r} \qquad T_1 = \sum_{j=1}^{r} X_{ij} \qquad \text{etc.}$$

Suppose X_{ij} is $N(\mu_i, \sigma^2)$, where μ_i is the mean effect of the ith treatment. Three pertinent independent linear forms are: $L_1 = 3T_0 - T_1 - T_2 - T_3$, for determining the effect of nitrogen; $L_2 = T_3 - T_1$, for determining the linear effect; $L_3 = T_1 - 2T_2 + T_3$, for determining the quadratic effect. Find a fourth linear form to complete the set, and determine the variance of each.

6.9 (a) Set up the joint probability distribution of the random sample of size n, $(X_{1i}, X_{2i})'$, i = 1, ..., n, from a bivariate normal distribution with means μ_1 and μ_2, variances σ_1^2 and σ_2^2, and correlation ρ. (b) Show that the joint density function of $\bar{X}_1$ and $\bar{X}_2$ is

$$\frac{n}{2\pi\sigma_1\sigma_2\sqrt{1 - \rho^2}} e^{-\theta}$$

where

$$\theta = \frac{n}{2(1 - \rho^2)}\left[\frac{(\bar{x}_1 - \mu_1)^2}{\sigma_1^2} + \frac{(\bar{x}_2 - \mu_2)^2}{\sigma_2^2} - \frac{2\rho(\bar{x}_1 - \mu_1)(\bar{x}_2 - \mu_2)}{\sigma_1\sigma_2}\right]$$

$$-\infty < \bar{x}_1 < \infty \qquad \infty < \bar{x}_2 < \infty$$

6.10 Given a normal population with $\sigma^2 = 25$, a random sample of 12 produced the following values: 24, 30, 26, 33, 21, 24, 20, 32, 24, 30, 33, 34. The observed sample mean is 27.58. Assuming that the population mean is 24, find $P(\bar{X} > 27.58)$.

6.11 Letting the random variable X have a χ^2 distribution with ν degrees of freedom, find the distribution of

$$Y = \frac{X}{1 + X}$$

6.12 Given that $X_1 \sim \chi^2(\nu_1)$ and $X_2 \sim \chi^2(\nu_2)$ and the two functions

$$U = a_1X_1 + a_2X_2$$

$$V = X_1 + X_2$$

with X_1 and X_2 independently distributed, find the distribution of $W = U/V$. What effect would the provision $a_1 > a_2 > 0$ have on the results?

6.13 If X_1 and X_2 are NID(0,1), show that the variables r and θ, defined as follows,

$$X_1 = r \sin \theta \qquad X_2 = r \cos \theta$$

are independently distributed and that r^2 is distributed as χ^2 with 2 degrees of freedom.

6.14 It is shown in Example 6.7 for large n that the binomial distribution may be approximated by the normal distribution with $\mu = np$ and $\sigma^2 = npq$. If 20 coins are tossed, obtain the approximate probability of obtaining 8 or more heads.

6.15 Given the bivariate normal distribution with means zero and variances σ_1^2 and σ_2^2 and correlation ρ, show that

$$U = \frac{1}{1 - \rho^2}\left(\frac{X_1^2}{\sigma_1^2} + \frac{X_2^2}{\sigma_2^2} - \frac{2\rho X_1X_2}{\sigma_1\sigma_2}\right)$$

is distributed as chi-square with 2 degrees of freedom. *Hint:* Show that the moment-generating function of U is the same as that of χ^2 with 2 degrees of freedom.

6.16 Using the result in Exercise 6.15 find the probability that $U \leq c^2$.

6.17 In Exercise 6.15 let $\sigma_1^2 = \sigma_2^2 = 1$ and $\rho = 0$. Find the equation of the circle with center at the origin which would contain 99% of a large number of samples of X_1 and X_2.

6.18 A random sample of size n is drawn from a N(0,1) population. Suppose the sample is subdivided into r subclasses with two or more in each ($n_1, n_2, \ldots, n_r$ being the numbers in the subclasses). The sum of squares $Q_i (i = 1, 2, \ldots, r)$ of deviations from the subclass mean is computed for each subclass. What is the distribution of $\Sigma_{i=1}^{r} Q_i$?

6.19 Compute the 1% and 5% tabular values of χ^2 for $\nu = 1, 2$, and 4. Check with the results in Table C.2 (Appendix C).

6.20 Given $S^2/\sigma^2 = 2$ with $n = 5$, what is the probability of obtaining a value this large or larger? Check your results by interpolating with the tabled values.

6.21 Use the data in Exercise 6.10 and assume that $\mu = 30$ and $\sigma^2 = 25$. Check that $\Sigma(x_i - \mu)^2 = 343$ and compute $P(\Sigma(X_i - \mu)^2 > 343)$.

6.22 Repeat Exercise 6.21 when no assumption is made concerning μ. Check that the sample variance is equal to 24.81 and compute $P(S^2 > 24.81)$.

6.23 Show that the distribution of $S^2 = \Sigma_{i=1}^{n} (X_i - \bar{X})/(n - 1)$ approaches normality for large n. What are the skewness and kurtosis of S^2?

6.24 Find the moment-generating function of log S, where S^2 is the usual unbiased estimator of variance. Show that the distribution of log S is independent of σ^2 apart from its mean value.

6.25 The t distribution with 1 degree of freedom is the so-called Cauchy distribution. What about its mean and variance?

6.26 Show that the t distribution approaches normality for large n.

6.27 The distribution of the correlation coefficient r in samples of n from a normal bivariate population with $\rho = 0$ is

$$f(r) = \begin{cases} \dfrac{1}{B((n-2)/2,\ 1/2)} (1 - r^2)^{(n-4)/2} & -1 < r < 1 \\ 0 & \text{elsewhere} \end{cases}$$

Show that

$$T = r\sqrt{\frac{n-2}{1-r^2}}$$

is distributed as t with n - 2 degrees of freedom.

6.28 In Exercise 6.27 (a) compute the 95% point of T for 2 degrees of freedom. (b) What is the probability that $|T| > 3$?

6.29 In a certain test, given to the 45 members of a class in statistics, 20 women students had an average score of 40 with a variance of 16, while 25 men students had an average of 46 with a variance of 16. What is the probability of obtaining such results if both groups were equally well prepared for the test? What assumptions are made in obtaining this probability value? On the basis of this sample and the validity of the assumptions is there much evidence that both groups are not equally well prepared?

6.30 Show that $\int_0^\infty g(f)\, df = 1$ for the F distribution.

6.31 When $n_1 = 1$, show that $F = t^2$.

6.32 Let F have the $F(n_1,n_2)$ distribution with n_1 degrees of freedom for the numerator and n_2 for the denominator. (a) Derive the distribution of $F' = 1/F$ and show that F' is distributed as $F(n_2,n_1)$. (b) Hence show that $P(F' < 1/c) = P(F > c)$.

6.33 Show that $P(F > F_\alpha)$ is an incomplete beta function. Set up this function. *Hint:* Let

$$X = \frac{n_1F/n_2}{1 + n_1F/n_2}$$

6.34 Use the result in Exercise 6.33 to determine a general formula for μ_i of F.

6.35 Determine the 5% tabular value for F if (a) $n_1 = 2$, $n_2 = 4$, and (b) $n_1 = 4$, $n_2 = 4$.

6.36 What is $P(F > 4)$ for $n_1 = 4$, $n_2 = 4$?

6.37 Given $s_1^2 = 40$ and $s_2^2 = 10$, each based on 4 degrees of freedom. Determine the probability that sample variances as divergent as these could be estimates of the sample population variance when $\sigma_1^2 = \sigma_2^2$.

6.38 Given that one mean square, 80, with 4 degrees of freedom, is an estimate of $\sigma_2^2 + 5\sigma_1^2$, while another mean square, 10, with 20 degrees of freedom, is an estimate of σ_2^2. The ratio of the two mean squares has an F distribtuion when $\sigma_1^2 = 0$. What is the probability that the ratio exceeds 80/10?

6.39 R. A. Fisher first considered the problem of the ratio of two variances as the difference between the logarithms of the two variances. If we let $Z = (1/2) \log F$, we obtain his z distribution, which is more nearly normal than the F distribution. Show that

$$f(z) = 2ke^{n_1 z}\left(1 + \frac{n_1}{n_2} e^{2z}\right)^{-(n_1+n_2)/2} \qquad -\infty < z < \infty$$

where

$$k = \frac{(n_1/n_2)^{n_1/2}}{B(n_1/2,\ n_2/2)}$$

6.40 Show that $f(z)$ is symmetrical if $n_1 = n_2$.

6.41 Let $X_{(1)}, X_{(2)}, \ldots, X_{(n)}$ be the order statistics for a random sample of size n from an arbitrary population. Define $Y = X_{(1)}$ and $Z = X_{(n)}$, show that the joint cumulative distribution function of Y and Z is

$$G(y,z) = \begin{cases} [F(z)]^n - [F(z) - F(y)]^n & \text{if } y \le z \\ [F(z)]^n & \text{if } y > z \end{cases}$$

Hence show that the joint density function is

$$g(y,z) = \begin{cases} n(n-1)[F(z) - F(y)]^{n-2} f(z)f(y) & \text{if } y \le z \\ 0 & \text{if } y > z \end{cases}$$

6.42 In Exercise 6.41, if the random sample is taken from the uniform distribution with $f(x) = 1/\theta$, $0 < x < \theta$, find $G(y,z)$ and $g(y,z)$. Find the density function of the *sample range* $R = X_{(n)} - X_{(i)} = Z - Y$.

REFERENCES

Bancroft, T. A. (1950). Probability values for the common tests of hypotheses. J. Amer. Stat. Assoc. 45:211-217.

David, H. A. (1981). *Order Statistics*, 2nd ed. Wiley, New York.

Dixon, W. J., and Massey, F. J. (1969). *Introduction to Statistical Analysis*, 3rd ed. McGraw-Hill, New York.

Fisher, R. A. (1974). *Design of Experiments*, 9th ed. Oliver & Boyd, Edinburgh.

Fisher, R. A., and Yates, F. (1963). *Statistical Tables for Biological, Agricultural, and Medical Research*, 6th ed. Oliver & Boyd, Edinburgh.

Lancaster, H. O. (1969). *The Chi-Squared Distribution*. Wiley, New York.

Neyman, J. (1935). On the problem of confidence intervals. Ann. Math. Stat. 6:111-116.

Pearson, K. (1922). *Tables of the Incomplete Γ Function*. H. M. Stationery Office, London.

Snedecor, G. W. (1937). *Statistical Methods*. Iowa State University Press, Ames.

Thompson, C. M. (1941). Tables of percentage points of the incomplete beta function and the chi-squared distribution. Biometrika 32: 151-181, 187-191.

7

POINT ESTIMATION

7.1 INTRODUCTION: A worker in animal husbandry may wish to estimate the mean gain in weight of swine of the same breed, sex, and age, fed on the same ration, and managed similarly for a period of 20 days. To this end the following sample set of gains in pounds was obtained: 27, 28, 28, 29, 29, 29, 30, 30, 30, 30, 31, 31, 31, 32, 32, 33. Now, if it is assumed that these observations are a random sample from a population of such gains which are normally distributed with mean μ, then we must decide what function of the sample observations must be used to obtain a "good" or "best" estimate of μ. We must, of course, define "good" or "best" in some technical sense. As will be shown later, it turns out that the sample mean, $\bar{X} = 30$, is a "good" estimate of μ. Our investigation concerns itself, then, with drawing a sample from some population of specified mathematical form for the purpose of estimating a parameter of the population. In subsequent discussions the function of the observations chosen to estimate the population parameter, in this case $\bar{X}$, will be called an *estimator*, while the particular numerical value obtained in an application will be called an *estimate*.

In Sections 7.2 to 7.8, we discuss properties of estimators and criteria for judging the goodness of estimators. Methods of finding "good" estimators are given in Sections 7.9 to 7.14. The last section (Section 7.15) discusses robust estimation.

7.2 PROBLEM OF POINT ESTIMATION: A sample $X_1, X_2, \ldots, X_n$ is specified as having been drawn from a common population with density function $f(x; \theta_1, \theta_2, \ldots, \theta_k)$, where X is the variate and $\theta_1, \theta_2, \ldots, \theta_k$ are population parameters. We wish to find functions of the observations, say $\hat{\theta}_1(X_1,X_2,\ldots,X_n)$, $\hat{\theta}_2(X_1,X_2,\ldots,X_n)$, ..., $\hat{\theta}_k(X_1,X_2,\ldots,X_n)$, such that the distribution of these functions will be concentrated closely about the respective true values of the parameters. By saying that the distribution of $\hat{\theta}_i(X_1,X_2,\ldots,X_n)$ will be concentrated closely about the true value we can mean one of several conditions, such as the following:

a. The probability that the estimator falls within a short distance of the true value shall be large regardless of the fact that this requirement may be satisfied only by an estimator which is distributed in such a way that there is a possibility (though small) of a very large deviation from the parameter.

b. The probability that the estimator falls more than a specified distance from the parameter shall be negligible, or even zero, regardless of how the estimator is distributed inside this region.

c. Or we may be willing to have the estimator deviate in one direction from the parameter but may wish to minimize the chance of large deviations in the other direction.

These three conditions may be represented by estimators which are distributed as shown in Fig. 7.1 (θ is the parameter).

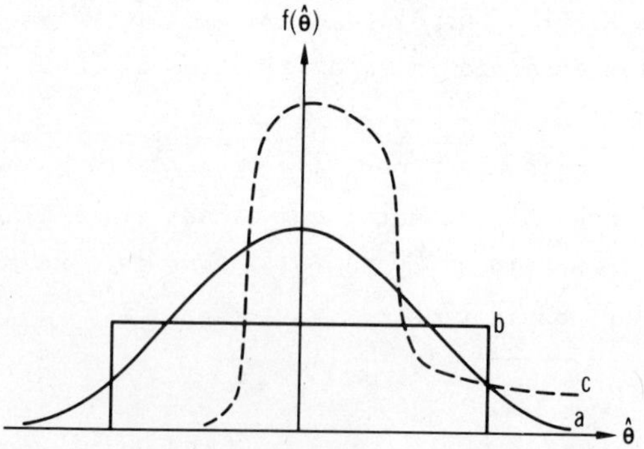

FIG. 7.1 Distributions of estimators.

There are other considerations which might influence one in deciding whether or not a given estimator is a "good" one. For example, is the estimator appropriate for small samples as well as large? It also might be possible to set down a cost of a given deviation of an estimator from the parameter, the cost presumably being an increasing function of the size of the deviation. Then, if we could determine this cost function for various proposed estimators, it would be possible to estimate the total cost for each and adopt that estimator which produced a minimum cost.

Most of these problems of estimation are resolved if all the estimators are normally distributed. In this case, the best estimator can reasonably be assumed to be that one which has the minimum variance. However, it is difficult to say which estimator is superior if one is distributed normally and the other uniformly, for example. Although most common estimators are asymptotically normally distributed (n large), few are normal for small samples and many are far from normal in this case. On the whole we can advance only a few guiding principles to be used in deciding whether an estimator is "good" or not. While these principles have yielded fruitful results, much work remains to be done, especially regarding nonnormally distributed estimators and nonrandom samples.

In the following sections we shall define certain characteristics of a "good" estimator and introduce a method which sometimes yields an estimator which satisfies all of these characteristics. Much of the philosophy and theory of estimation are a result of the work of R. A. Fisher.

7.3 UNBIASEDNESS: An estimator or statistic is a function of the random variables and it has a certain distribution. It is desirable that the mean value of this distribution be equal to the true value θ.

Definition. An estimator $\hat{\theta}$ is said to be unbiased if $E(\hat{\theta}) = \theta$. It is said to be positively or negatively biased, respectively, according as to whether $E(\hat{\theta}) > \theta$ or $E(\hat{\theta}) < \theta$.

Unbiasedness is a desirable but not necessarily an indispensable property of a "good" estimator. If the amount of bias is small compared with the standard deviation of the estimator (say less than 10%), the estimator, though biased, may be entirely satisfactory.

7.4 CONSISTENCY: When the sample size increases, a good estimator should possess the property that it moves closer and closer to the true value θ. This is the property of consistency.

Definition. An estimator $\hat{\theta}$ is said to be a consistent estimator of θ if $\hat{\theta}$ converges stochastically to θ as n approaches ∞. Symbolically, $\hat{\theta}$ will converge stochastically to θ as n approaches ∞, if for two arbitrarily small positive numbers ε and η, a large enough sample can be taken so that

$$P(|\theta - \hat{\theta}| \geq \varepsilon) \leq \eta$$

A useful relation for proving consistency is furnished by *Tchebysheff's inequality:* Given a random variable X with density function f(x), mean μ, and variance σ^2 assumed to exist, then for a given δ (> 0),

$$P(|X - \mu| \geq \delta\sigma) \leq \frac{1}{\delta^2} \tag{7.1}$$

The *inequality* may be established from Fig. 7.2. It follows that

$$\sigma^2 = \int_{-\infty}^{\infty} (x - \mu)^2 f(x)\, dx$$

$$= \int_{-\infty}^{\mu-\delta\sigma} (x - \mu)^2 f(x)\, dx + \int_{\mu-\delta\sigma}^{\mu+\delta\sigma} (x - \mu)^2 f(x)\, dx$$

$$+ \int_{\mu+\delta\sigma}^{\infty} (x - \mu)^2 f(x)\, dx$$

Also, since the second integral is nonnegative,

$$\sigma^2 \geq \int_{-\infty}^{\mu-\delta\sigma} (x - \mu)^2 f(x)\, dx + \int_{\mu+\delta\sigma}^{\infty} (x - \mu)^2 f(x)\, dx$$

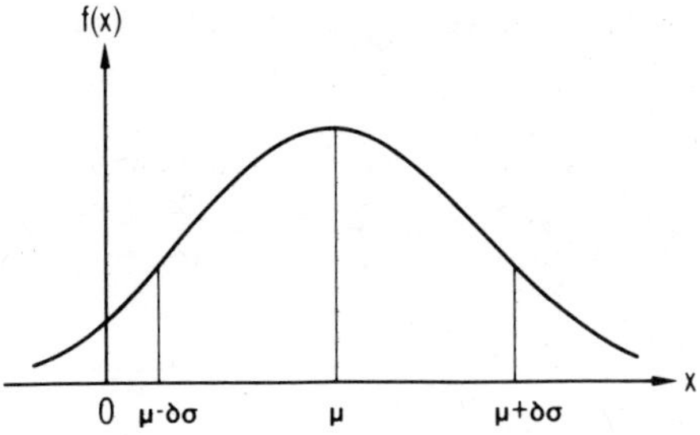

FIG. 7.2

Since, in the range of integration, $|x - \mu| \geq \delta\sigma$, the factor $(x - \mu)^2$ may be replaced by $\delta^2\sigma^2$ and

$$\sigma^2 \geq \delta^2\sigma^2 \int_{-\infty}^{\mu-\delta\sigma} f(x)\ dx + \delta^2\sigma^2 \int_{\mu+\delta\sigma}^{\infty} f(x)\ dx$$

It is now evident that

$$\frac{1}{\delta^2} \geq P(|X - \mu| \geq \delta\sigma)$$

or

$$P(|X - \mu| \geq \delta\sigma) \leq \frac{1}{\delta^2}$$

It should be noted that this inequality holds for any distribution but is not very efficient for a normal distribution. For example, if X is distributed as $N(\mu,\sigma^2)$, then the exact probability α for $\delta = 2$ is .046, while Tchebysheff's inequality gives $\alpha \leq .25$.

In the limit a consistent estimator will necessarily be unbiased although for finite sample sizes the consistent estimator may be biased. An unbiased estimator may or may not be consistent. It can be proved (Cramer, 1946) that an unbiased estimator will be consistent if $\sigma_{\hat{\theta}}^2 \to 0$ as $n \to \infty$. Note that consistency is a large sample criterion, while bias is applied to small samples as well.

There are usually many consistent estimators of θ; hence, other criteria are needed to select the "best" from among those with the property of consistency. In general, consistency is a desirable property of an estimator.

EXAMPLE 7.1 Since $E(\bar{X}) = \mu$, the sample mean $\bar{X}$ is an unbiased estimator of the population mean μ. Also, if σ^2 exists, $\sigma_{\bar{X}}^2 = \sigma^2/n$, and then, by Tchebysheff's inequality,

$$P\left(|\bar{X} - \mu| \geq \frac{\delta\sigma}{\sqrt{n}}\right) \leq \frac{1}{\delta^2}$$

Now, if we set $\varepsilon = \delta\sigma/\sqrt{n}$ and $\eta = 1/\delta^2$, then $\bar{X}$ will converge stochastically to μ if we choose $n \geq \delta^2\sigma^2/\varepsilon^2$, i.e., the probability that the sample mean differs from the population mean can be made arbitrarily small if we select a large enough sample. Hence $\bar{X}$ is a consistent estimator for μ.

EXAMPLE 7.2 It can be shown that the distribution of the median M for a random sample of n from $N(\mu,\sigma^2)$ approaches $N(\mu,\sigma_M^2)$ for n large, with $\sigma_M^2 = \pi\sigma^2/2n$. Hence, for large n, M may be taken as unbiased and is also a consistent estimator of μ for normally distributed data. [It can be shown in general that $\sigma_M^2 = 1/4nf^2(m)$ if $f(m) \neq 0$. σ_M^2 is the asymptotic variance.]

7.5 *EFFICIENCY:*

Since there are usually many consistent estimators of a given parameter θ, we require some additional criterion to use in deciding which of these consistent estimators is the "best." Also, there might be cases in which it is desirable to use an estimator for which the limiting value deviates by a small amount from the parameter, that is, an inconsistent estimator, if such an estimator were superior in other respects. Another criterion advanced by R. A. Fisher is that the estimator shall have a minimum variance in large samples. An estimator possessing this property is said to be *efficient*. (Efficiency for small samples will be discussed in Section 7.8.)

Definition. $\hat{\theta}$ is an efficient estimator of θ if (1) $\sqrt{n}(\hat{\theta} - \theta)$ approaches $N(0,\sigma^2)$ as $n \to \infty$; (2) for any other estimator θ' for which $\sqrt{n}(\theta' - \theta)$ approaches $N(0,\sigma'^2)$, $\sigma'^2 \geq \sigma^2$.

The efficiency of θ' is given by $e = \sigma^2/\sigma'^2$. We shall consider later an estimation method which provides an estimator with minimum variance.

EXAMPLE 7.3 Consider two consistent estimators of μ furnished by a random sample of n taken from $N(\mu,\sigma^2)$, that is, the mean $\bar{X}$ and the median M. We know that $\sigma_{\bar{X}}^2 = \sigma^2/n$ and also for large samples that $\sigma_M^2 = \pi\sigma^2/2n$. The efficiency of the median relative to the mean is given by

$$e = \frac{\sigma^2/n}{\pi\sigma^2/2n} = .64$$

EXAMPLE 7.4 Consider the midrange (MR) as an estimator of μ. MR is the average of the largest X and the smallest X in the sample. It can be shown that if the n members of the sample $\{X_i\}$ are $N(\mu,\sigma^2)$, the $E(MR) = \mu$ and

$$\sigma_{MR}^2 = \frac{\pi^2\sigma^2}{24 \log n} + 0\left(\frac{1}{\log^2 n}\right)$$

Hence, the efficiency of MR using the first term only of σ_{MR}^2 relatively to the mean $\bar{X}$ is given by

$$e = \frac{\sigma^2/n}{\pi^2\sigma^2/24 \log n}$$

But $\lim_{n\to\infty} e = 0$. Hence, although MR is an unbiased and consistent estimator for μ, it is decidedly *inefficient* in large samples as compared with the mean $\bar{X}$.

EXAMPLE 7.5 Consider the efficiencies of $\bar{X}$, M, and MR as estimators of μ from a sample of n from the rectangular distribution: $f(x) = 1/\theta$, $0 < x < \theta$, $\theta < \infty$, $\mu = \theta/2$, and $\sigma^2 = \theta^2/12$. All three estimators are unbiased. The variances are

$$\sigma_{\bar{X}}^2 = \frac{\sigma^2}{n} = \frac{\theta^2}{12n} \qquad \sigma_M^2 = \frac{3\sigma^2}{n} \qquad \sigma_{MR}^2 = \frac{6\sigma^2}{(n + 1)(n + 2)}$$

Hence each estimator is consistent. If MR were asymptotically normally distributed, we could infer that the efficiency of the mean or the median relative to midrange would be zero. However, MR is not so distributed; hence, we cannot use the criterion of efficiency in making such a comparison. The efficiency of M relative to $\bar{X}$ is 1/3.

7.6 SUFFICIENCY: Another criterion which is useful for small samples is *sufficiency*. In the language of R. A. Fisher, a sufficient estimator is one which exhausts all of the information in the sample. When a sample of size n is taken, we observe n values. A question of interest is that of whether it is sufficient to study a smaller number of values (which are functions of the n observations) without loss of any information about the parameter contained in the sample. This is interesting because, in general, it is more convenient to study, say, a single estimator than the n values in the sample.

Definition. $\hat{\theta}$ is said to be a sufficient estimator for estimating θ if the joint density function of the sample $\{X_i\}$ of n observations may be put in the form

$$f(x_1,\ldots,x_n;\theta) = g(x_1,x_2,\ldots,x_n|\hat{\theta})\cdot h(\hat{\theta};\theta) \tag{7.2}$$

where $g(x_1,x_2,\ldots,x_n|\hat{\theta})$ does not involve θ.

In words, it must be possible to subdivide the joint distribution of the sample into the conditional distribution of $(X_1,X_2,\ldots,X_n)$ given $\hat{\theta}$ multiplied by the distribution of $\hat{\theta}$ (which depends on θ) in such a way that the conditional distribution does not involve θ. In practice, it suffices to show that g is a function of the X's only and h is a function of $\hat{\theta}$ and θ.

In the factored form it is easy to see that no other estimator, say θ', can provide any information about θ. For the distribution of θ' for a fixed $\hat{\theta}$ will be determined by $g(x_1,x_2,\ldots,x_n|\hat{\theta})$ and will involve $\hat{\theta}$ but not θ. Hence θ' will provide information about $\hat{\theta}$ but not θ. But, for any given problem, $\hat{\theta}$ is known, so that this information is of no value.

Although the above discussion implies that the sufficient statistic $\hat{\theta}$ is a single variate, there exist other sufficient statistics which have a number of variates in the sufficient set. For example, the order statistics, $X_{(1)} \leq X_{(2)} \leq \cdots \leq X_{(n)}$, may be used as a set of sufficient statistics which has n elements. Usually we would like to have a sufficient set which consists of the smallest possible number of variates. This leads to the consideration of minimal sufficient statistics. The discussion of minimal sufficient statistics is beyond the scope of this book. Fortunately for most distributions used commonly in statistics, the factorization criterion usually leads to a minimal sufficient statistic.

It should be emphasized that sufficiency is not an asymptotic property; it does not require that n be increased without limit or that $\hat{\theta}$ be distributed normally for large n. In view of the above discussion it would appear appropriate to consider a sufficient estimator as the "best" estimator. Certainly if an estimator of θ is not a function of $\hat{\theta}$, it can be of no use in estimating θ. Conversely, if an estimator exhausts all the information in a sample, it seems useless to consider any other estimator. We have discussed a single-parameter case; when there are several parameters, a number of statistics may be found to be jointly sufficient for these parameters. Then $\hat{\theta}_1, \hat{\theta}_2, \ldots, \hat{\theta}_k$ are said to be jointly sufficient for estimating $\theta_1, \theta_2, \ldots, \theta_k$ if $f(x_1,\ldots,x_n;\theta_1,\ldots,\theta_k) = g(x_1,\ldots,x_k)h(\hat{\theta}_1,\ldots,\hat{\theta}_k;\theta_1,\ldots,\theta_k)$.

7.7 *CRAMER-RAO INEQUALITY:* In discussing efficiency, we wish to use an estimator with the smallest possible variance. Then how small can the variance be? This question may be answered by the use of the Cramer-Rao inequality. Let $X_1, X_2, \ldots, X_n$ be a random sample from a population with probability density function $f(x,\theta)$. The joint density function is $f(\underline{x};\theta) = \Pi_{i=1}^{n} f(x_i;\theta)$, where $\underline{x} = (x_1,x_2,\ldots,x_n)$. Let T be an unbiased estimator of θ based on a random sample of size n and σ_T^2 the variance of T. Then the following inequality holds:

$$\sigma_T^2 \geq \frac{1}{nE[(\partial \log f)/\partial\theta]^2}$$

where $f(X;\theta)$ has been abbreviated to f. This inequality is usually referred to as the *Cramer-Rao inequality*.

We now prove this inequality when X is a continuous random variable (a similar proof can be established for a discrete random variable by using summation instead of integration). We assume that we may differentiate under the integral or summation signs with respect to the parameter θ. In such a case, the range of x must not depend on θ. Define the random variable Y by

$$Y = \frac{\partial}{\partial\theta} \log f(\underline{X};\theta)$$

$$= \frac{1}{f(\underline{X};\theta)} \frac{\partial}{\partial\theta} f(\underline{X};\theta)$$

Then

$$E(Y) = E\left(\frac{1}{f(\underline{X};\theta)} \frac{\partial}{\partial\theta} f(\underline{X};\theta)\right)$$

$$= \int_{-\infty}^{\infty} \cdots \int_{-\infty}^{\infty} \frac{\partial}{\partial\theta} f(\underline{x};\theta) \prod_{i=1}^{n} dx_i$$

$$= \frac{\partial}{\partial\theta} \int_{-\infty}^{\infty} \cdots \int_{-\infty}^{\infty} f(\underline{x};\theta) \prod_{i=1}^{n} dx_i = 0$$

$$\sigma_Y^2 = E(Y^2) = E\left(\left[\sum_{i=1}^{n} \frac{\partial}{\partial\theta}\log f(X_i;\theta)\right]^2\right)$$

$$= nE\left(\left(\frac{\partial \log f}{\partial\theta}\right)^2\right)$$

The covariance between T and Y is

$$\sigma_{TY} = E(TY)$$

$$= \int_{-\infty}^{\infty} \cdots \int_{-\infty}^{\infty} t \frac{\partial}{\partial\theta} f(\underline{x};\theta) \prod_{i=1}^{n} dx_i$$

$$= \frac{\partial}{\partial\theta} \int_{-\infty}^{\infty} \cdots \int_{-\infty}^{\infty} tf(\underline{x};\theta) \prod_{i=1}^{n} dx_i$$

$$= \frac{\partial}{\partial\theta} \theta = 1$$

since T is an unbiased estimator of θ. Now the correlation coefficient between T and Y lies between -1 and 1; hence

$$\frac{\sigma_{TY}^2}{\sigma_T^2 \sigma_Y^2} \leq 1$$

so we have

$$\sigma_T^2 \geq \frac{\sigma_{TY}^2}{\sigma_Y^2}$$

or

$$\sigma_T^2 \geq \frac{1}{nE[(\partial \log f)/\partial\theta]^2}$$

which establishes the inequality.

When the lower bound is achieved by an unbiased estimator, say $\hat{\theta}$, then $\hat{\theta}$ has the minimum variance. It should be noted that the lower bound is not always achievable. That is, in some situations, there does not exist a $\hat{\theta}$ which achieves the lower bound.

7.8 AMOUNT OF INFORMATION AND A MEASURE OF EFFICIENCY FOR SMALL SAMPLES: The minimum variance for $\hat{\theta}$ is given by

$$\sigma_{\hat{\theta},\min}^2 = \frac{1}{nE[(\partial \log f)/\partial\theta]^2}$$

This minimum variance can be achieved only if $\hat{\theta}$ is an unbiased sufficient estimator of θ and if

$$\frac{\partial \log h(\hat{\theta};\theta)}{\partial\theta} = (\hat{\theta} - \theta)k(\theta) \tag{7.3}$$

where k may be a function of θ but is independent of $\hat{\theta}$, and $h(\hat{\theta};\theta) > 0$.

R. A. Fisher has designated the reciprocal of $\sigma_{\hat{\theta},\min}^2$ the *amount of information in the sample of n*, that is,

$$I = n\int_{-\infty}^{\infty} \left(\frac{\partial \log f}{\partial\theta}\right)^2 f(x;\theta)\ dx \tag{7.4}$$

Now, if $\hat{\theta}$ is the estimator with minimum variance, then the efficiency of any other estimator will be measured by

$$e = \frac{1}{I\sigma^2_{\hat{\theta}}}, \tag{7.5}$$

If the estimator $\hat{\theta}$ is asymptotically normally distributed so that

$$\log h(\hat{\theta};\theta) = \text{constant} - \frac{(\hat{\theta} - \theta)^2}{2\sigma^2_{\hat{\theta}}}$$

then

$$\frac{\partial \log h}{\partial \theta} = \frac{\hat{\theta} - \theta}{\sigma^2_{\hat{\theta}}}$$

Hence, in this case the second requirement for minimum variance is met with $k = 1/\sigma^2_{\hat{\theta}}$, provided $\sigma^2_{\hat{\theta}}$ exists.

We can write the integrand of I in several additional ways. Since

$$\frac{\partial \log f}{\partial \theta} = \frac{1}{f}\frac{\partial f}{\partial \theta}$$

the integrand may be written as

$$\frac{1}{f}\left(\frac{\partial f}{\partial \theta}\right)^2$$

Also,

$$\frac{\partial^2 \log f}{\partial \theta^2} = \frac{1}{f}\frac{\partial^2 f}{\partial \theta^2} - \frac{1}{f^2}\left(\frac{\partial f}{\partial \theta}\right)^2$$

Hence

$$-n\int_{-\infty}^{\infty}\left(\frac{\partial^2 \log f}{\partial \theta^2}\right) f\,dx = n\int_{-\infty}^{\infty}\frac{1}{f}\left(\frac{\partial f}{\partial \theta}\right)^2 dx - n\int_{-\infty}^{\infty}\frac{\partial^2 f}{\partial \theta^2}\,dx$$

But the last integral on the right is zero; hence

$$I = -n\int_{-\infty}^{\infty}\left(\frac{\partial^2 \log f}{\partial \theta^2}\right) f\,dx = -nE\left(\frac{\partial^2 \log f}{\partial \theta^2}\right) \tag{7.6}$$

EXAMPLE 7.6 Consider a random sample of n from $N(\mu,\sigma^2)$. The joint density function is

$$\prod_{i=1}^{n} f(x_i;\mu,\sigma^2) = \frac{1}{(\sqrt{2\pi}\sigma)^n} e^{-\Sigma_{i=1}^{n}(x_i-\mu)^2/2\sigma^2}$$

Since $\Sigma_{i=1}^{n} (x_i - \mu)^2 = \Sigma_{i=1}^{n} (x_i - \bar{x})^2 + n(\bar{x} - \mu)^2$ then

$$\prod_{i=1}^{n} f(x_i;\mu,\sigma^2) = \frac{1}{(\sqrt{2\pi}\sigma)^n} e^{-n(\bar{x}-\mu)^2/2\sigma^2} e^{-\Sigma_{i=1}^{n}(x_i-\bar{x})^2/2\sigma^2}$$

Then we may make the identifications

$$h(\hat{\theta};\theta) = h(\bar{x};\mu) = e^{-n(\bar{x}-\mu)^2/2\sigma^2}$$

and

$$g(x_1,x_2,\ldots,x_n|\hat{\theta}) = g(x_1,x_2,\ldots,x_n|\bar{x}) = e^{-\Sigma_{i=1}^{n}(x_i-\bar{x})^2/2\sigma^2}$$

But the function g is independent of μ; hence, the estimator $\bar{X}$ is a sufficient estimator of μ.

Also, since

$$\log f(x;\mu) = \log(\sigma\sqrt{2\pi}) - \frac{(x - \mu)^2}{2\sigma^2}$$

then

$$I = n \int_{-\infty}^{\infty} \frac{(x - \mu)^2}{\sigma^4} f(x;\mu)\, dx = \frac{n}{\sigma^2}$$

Hence the minimum variance is σ^2/n, which is the variance of $\bar{X}$. So $\bar{X}$ is efficient.

EXAMPLE 7.7 Suppose that a random sample of size n is taken from a Poisson population so that

$$f(x;m) = e^{-m} \frac{m^x}{x!}$$

The probability of obtaining the sample, that is, the joint density function, is

$$\prod_{i=1}^{n} f(x_i;m) = e^{-nm}(m)^{\Sigma x_i} \frac{1}{\Pi_{i=1}^{n} x_i!}$$

Since we may set

$$h(\bar{x};m) = (m^{\bar{x}}e^{-m})^n$$

the estimator $\bar{X} = \Sigma_{i=1}^{n} X_i/n$ is a sufficient estimator for m. Also, $\bar{X}$ is unbiased, since

$$E(\bar{X}) = \sum_{i=1}^{n} \frac{E(X_i)}{n} = m$$

Again, since

$$\log f(x;\lambda) = x \log m - m - \log x!$$

then

$$I = n \sum_{x=0}^{\infty} \frac{(x - m)^2}{m^2} f(x;m) = \frac{n}{m}$$

Hence the minimum variance is m/n, which is the variance of $\bar{X}$.

7.9 PRINCIPLES OF POINT ESTIMATION: In the previous section, we have discussed some characteristics of "good" estimators. In order to obtain such "good" estimators, several principles of estimation, leading to routine mathematical procedures, have been proposed. These include the *method of moments, the principle of maximum likelihood, the Bayesian principle, and the method of least squares*. The application of these principles in particular cases will lead to estimators which may differ and hence possess different attributes of "goodness." The method of least squares will be discussed in Chapter 13.

7.10 METHOD OF MOMENTS: The method of moments determines the estimators of the unknown parameters by equating the sample moments to the corresponding population moments. Let $X_1, X_2, \ldots, X_n$ denote a random sample from a distribution with density function $f(x;\theta_1,\ldots,\theta_k)$. Let M_r be the rth sample moment, i.e.,

$$M_r = \frac{1}{n} \sum_{i=1}^{n} X_i^r \tag{7.7}$$

Then equating the first k sample moments to the corresponding population moments yields

$$M_r = \mu'_r \qquad r = 1, \ldots, k \tag{7.8}$$

where μ'_r is a function of $\theta_1, \ldots, \theta_k$. Hence, there are k unknown parameters in k equations. Solving the equations for the θ's, we obtain estimators $\hat{\theta}_1, \ldots, \hat{\theta}_k$.

EXAMPLE 7.8 Consider a random sample of n from $N(\mu,\sigma^2)$. There are two parameters; therefore, we equate the first two moments. Now $\mu'_1 = \mu$ and $\mu'_2 = \sigma^2 + \mu_1^2$. Hence

$$\bar{X} = \hat{\mu}$$

$$\frac{1}{n} \sum_{i=1}^{n} X_i^2 = \hat{\sigma}^2 + \hat{\mu}^2$$

Solving these two equations for $\hat{\mu}$ and $\hat{\sigma}^2$, we obtain

$$\hat{\mu} = \bar{X}$$

$$\hat{\sigma}^2 = \frac{1}{n} \sum_{i=1}^{n} X_i^2 - \bar{X}^2 = \frac{1}{n} \sum_{i=1}^{n} (X_i - \bar{X})^2$$

The method of moments is simple to apply. However, it does not give the most efficient estimators in general.

7.11 PRINCIPLE OF MAXIMUM LIKELIHOOD: The procedure for determining the maximum-likelihood (ML) estimator for a population parameter θ is as follows:

1. Determine the joint density function of the sample $f(x_1, x_2, \ldots, x_n; \theta)$. This is the probability of obtaining the particular sample for discrete variates and is the probability without the differentials for continuous variates. R. A. Fisher has called this the *likelihood* of the sample. The likelihood is a function of θ; we wish to find the value of θ which maximizes the likelihood. If such a value exists, we have obtained the value of θ from which the sample is most likely to have come.

2. Determine $L = f(X_1, X_2, \ldots, X_n; \theta)$ or $\log L$.

3. Determine the value of θ, say $\hat{\theta}$, which will maximize L or $\log L$. If L or $\log L$ is differentiable with respect to θ, one can obtain $\hat{\theta}$ by solving the equation

$$\frac{\partial L}{\partial \theta} = 0 \qquad \text{or} \qquad \frac{\partial \log L}{\partial \theta} = 0 \tag{7.9}$$

The latter will also maximize the likelihood. When $\partial L/\partial \theta$ does not exist, one may determine $\hat{\theta}$ by the graphical method.

From Section 7.6, we recall that a sufficient estimator for θ exists when the joint distribution of the sample may be put in the form

$$f(X_1, X_2, \ldots, X_n; \theta) = g(X_1, X_2, \ldots, X_n | \hat{\theta}) h(\hat{\theta}; \theta)$$

where $g(X_1, X_2, \ldots, X_n | \hat{\theta})$ does not involve θ. Hence the ML equation reduces to

$$\frac{\partial \log h(\hat{\theta}; \theta)}{\partial \theta} = 0$$

But any solution of this equation can depend only on $\hat{\theta}$; hence, this ML solution must be the sufficient estimator, as it depends on no other estimator.

7.12 MAXIMUM-LIKELIHOOD AND EFFICIENT ESTIMATORS IN SMALL SAMPLES: If an efficient estimator for small samples exists (has minimum variance), the ML estimator, adjusted for bias if necessary, will be efficient.

In order that $\hat{\theta}$ have minimum variance, we know that $\hat{\theta}$ must be an unbiased sufficient estimator and also that

$$\frac{\partial \log h(\hat{\theta};\theta)}{\partial\theta} = (\hat{\theta} - \theta)k(\theta)$$

But to obtain the ML estimator we set this, or what amounts to the same thing,

$$\frac{\partial \log h(\hat{\theta};\theta)}{\partial\theta} = 0$$

Hence, $\hat{\theta}$, the unbiased ML estimator, is an efficient estimator in small samples. It follows that the minimum variance that an *unbiased* ML estimator can have is $1/I(\theta)$. If the ML estimator $\hat{\theta}$ is biased in such a way that $E(\hat{\theta}) = \theta + b(\theta)$, the minimum variance is

$$\frac{\{1 + d[b(\theta)]/d\theta\}^2}{I(\theta)}$$

If the ML estimator is biased, we may adjust it for bias and the minimum variance of the adjusted $\hat{\theta}$ will be $1/I(\theta)$.

7.13 MAXIMUM-LIKELIHOOD ESTIMATORS FOR TWO OR MORE PARAMETERS: If there are k parameters for which estimators are desired, we solve the set of k equations

$$\frac{\partial L}{\partial\theta_i} = 0 \qquad i = 1, 2, \ldots, k \tag{7.10}$$

In the case of two parameters θ_1 and θ_2, Cramer (1946) has shown that the minimum variance of θ_i is

$$\frac{1}{1 - \rho^2}\,\frac{1}{I(\theta_i)}$$

where

$$\rho^2 = n^2\left[E\left(\frac{\partial \log f}{\partial\theta_1} \cdot \frac{\partial \log f}{\partial\theta_2}\right)\right]^2 \Big/ I(\theta_1)I(\theta_2)$$

The joint efficiency in small samples is given by Cramer as

$$\frac{1}{(1-\rho^2)^2} \quad \frac{1}{I(\theta_1)I(\theta_2)\sigma_1^2\sigma_2^2}$$

The asymptotic efficiency is given by the limiting value of this joint small-sample efficiency as $n \to \infty$. Similar formulas may be derived for $k > 2$.

EXAMPLE 7.9 Consider a random sample of n for $N(\mu,\sigma^2)$. In this case

$$\log L = \log f(X_1, X_2, \ldots, X_n; \mu, \sigma^2)$$

$$= -\frac{1}{2}\left[n(\log 2\pi + \log \sigma^2) + \frac{\Sigma_{i=1}^n (X_i - \mu)^2}{\sigma^2}\right]$$

Let us study the following three cases:

(a) σ^2 *known*, μ *unknown*

$$\frac{\partial \log L}{\partial \mu} = 0: \quad \sum_{i=1}^{n} (X_i - \hat{\mu}) = 0$$

Therefore, the maximum-likelihood estimator of μ is

$$\hat{\mu} = \frac{\Sigma_{i=1}^n X_i}{n} = \bar{X}$$

and

$$\sigma_{\bar{X}}^2 = \frac{\sigma^2}{n} \qquad \text{(the minimum variance; see Example 7.6)}$$

We also note that $(\partial^2 \log L)/\partial\mu^2$ evaluated at $\bar{x}$ is negative to insure it is a maximum.

(b) μ *known*, σ^2 *unknown*

$$\frac{\partial \log L}{\partial \sigma^2} = 0: \quad \frac{n}{\hat{\sigma}^2} = \frac{\Sigma_{i=1}^n (X_i - \mu)^2}{(\hat{\sigma}^2)^2}$$

Therefore, the maximum-likelihood estimator of σ^2 is

$$\hat{\sigma}^2 = \frac{\Sigma_{i=1}^{n} (X_i - \mu)^2}{n}$$

[The second derivation $\partial^2 \log L/\partial(\sigma^2)^2$ evaluated at $\hat{\sigma}^2$ is negative.] The variance of $\hat{\sigma}^2$ is

$$\sigma^2_{\hat{\sigma}^2} = \frac{2\sigma^4}{n}$$

The lower bound or the minimum variance is

$$\frac{1}{I} = -\left\{nE\left[\frac{\partial^2 \log f}{\partial(\sigma^2)^2}\right]\right\}^{-1}$$

Now

$$\log f = -\frac{1}{2} \log 2\pi - \frac{1}{2} \log \sigma^2 - \frac{1}{2\sigma^2} (X - \mu)^2$$

$$\frac{\partial \log f}{\partial(\sigma^2)} = -\frac{1}{2\sigma^2} + \frac{1}{2(\sigma^2)^2} (X - \mu)^2$$

$$\frac{\partial^2 \log f}{\partial(\sigma^2)^2} = \frac{1}{2(\sigma^2)^2} - \frac{1}{(\sigma^2)^3} (X - \mu)^2$$

Hence the minimum variance is

$$-\left[\frac{n}{2\sigma^4} - \frac{n}{\sigma^6}E(X - \mu)^2\right]^{-1} = \frac{2\sigma^4}{n}$$

Since $\hat{\sigma}^2$ achieves the lower bound, $\hat{\sigma}^2$ is efficient. Note that

$$s^2 = \frac{\Sigma_{i=1}^{n} (X_i - \bar{X})^2}{n - 1}$$

in this case is not completely efficient in small samples since

$$\sigma^2_{s^2} = \frac{2\sigma^4}{n - 1}$$

(c) *Both* μ *and* σ^2 *unknown* The joint maximum likelihood estimators of μ and σ^2 are

$$\hat{\mu} = \bar{X} \qquad \hat{\sigma}^2 = \frac{\Sigma_{i=1}^{n} (X_i - \bar{X})^2}{n}$$

In case (c), $\hat{\sigma}^2$ is biased, and we would use S^2 as the unbiased estimator for σ^2. It has previously been shown that $\bar{X}$ and S^2 are independently distributed. The two estimators $\bar{X}$ and S^2 are jointly sufficient since $f(x_1,x_2,\ldots,x_n;\mu,\sigma^2) = g(x_1,x_2,\ldots,x_n)h_1(\bar{x};\mu,\sigma^2)$ $h_2(s^2;\mu,\sigma^2)$ where $g(x_1,x_2,\ldots,x_n)$ is independent of μ and σ^2 (in this case, also independent of $\bar{x}$ and s^2).

We see that $\bar{X}$ is still efficient in small samples with $\sigma_{\bar{X}}^2 = \sigma^2/n$. However, S^2 is not quite efficient, since $\sigma_{S^2}^2 = 2\sigma^4/(n - 1)$ and the minimum variance in small samples is still $2\sigma^4/n$ ($\bar{X}$ and S^2 are independently distributed). Hence the efficiency of S^2 is $(n - 1)/n$, which approaches 1 as $n \to \infty$. The *joint* small-sample efficiency is also $(n - 1)/n$.

EXAMPLE 7.10 Given N random samples consisting of n trials each from a population assumed binomially distributed with constant probability p so that

$$f(x;p) = \binom{n}{x} p^x (1 - p)^{n-x}$$

Then

$$f(X_1,X_2,\ldots,X_N;p) = \left[\prod_{i=1}^{N} \binom{n}{X_i}\right] p^{\Sigma_{i=1}^{N} X_i} (1 - p)^{\Sigma_{i=1}^{N}(n-X_i)}$$

and

$$\log L = \sum_{i=1}^{N} \log\binom{n}{X_i} + \left(\sum_{i=1}^{N} X_i\right) \log p + \left[\sum_{i=1}^{N} (n - X_i)\right] \log(1 - p)$$

Hence

$$\frac{\partial \log L}{\partial p} = 0: \qquad \hat{p} = \frac{\Sigma_{i=1}^{N} X_i}{\Sigma_{i=1}^{N}(n - X_i) + \Sigma_{i=1}^{N} X_i} = \frac{\Sigma_{i=1}^{N} X_i}{Nn} = \bar{X}$$

EXAMPLE 7.11 A random sample of size n is taken from the uniform distribution

$$f(x) = \begin{cases} \frac{1}{\theta} & 0 < x < \theta \\ 0 & \text{otherwise} \end{cases}$$

The likelihood function is

$$L = \begin{cases} \frac{1}{\theta^n} & \text{if all } X_i < \theta \\ 0 & \text{otherwise} \end{cases}$$

In this case the maximum-likelihood estimator cannot be obtained by differentiating L with respect to θ since the range of X depends on θ. The likelihood function is depicted in Fig. 7.3. It is clear that the maximum-likelihood estimator is $X_{(n)}$, the largest observation.

7.14 BAYESIAN PRINCIPLE: The Bayesian principle is based on the subjective interpretation of probability and the use of Bayes' formula. The unknown parameter θ is treated as a random variable. The distribution of θ before sampling is called the prior distribution. After the experimenter or investigator collects the data, the joint distribution of $(X_1, X_2, \ldots, X_n)$ is combined with the prior distribution by multiplying the two together. When the result is normalized,

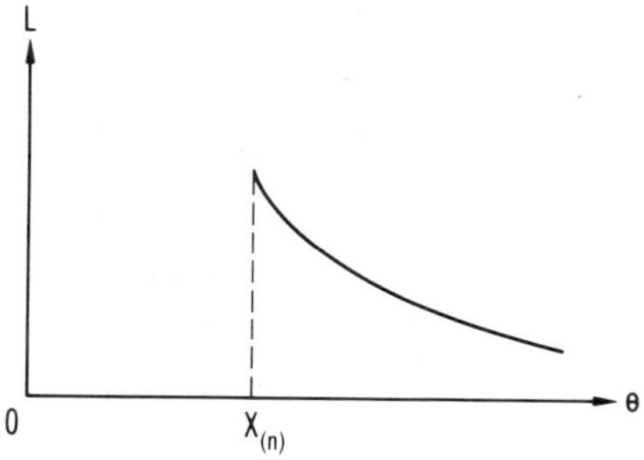

FIG. 7.3 Likelihood function.

it gives the posterior distribution of θ given the observed data. Therefore, the posterior distribution is obtained by modifying the prior distribution by the data.

The Bayesian inference is made through the posterior distribution. In this section, we consider point estimation of θ with respect to a prior distribution. Let $g(\theta)$ be the prior density function of θ and $f(x_1,x_2,\ldots,x_n|\theta)$ the joint probability density function of $(X_1,X_2,\ldots,X_n)$. By using Bayes' formula the posterior density function of θ is

$$h(\theta|x_1,\ldots,x_n) = \frac{g(\theta)f(x_1,\ldots,x_n|\theta)}{\int_{-\infty}^{\infty} g(\theta)f(x_1,\ldots,x_n|\theta)\, d\theta} \tag{7.11}$$

if θ is a continuous variate. When θ is a discrete variate, the posterior density function of θ is

$$h(\theta|x_1,\ldots,x_n) = \frac{g(\theta)f(x_1,\ldots,x_n|\theta)}{\Sigma_\theta\, g(\theta)f(x_1,\ldots,x_n|\theta)} \tag{7.12}$$

The denominator in the posterior density function is a normalizing constant which ensures the total probability to be unity.

The posterior distribution is a conditional distribution of θ given $(X_1,\ldots,X_n) = (x_1,\ldots,x_n)$. The mean of the posterior distribution is usually used as an estimator of θ because if we wish to determine an estimator, say $\tilde{\theta}$, such that $E((\theta - \tilde{\theta})^2)$ is minimized, then $\tilde{\theta}$ would be the mean value. The function $(\theta - \tilde{\theta})^2$ is called a squared error loss function. Therefore, under the squared error loss function, the Bayes estimator is the mean of the posterior distribution. It should be noted that the Bayes estimate depends on the form of the loss function. Although the squared error loss function is a reasonable loss function, the experimenter may well select other functions to express the loss incurred for an estimator. For example, the absolute error loss function $|\theta - \tilde{\theta}|$ may be used. In such a case a Bayes estimator which minimizes $E|\theta - \tilde{\theta}|$ is a median of the posterior distribution.

When there are several parameters, a joint prior distribution of the parameters is first obtained. After the collection of data, the joint posterior distribution of the parameters can be derived. Then a Bayes estimate would be obtained from this joint posterior distribution.

EXAMPLE 7.12 Consider estimating the mean of a normal distribution $N(\mu,\sigma^2)$ where σ^2 is known. Suppose the prior distribution of μ is $N(\nu,\tau^2)$ and let $X_1, \ldots, X_n$ be a random sample from $N(\mu,\sigma^2)$. The posterior distribution is found as follows:

$$g(\mu)f(x_1,\ldots,x_n|\mu) = \frac{1}{\sqrt{2\pi}\tau} e^{-(1/2\tau^2)(\mu-\nu)^2}\left(\frac{1}{\sqrt{2\pi}\sigma}\right)^n$$

$$\times\ e^{-(1/2\sigma^2)\Sigma_{i=1}^n(x_i-\mu)^2}$$

$$= k\ \exp\left\{-\frac{1}{2}\left(\frac{1}{\tau^2}+\frac{n}{\sigma^2}\right)\left[\mu - \left(\frac{1}{\tau^2}+\frac{n}{\sigma^2}\right)^{-1}\right.\right.$$

$$\left.\left.\times\left(\frac{1}{\tau^2}\nu + \frac{n}{\sigma^2}\bar{x}\right)\right]^2\right\}$$

where k is a multiplier which does not involve μ. The integral with respect to μ yields

$$\int_{-\infty}^{\infty} g(\mu)f(x_1,\ldots,x_n|\mu)\ d\mu = k\sqrt{2\pi}\left(\frac{1}{\tau^2}+\frac{n}{\sigma^2}\right)^{-1/2}$$

Therefore, the posterior density function is

$$h(\mu|x_1,\ldots,x_n) = \frac{1}{\sqrt{2\pi}}\left(\frac{1}{\tau^2}+\frac{n}{\sigma^2}\right)^{1/2}$$

$$\times\ \exp\left\{-\frac{1}{2}\left(\frac{1}{\tau^2}+\frac{n}{\sigma^2}\right)\left[\mu - \left(\frac{1}{\tau^2}+\frac{n}{\sigma^2}\right)^{-1}\left(\frac{1}{\tau^2}\nu+\frac{n}{\sigma^2}\bar{x}\right)\right]^2\right\}$$

So the posterior distribution is also a normal distribution.

Under the squared error loss function, the Bayes estimate of μ is the mean

$$\frac{(1/\tau^2)\nu + (n/\sigma^2)\bar{x}}{1/\tau^2 + n/\sigma^2} \tag{7.13}$$

This is a weighted average of the sample mean and the mean of the prior distribution with the weights proportional to the inverses of their variances. It should be pointed out that

$$\lim_{n\to\infty} \frac{(1/\tau^2)\nu + (n/\sigma^2)\bar{x}}{1/\tau^2 + n/\sigma^2} = \bar{x}$$

Therefore, when the experimenter has a lot of data, the Bayes estimate tends toward the sample mean. In such a case, it is seen that the data dominates the prior.

The Bayes estimator is obtained from the posterior distribution of θ. Strictly speaking, θ is a random variable from the Bayesian viewpoint. The Bayes estimator should be considered as a predictor of the random variable, i.e., a value which would be most likely to be obtained for the next observation. The goodness of the predictor depends on the prior distribution which has a subjective interpretation. Hence the measures of the goodness of estimators such as unbiasedness and efficiency cannot be judged objectively for Bayes estimators.

7.15 ROBUST ESTIMATION: In applied statistics, particularly data analysis, the investigator often questions the validity of the mathematical-statistical model which underlies the assumptions that he or she imposes. A posterior statistical test for the validity of models is available and will be studied in Chapter 11. Again, in Chapter 12 a discussion is given of the use of inference procedures based on conditionally specified models which make use of the data from the same investigation to resolve certain uncertainties regarding a proposed model and also provides final inferences. In addition to the detection of the departure of assumptions, modern statistical estimation theory has dealt with the possible selection of an estimator which is insensitive to the violation of certain assumptions of the model. An estimator which performs well under a variety of underlying conditions is called a *robust estimator*.

It was mentioned in Section 3.10 that data arising from many different measurements taken on plants and animals are specified as following the normal distribution. As demonstrated previously, the normal distribution has many desirable properties such as symmetry and mathematical convenience. But it should not be assumed that every continuous distribution representing actual data should be normal. Data collected in an investigation might well be asymmetric, heavy-tailed, or may contain some "wild" observations such as outliers. When a data set contains unusually large (or small) observations, they are suspected as outliers. Hence an outlier is an observation whose value is comparatively markedly different from the other observations in the sample. When outlying observations are present, the investigator may wish to use, for example, the sample median rather than the sample mean as the estimator of the population average

EXAMPLE 7.13 A random sample of 9 facotry workers is taken and their annual income in thousand dollars are

10.4, 11.5, 11.9, 12.1, 12.5, 13.4, 14.6, 14.7, 35.7

The sample mean is 15.2 and the sample median is 12.5 as estimates of the average annual income of the factory workers. It should be noted that the sample mean is larger than 8 of the 9 observations. In this case the median is a preferable estimator. A subsequent investigation reveals that the value 35.7 is the manager's income, which is an outlier caused by mistakenly including the manager's income with those of the factory workers. It is advisable to examine the causes when outlying observations are located. Appropriate actions should then be taken to make the necessary adjustments.

The median is not the only robust estimator for the location parameter. Some robust estimators are constructed by trimming the observations from the upper and the lower tail of the sample. Let $X_{(1)} \leq X_{(2)} \leq \cdots \leq X_{(n)}$ denote the order statistics. The 100α% symmetrically trimmed mean is defined as

$$\bar{X}_\alpha = \frac{pX_{([\alpha n + 1])} + X_{[\alpha n + 2]} + \cdots + pX_{(n - [\alpha n])}}{n(1 - 2\alpha)} \qquad (7.14)$$

where $0 \leq \alpha < .50$, $p = 1 + [\alpha n] - \alpha n$, and $[\alpha n]$ denotes the largest integer less than or equal to αn. The trimmed mean uses only the middle $100(1 - 2\alpha)\%$ observations and the observations located at the two tails are eliminated. It is noted that the 50% trimmed mean is the median.

Instead of eliminating the observations at the tails, we may replace their values by their adjacent "milder" values. This type of adjustment of the extreme observations is called *Winsorization*, which is coined in honor of Charles P. Winsor. The g-times symmetrically Winsorized sample is given by

$$\underbrace{X_{(g+1)}, \ldots, X_{(g+1)}}_{g \text{ times}}, X_{(g+1)}, X_{(g+2)}, \ldots, X_{(n-g-1)}, X_{(n-g)}$$

$$\underbrace{X_{(n-g)}, \ldots, X_{(n-g)}}_{g \text{ times}}$$

The g-times Winsorized mean is defined as

$$\bar{X}_{Wg} = \frac{1}{n}[(g + 1)X_{(g+1)} + X_{(g+2)} + \cdots + X_{(n-g-1)} + (g + 1)X_{(n-g)}] \tag{7.15}$$

EXAMPLE 7.14 Dixon and Tukey (1968) gave the following coded values of percentage Al_2O_3 found in a synthetic granite glass by 12 laboratories as reported by Fairbairn and Schairer (1952):

% Al_2O_3	Coded Value [= $100(\%Al_2O_3 - 16.15)$]
16.78	63
16.39	24
16.34	19
16.27	12
16.20	5
16.14	- 1
16.12	- 3
16.12	- 3
15.98	-17
15.91	-24
15.85	-30
15.79	-36

(a) The sample mean for the coded data is 0.75.

(b) The 10% trimmed mean is

$$\bar{x}_{.10} = \frac{(.8)(-30) - 24 - 17 - 3 - 3 - 1 + 5 + 12 + 19 + (.8)24}{12(1 - .2)}$$

$$= -1.75$$

(c) The 25% trimmed mean is

$$\bar{x}_{.25} = \frac{-17 - 3 - 3 - 1 + 5 + 12}{12(1 - .5)} = -1.17$$

(d) The once-Winsorized mean is

$$\bar{x}_{W1} = \frac{1}{12}[2(-30) - 24 - 17 - 3 - 3 - 1 + 5 + 12 + 19 + 2(24)]$$

$$= -2.00$$

(e) The twice-Winsorized mean is

$$\bar{x}_{W2} = \frac{1}{12}[3(-24) - 17 - 3 - 3 - 1 + 5 + 12 + 3(19)] = -1.83$$

The sample mean is large due to one large observation.

We have discussed some robust estimators in this section. For other types of robust estimators, the readers are referred to the Princeton study by Andrews et al. (1972). In their study, the robust estimators are evaluated by simulations. Stigler (1977) evaluated several robust estimators with real data.

EXERCISES

7.1 Show that $\Sigma(X_i - \bar{X})^2)/n$ is a biased estimator for σ^2 in a random sample of size n, assuming σ^2 finite.

7.2 Given a random sample of size n from $N(0,\sigma^2)$.

(a) Show that $S^2 = \Sigma X_i^2/n$ is an unbiased estimator for σ^2. *Hint:* $S^2 = (\sigma^2/n)\chi^2$. (b) Show that $\sigma^2_{S2} = 2\sigma^4/n$. *Hint:* $\sigma^2_{S2} = E((S^2 - \sigma^2)^2) = E((S^2)^2) - (\sigma^2)^2$. Use $E((S^2)^2) = (\sigma^4/n^2)E((\chi^2)^2)$ to complete the demonstration. (c) Is S^2 also consistent? *Hint:* Use the results of (b) and Tchebysheff's inequality. (d) Using the concept of information I, show that the minimum variance possible for an unbiased estimator for σ^2 is $2\sigma^4/n$. Is S^2 efficient in small samples? (e) Is S^2 a sufficient estimator for σ^2?

7.3 Determine the ML estimator for m in the Poisson distribution in Example 7.7.

7.4 (a) Show that $h(\bar{X};m)$ for a random sample from a Poisson distribution meets the conditions given for minimum variance. *Hint:* From Example 7.7, $h(\bar{X};m) = (m^{\bar{X}}e^{-m})^n$. Now, show that $\partial \log [h(\bar{X};m)]/\partial m = (n/m)(\bar{X} - m)$. (b) Is $\bar{X}$ consistent in this case?

7.5 Given a random sample of size n from $f(x) = 3x^2/\theta^3$, $0 < x < \theta$, and 0 otherwise, (a) use the method of moments to find an estimator for θ; (b) discuss this estimator as to sufficiency, bias, and consistency.

7.6 Let $X_1, X_2, \ldots, X_n$ be independent Bernoulli trials with $P(X_i = 1) = p$ and $P(X_i = 0) = 1 - p$. (a) Show that the sample mean $\bar{X}$ is the maximum-likelihood estimator of p. (b) Find the Cramer-Rao lower bound for estimating p. (c) Show that $\bar{X}$ is unbiased, consistent, and efficient.

7.7 Suppose N independent experiments consisting of n trials each are conducted with data assumed binomially distributed with constant probability p. That is,

$$f(x;p) = \binom{n}{x} p^x q^{n-x} \qquad x = 0, 1, \ldots, n$$

We estimate p by $\bar{X} = \Sigma_{i=1}^{N} X_i/Nn$. Discuss this estimator as to (a) Sufficiency, by showing that $h(\bar{X};p) = [p^{\bar{X}}(1 - p)^{1-\bar{X}}]^{Nn}$; (b) Bias; (c) Information and Cramer-Rao lower bound, by showing $I = Nn/pq$; (d) Consistency, by using $\sigma_{\bar{X}}^2 = pq/Nn$ and Tchebysheff's inequality.

7.8 Given a random sample of size n taken from the Cauchy distribution

$$f(x;\mu) = \frac{1}{\pi[1 + (x - \mu)^2]} \qquad -\infty < x < \infty$$

(a) It can be shown that $\bar{X}$ has the same distribution as X. Hence what can be said about the usefulness of $\bar{X}$ as an estimate of μ in this case? (b) Using the concept of information, show that the minimum variance of $\hat{\mu}$ is $2/n$. (c) Show that the asymptotic variance of the median for this distribution is $\pi^2/4n$. What is its asymptotic efficiency? (d) Is the median consistent in this case?

7.9 R. A. Fisher has shown that

$$\sigma^2_{S^2} = \frac{1}{n}\left(\mu_4 - \frac{n-3}{n-1}\sigma^4\right)$$

where $S^2 = \Sigma_{i=1}^{n} (X_i - \bar{X})^2/(n - 1)$ and the sample is from an arbitrary universe. Assume that μ_4 and μ_2 exist. Show that S^2 is a consistent estimator for σ^2.

7.10 Show that the maximum-likelihood estimator in Example 7.11 is biased. *Hint:* Consider the distribution of $X_{(n)}$; see Section 6.13.

7.11 Given a random sample of size n from

$$f(x) = \begin{cases} e^{-(x-\theta)} & x > \theta \\ 0 & \text{otherwise} \end{cases}$$

Find the maximum likelihood estimator of θ.

7.12 Given the Pearson Type III distribution

$$f(x) = \begin{cases} \dfrac{1}{\beta^{\alpha}\Gamma(\alpha)} x^{\alpha-1} e^{-x/\beta} & 0 < x < \infty \\ 0 & \text{otherwise} \end{cases}$$

where α is some fixed positive constant. (a) Show that the ML estimator for β for a random sample of n is $\hat{\beta} = \bar{X}/\alpha$. (b) Show that $E(X) = \alpha\beta$, and hence that $E(\hat{\beta}) = \beta$ and $\hat{\beta}$ is unbiased. (c) It can be shown that $\sigma^2_{\hat{\beta}} = \beta^2/n\alpha$. Using the concept of information, show that the minimum variance is also given by $\beta^2/n\alpha$. (d) Is the estimator efficient in small samples? (e) Is the estimator consistent? (f) Show that $\hat{\beta}$ is a sufficient estimator for β.

7.13 Derive for random samples of size n the ML estimators $\hat{\alpha}$, $\hat{\beta}$, and $\hat{\sigma}^2$ of α, β, and σ^2 in the regression equation $Y = \alpha + \beta(x - \bar{x}) + \varepsilon$, where ε is $N(0, \sigma^2)$, and the x's are fixed known constants. (a) Show that $\hat{\alpha}$, $\hat{\beta}$, and $\hat{\sigma}^2$ are jointly sufficient. (b) What are the variances of $\hat{\alpha}$ and $\hat{\beta}$? (c) Is the ML estimator for σ^2 unbiased?

7.14 Given the bivariate normal distribution with parameters μ_x, μ_Y, σ_X^2, σ_Y^2, ρ_{XY}: (a) Find the joint ML estimators for the parameters for a random sample of size n. (b) Show that $(\hat{\rho})^2 = (\hat{\beta})^2(\hat{\sigma}_X^2/\hat{\sigma}_Y^2)$ where $\hat{\beta}$ is obtained from Exercise 7.13. (c) Is there any difference between $\hat{\sigma}_Y^2$ above and the $\hat{\sigma}^2$ of Exercise 7.13?

7.15 Suppose $\begin{pmatrix} X_{1j} \\ X_{2j} \end{pmatrix}$, j = 1, 2, ..., n, is a random sample from a bivariate normal distribution with mean vector $\begin{pmatrix} \mu_1 \\ \mu_2 \end{pmatrix}$ and covariance matrix $\begin{pmatrix} \sigma_1^2 & \sigma_{12} \\ \sigma_{12} & \sigma_2^2 \end{pmatrix}$. Find the distribution of $X_{1j} - X_{2j}$ and use this distribution to determine the maximum-likelihood estimator of $\mu_1 - \mu_2$.

7.16 Consider estimating the parameter p in the binomial distribution $f(x) = \binom{n}{x} p^x (1 - p)^{n-x}$, x = 0, 1, ..., n. If the prior distribution of p is a beta distribution $g(p) = [1/B(a,b)]p^{a-1}(1 - p)^{b-1}$, $0 < p < 1$, $a > 0$, $b > 0$, and a random sample of size N is available, obtain the posterior distribution of p and the Bayes estimate under a quadratic loss function.

7.17 Consider the estimation of the parameter m in the Poisson distribution $f(x) = e^{-m}m^x/x!$, x = 0, 1, 2, ... If the prior distribution of m is a gamma distribution $g(m) = [1/\Gamma(\alpha)\beta^\alpha]m^{\alpha-1} e^{-m/\beta}$, $0 < m < \infty$, $\alpha > 0$, $\beta > 0$, and a random sample of size n is available, find the posterior distribution of m and the Bayes estimate under a quadratic loss function.

7.18 A random sample of 20 households in a town was taken and their consumptions of electricity in a month were recorded (in kW-hr used):

1	1	9	10	12	12	14	14	14	15
16	20	22	22	23	25	25	28	32	96

(The two smallest values were obtained for the two families on vacation and the largest value was for a small family business.) Compute (a) the sample mean; (b) the median; (c) the 5% and 10% symmetrically trimmed means; (d) the once and twice symmetrically Winsorized means.

REFERENCES

Andrew, D. F., Bickel, P. J., Hampel, F. R., Huber, P. J., Rogers, W. H., and Tukey, J. W. (1972). *Robust Estimates of Location*. Princeton University Press, Princeton, N.J.

Cramer, H. (1946). *Mathematical Methods of Statistics*. Princeton University Press, Princeton, N.J.

Dixon, W. J., and Tukey, J. W. (1968). Approximate behavior of the distribution of Winsorized t (trimming/Winsorization 2). Technometrics 10:93-98.

Fairbairn, H. W., and Schairer, J. E. (1952). A test of the accuracy of chemical analysis of silicate rocks. Amer. Mineralogist 37:744-757.

Stigler, S. (1977). Do robust estimators work with real data? Ann. Stat. 5:1055-1078.

8

SAMPLING FROM FINITE POPULATIONS

8.1 INTRODUCTION: In the previous chapters, random samples are assumed to be taken from infinite populations. Sometimes the infinite population may be a hypothetical population. For example, in order to estimate the mean gain in weight of swine, a random sample is taken from the population of swine of the same breed, sex, and age, fed on the same ration and managed similarly for a period of 20 days. Such a population is a hypothetical one since in practice it is difficult to feed *all* swine on the same ration and managed similarly for a period of 20 days. But conceptually this could be done. In other real-world problems, the population may consist of finite number of units. For example, in order to estimate the average price of beef in a city, a simple random sample is taken from the population of grocery stores. This population is clearly finite. In this chapter we consider sampling from finite populations and discuss various simple basic designs in sample surveys.

8.2 SIMPLE RANDOM SAMPLING AND SYSTEMATIC SAMPLING: Let the finite population consist of N units. The value of the ith unit is denoted by y_i, i = 1, ..., N. The population mean and variance, denoted by μ and σ^2, respectively, are defined as

$$\mu = \frac{1}{N} \sum_{i=1}^{N} y_i \tag{8.1}$$

$$\sigma^2 = \frac{1}{N} \sum_{i=1}^{N} (y_i - \mu)^2 \tag{8.2}$$

Further define

$$S^2 = \frac{1}{N - 1} \sum_{i=1}^{N} (y_i - \mu)^2 = \frac{N}{N - 1} \sigma^2 \qquad (8.3)$$

In most sample surveys, the objective is to estimate μ (or equivalently to estimate the population total $\Sigma_{i=1}^{N} y_i$). We first consider two basic sampling designs: simple random sampling and systematic sampling.

Simple random sampling is a sampling procedure such that every possible sample of fixed size n has an equal chance of being chosen. Simple random sampling may be either with replacement or without replacement. When a simple random sample is taken with replacement, the sampled unit is returned to the population at each draw before the next unit is selected. When the unit is not returned to the population after it is drawn, the sample is a simple random sample without replacement. Since the latter is used most frequently in practice, we shall only consider sampling without replacement in this chapter.

Since there are $\binom{N}{n}$ possible samples in simple random sampling without replacement, the probability of a particular sample being chosen is $1/\binom{N}{n}$. Let $y_1, y_2, \ldots, y_n$ be the values of units in the sample (note that these need not be the first n units in the population). The sample mean is

$$\bar{y} = \frac{1}{n} \sum_{i=1}^{n} y_i \qquad (8.4)$$

In order to discuss the properties of $\bar{y}$, we define the random variable

$$T_i = \begin{cases} 1 & \text{if the ith unit is in the sample} \\ 0 & \text{otherwise} \end{cases} \qquad (8.5)$$

Then $\bar{y}$ can be written as

$$\bar{y} = \frac{1}{n} \sum_{i=1}^{N} y_i T_i$$

Hence $\bar{y}$ is a function of the random variables T_i and may be written as $\bar{y}(T_1,\ldots,T_N)$. But we write it as $\bar{y}$ for simplicity. This remark also applies to other estimators in this chapter.

Now $E(T_i) = P(T_i = 1) = \binom{N-1}{n-1}\Big/\binom{N}{n} = n/N$, so we have

$$E(\bar{y}) = \frac{1}{n}\sum_{i=1}^{N} y_i E(T_i) = \frac{1}{n}\sum_{i=1}^{N} y_i P(T_i = 1) = \frac{1}{N}\sum_{i=1}^{N} y_i = \mu \quad (8.6)$$

Hence $\bar{y}$ is an unbiased estimator of μ. Also $E(T_i^2) = n/N$ and $E(T_iT_j) = n(n-1)/N(N-1)$; hence

$$V(T_i) = \frac{n}{N}\left(1 - \frac{n}{N}\right)$$

and

$$\mathrm{Cov}(T_i,T_j) = -\frac{1}{N-1}\frac{n}{N}\left(1 - \frac{n}{N}\right) \qquad i \neq j$$

So

$$V(\bar{y}) = \frac{1}{n^2}V\left(\sum_{i=1}^{N} y_iT_i\right)$$

$$= \frac{1}{n^2}\left[\sum_{i=1}^{N} y_i^2V(T_i) + 2\sum_{i<j}\sum y_iy_j\mathrm{Cov}(T_i,T_j)\right]$$

$$= \frac{(1-f)S^2}{n} \quad (8.7)$$

where $f = n/N$ is called the *sampling fraction*. The factor $1 - f$ is called the *finite population correction*. It is noted that for fixed n, $f \to 0$ as $N \to \infty$, further $S^2 \to \sigma^2$. So that $V(\bar{y})$ coincides with the variance of the sample mean from an infinite population. When $V(\bar{y})$ is unknown, an unbiased estimator is $(1 - f)\ \Sigma_{i=1}^{n}\ (y_i - \bar{y})^2/[n(n-1)]$.

To estimate the population total. The estimator $N\bar{y}$ is unbiased provided N is known. The variance of $N\bar{y}$ is $N^2(1 - f)S^2/n$.

Next let us consider systematic sampling. Because the population size is finite, the N units may be labled from 1 to N. Suppose $N = nk$, where k is an integer. A systematic sample is obtained by selecting a random integer i, i = 1, 2, ..., k; then the units y_i, y_{k+i}, y_{2k+i}, ..., $y_{(n-1)k+i}$ will comprise a particular systematic sample for each value of i. Therefore, there are k possible samples and each sample has probability 1/k to be selected.

We may rearrange the units and let y_{ij} be the jth unit in the ith sample, j = 1, ..., n, i = 1, ..., k. Denote the systematic sample mean by $\bar{y}_{sy}$ (sy for systematic), then

$$\bar{y}_{sy} = \bar{y}_i = \frac{1}{n} \sum_{j=1}^{n} y_{ij} \tag{8.8}$$

when the integer i is chosen. It can be seen that $\bar{y}_{sy}$ is unbiased. Define

$$Z_i = \begin{cases} 1 & \text{if the integer i is chosen} \\ 0 & \text{otherwise} \end{cases} \tag{8.9}$$

then $P(Z_i = 1) = 1/k$. We can write

$$\bar{y}_{sy} = \sum_{i=1}^{k} \bar{y}_i Z_i$$

Hence $\bar{y}_{sy}$ is a function of the random variables Z_i. The expectation of $\bar{y}_{sy}$ is

$$E(\bar{y}_{sy}) = \sum_{i=1}^{k} \bar{y}_i E(Z_i) = \frac{1}{k} \sum_{i=1}^{k} \bar{y}_i = \frac{1}{kn} \sum_{i=1}^{k} \sum_{j=1}^{n} y_{ij} = \mu \tag{8.10}$$

The variance of $\bar{y}_{sy}$ is

$$V(\bar{y}_{sy}) = \frac{1}{k} \sum_{i=1}^{k} (\bar{y}_i - \mu)^2 \tag{8.11}$$

Since the sample is taken following a particular path through the population units, the arrangements of the units plays an essential role to the performance of the estimator. If the units are arranged in such a way that the variation among $\bar{y}_i$ is small, from the expression of $V(\bar{y}_{sy})$ above, we see that $\bar{y}_{sy}$ would be a very efficient estimator. If the units are arranged at random, we would expect that systematic sampling and simple random sampling are equivalent. When a particular pattern exists in the arrangements such as periodic variation, precaution must be made before drawing a systematic sample to guard against potential loss of efficiency. See Cochran (1977) and Sukhatme and Sukhatme (1970) for further discussion.

EXAMPLE Suppose the population consists of nine families and their annual income in terms of thousand dollars are 6, 7, 8, 15, 16, 17, 24, 25, 26, respectively. The first three families may be called low-income families, the middle three families middle-income, and the last three high-income families among these nine families. A systematic sample of size 3 is taken to estimate the population mean. If the units are arranged as above, the three systematic samples are

$$s_1: \quad 6, 15, 24 \qquad \bar{y}_1 = 15$$

$$s_2: \quad 7, 16, 25 \qquad \bar{y}_2 = 16$$

$$s_3: \quad 8, 17, 26 \qquad \bar{y}_3 = 17$$

$$V(\bar{y}_{sy}) = \frac{1}{3} \Sigma (\bar{y}_i - \mu)^2 = \frac{2}{3}$$

However, if the units are arranged in such a way that the first sample includes all three low-income families, the second sample middle-income, and the third sample high-income families, then the three systematic samples are

$$s_1': \quad 6, 7, 8 \qquad \bar{y}_1 = 7$$

$$s_2': \quad 15, 16, 17 \qquad \bar{y}_2 = 16$$

$$s_3': \quad 24, 25, 26 \qquad \bar{y}_3 = 25$$

$$V(\bar{y}_{sy}) = \frac{1}{3} \Sigma (\bar{y}_i - \mu)^2 = 54$$

Hence the variance increased substantially.

8.3 STRATIFIED RANDOM SAMPLING: The objective of designing a sample survey is to obtain estimators with high efficiency. Stratified random sampling is a procedure widely used to achieve such an objective. In order to take a stratified random sample, the units in the population are divided into k groups called *strata*. The ith stratum consists of N_i units, $\Sigma_{i=1}^{k} N_i = N$. When a simple random sample is taken independently from each stratum, the sampling procedure is called stratified random sampling. Let n_i be the sample size from the ith stratum, then the total sample size is $\Sigma_{i=1}^{k} n_i = n$.

Denote by y_{ij} the value of the jth unit in the ith stratum, the population stratum mean is $\mu_i = \Sigma_{j=1}^{N_i} y_{ij}/N_i$. The mean for the whole population is $\mu = \Sigma_{i=1}^{k} N_i\mu_i/N = \Sigma W_i\mu_i$, where $W_i = N_i/N$ is called the ith *stratum weight*. Because a simple random sample of size n_i is taken from each stratum, the estimator for μ_i is $\bar{y}_i = \Sigma_{j=1}^{n_i} y_{ij}/n_i$. Hence the estimator used in stratified random sampling is

$$\bar{y}_{st} = \sum_{i=1}^{k} W_i\bar{y}_i \tag{8.12}$$

The subscript *st* stands for *stratified*. The estimator $\bar{y}_{st}$ is unbiased since each $\bar{y}_i$ is unbiased for μ_i. In view of the independence of the $\bar{y}_i$ terms, we have

$$V(\bar{y}_{st}) = \sum_{i=1}^{k} W_i^2 V(\bar{y}_i) \tag{8.13}$$

$$= \sum_{i=1}^{k} W_i^2 \frac{(1 - f_i)S_i^2}{n_i}$$

where

$$S_i^2 = \frac{1}{N_i - 1} \sum_{j=1}^{N_i} (y_{ij} - \mu_i)^2$$

and $f_i = n_i/N_i$ is the sampling fraction in the ith stratum. It is seen that for fixed W_i, $V(\bar{y}_{st})$ is small when all $V(\bar{y}_i)$ are small. Hence the strata should be constructed in such a way such that $V(\bar{y}_i)$ are small. In practice, the strata may be formed by administrative convenience. If units in the same administrative region are homogeneous, then $V(\bar{y}_{st})$ will be small.

Since simple random samples are taken from the strata, unbiased estimators for $V(\bar{y}_i)$ are available. An unbiased estimator of $V(\bar{y}_{st})$ is

$$\hat{V}(\bar{y}_{st}) = \sum_{i=1}^{k} W_i^2(1 - f_i) \frac{1}{n_i(n_i - 1)} \sum_{j=1}^{n_i} (y_{ij} - \bar{y}_i)^2 \qquad (8.14)$$

In this section, we discussed the case that simple random sampling is used in each stratum. The investigator could of course use systematic sampling instead of simple random sampling if systematic sampling is more appropriate. Then the sampling procedure is called stratified systematic sampling. The analysis of stratified systematic sampling will not be treated here.

8.4 ALLOCATION OF SAMPLE SIZES: In stratified random sampling, the sample size for the ith stratum is n_i which is to be determined. Usually the investigator has a limited budget, i.e., limited amount of money to spend on the survey. One way to determine n_i is to minimize $V(\bar{y}_{st})$ subject to budget or cost constraint. The cost function often has the form

$$C = c_0 + \sum_{i=1}^{k} c_i n_i \qquad (8.15)$$

where c_0 is the overhead cost and c_i is the cost of sampling one unit in the ith stratum. Letting λ be the Lagrange multiplier, the n_i values are determined by minimizing

$$V(\bar{y}_{st}) + \lambda(\Sigma c_i n_i + c_0 - C) = \Sigma\left(\frac{1}{n_i} - \frac{1}{N_i}\right)W_i^2 S_i^2$$
$$+ \lambda(\Sigma c_i n_i + c_0 - C)$$

Differentiating with respect to n_i, we obtain

$$-\frac{1}{n_i^2} W_i^2 S_i^2 + \lambda c_i = 0$$

or

$$n_i = \frac{W_i S_i}{\sqrt{\lambda c_i}}$$

Hence n_i is proportional to $W_iS_i/\sqrt{c_i}$. This implies that n_i is large when stratum size is large; within stratum variability is large and cost is small in the ith stratum.

By the relation that $\Sigma n_i = n$, $\sqrt{\lambda}$ is found to be

$$\sqrt{\lambda} = \frac{1}{n}\,\frac{\Sigma W_iS_i}{\sqrt{c_i}}$$

Therefore, in terms of the total sample size,

$$n_i = n\,\frac{W_iS_i/\sqrt{c_i}}{\Sigma\left(W_iS_i/\sqrt{c_i}\right)} \tag{8.16}$$

The total sample size is determined by the cost function. Substituting n_i in the cost function and solving for n, we obtain

$$n = (C - c_0)\,\frac{\Sigma\left(W_iS_i/\sqrt{c_i}\right)}{\Sigma W_iS_i\sqrt{c_i}} \tag{8.17}$$

In some cases, the investigator may wish to specify that $V(\bar{y}_{st})$ is not higher than a certain given value, say v. Then the n_i values are selected to minimize the cost subject to the variance constraint. The solution for n_i is the same as above. However, the total sample size is determined by using the variance function $V(\bar{y}_{st}) = v$ and we have

$$n = (\,W_iS_i\sqrt{c_i})\ \Sigma\,\frac{W_iS_i}{\sqrt{c_i}}\left(v + \Sigma\,\frac{W_i^2S_i^2}{N_i}\right)^{-1} \tag{8.18}$$

The allocation of the sample given in (8.16) is usually referred to as the optimum allocation because it minimizes either the cost or the variance. An important special case is that all c_i terms are equal. Then

$$n_i = \frac{nW_iS_i}{\Sigma W_iS_i} \tag{8.19}$$

This special allocation is called the Neyman allocation, after Neyman (1934); later it was found that Tschuprow (1923) also gave the same result.

Another useful allocation is the proportional allocation, which simply allocates the sample size proportional to the stratum size. Hence

$$n_i = nW_i$$

When c_i values are equal and S_i values are equal, the proportional allocation is the same as the optimum allocation.

It should be intuitively clear that the Neyman allocation is at least as good as the proportional allocation, since the former is optimum. Also because of stratification, the proportional allocation is at least as good as the simple random sampling if f_i is small. An analytic proof may be found, for example, in Cochran (1977) and in Sukhatme and Sukhatme (1970).

8.5 CLUSTER SAMPLING: When the units in the population are grouped into clusters, it may be more convenient and economical to sample the clusters. For example, rolls of paper towels are packed into cartons and cartons form clusters; it is easier to sample cartons than rolls. Let there be N clusters and each cluster has M elements (units in clusters). The value of the jth element in the ith cluster is denoted by y_{ij} and the population mean is

$$\mu = \frac{1}{NM} \sum_{i=1}^{N} \sum_{j=1}^{M} y_{ij} \tag{8.20}$$

We are interested in estimating this population mean.

In cluster sampling, a sample of clusters is selected at the first stage. If all the elements in the sampled clusters are enumerated, it is called a *single-stage cluster sampling*. If subsamples of elements are taken from the sampled clusters, it is called a *two-stage cluster sampling* or *subsampling*. The clusters selected at the first stage are also called *primary sampling units*. Let us first discuss the single-stage cluster sampling.

Suppose a simple random sample of size n is taken from N clusters and each element in the sampled cluster is enumerated. Define the ith cluster mean by

$$\bar{y}_i = \frac{1}{M} \sum_{j=1}^{M} y_{ij} = \mu_i$$

Then an estimator of μ is

$$\bar{\bar{y}} = \frac{1}{n} \sum_{i=1}^{n} \bar{y}_i \qquad (8.21)$$

which is the mean of cluster means. Since this is a simple random sample, $\bar{\bar{y}}$ is an unbiased estimator with variance

$$V(\bar{\bar{y}}) = (1 - f) \frac{1}{n} S_b^2 \qquad (8.22)$$

where

$$S_b^2 = \frac{1}{N - 1} \sum_{i=1}^{N} (\mu_i - \mu)^2$$

S_b^2 is the variation among cluster means. Therefore, $V(\bar{\bar{y}})$ is small when S_b^2 is small.

To compare the single-stage cluster sampling with the simple random sampling of size nM, we consider the relative efficiency of $\bar{\bar{y}}$ to the sample mean $\bar{y}$ of a simple random sample. The variance of $\bar{y}$ is

$$V(\bar{y}) = (1 - f) \frac{1}{nM} S^2$$

where

$$S^2 = \frac{1}{NM - 1} \sum_{i=1}^{N} \sum_{j=1}^{M} (y_{ij} - \mu)^2$$

which is the total variation in the population. The relative efficiency (RE) of $\bar{\bar{y}}$ to $\bar{y}$ is

$$RE = \frac{1/V(\bar{\bar{y}})}{1/V(\bar{y})} = \frac{S^2}{MS_b^2} \qquad (8.23)$$

Since S^2 is fixed, RE is large if S_b^2 is small. This shows that in order to have high efficiency for cluster sampling, the clusters should be formed in such a way that the variation among cluster means is small (or equivalently, the variation within clusters is large).

An unbiased estimator of $V(\bar{\bar{y}})$ is, from simple random sampling,

$$(1 - f) \frac{1}{n(n - 1)} \sum_{i=1}^{n} (\mu_i - \bar{\bar{y}})^2 \tag{8.24}$$

Let us now turn to two-stage sampling of clusters of equal size. The sampling procedure is to take a simple random sample of n primary sampling units at the first stage and a simple random sample of size m from each sampled primary sampling unit at the second stage. Denote the sample mean in the ith cluster by

$$\bar{y}_i = \frac{1}{m} \sum_{j=1}^{m} y_{ij}$$

An estimator of μ is

$$\bar{\bar{y}} = \frac{1}{n} \sum_{i=1}^{n} \bar{y}_i \tag{8.25}$$

This estimator is the same as that of the single-stage cluster sampling when m = M.

The estimator $\bar{\bar{y}}$ is unbiased since

$$E(\bar{\bar{y}}) = EE(\bar{y}_i|n) = E\left(\frac{1}{n} \sum_{i=1}^{n} \sum_{j=1}^{M} \frac{y_{ij}}{M}\right)$$

$$= \frac{1}{NM} \sum_{i=1}^{N} \sum_{j=1}^{M} y_{ij} = \mu \tag{8.26}$$

The second expectation in the above formula is the conditional expectation given the n primary sampling units. The variance of $\bar{\bar{y}}$ is derived by using the following result.

Let Z = g(X,Y) be a function of two random variables; then the variance of Z is

$$V(Z) = E(Z^2) - [E(Z)]^2$$
$$= E(E(Z^2|y)) - [EE(Z|y)]^2$$
$$= E(E(Z^2|y) - E(E(Z|y))^2 + E(E(Z|y))^2 - [EE(Z|y)]^2$$
$$= E(V(Z|y)) + V(E(Z|y)) \quad (8.27)$$

his result provides a way to evaluate V(Z) as the sum of two compo-ents: one component is the expected value of the conditional vari-nce given Y = y, the other is the variance of the conditional mean iven Y = y.

Applying this formula to find $V(\bar{\bar{y}})$, we have

$$V(\bar{\bar{y}}) = V(E(\bar{\bar{y}}|n)) + E(V(\bar{\bar{y}}|n))$$
$$= V\left(\frac{1}{n}\sum_{i=1}^{n}\sum_{j=1}^{M}\frac{y_{ij}}{M}\right) + E\left(\frac{1}{n^2}\sum_{i=1}^{n}\left(\frac{1}{m} - \frac{1}{M}\right)S_i^2\right)$$
$$= \left(\frac{1}{n} - \frac{1}{N}\right)S_b^2 + \frac{1}{n}\left(\frac{1}{m} - \frac{1}{M}\right)\bar{S}_W^2 \quad (8.28)$$

here

$$S_i^2 = \frac{1}{M-1}\sum_{j=1}^{M}(y_{ij} - \mu_i)^2$$

$$\bar{S}_W^2 = \frac{1}{N}\sum_{i=1}^{N}S_i^2$$

n unbiased estimator of $V(\bar{\bar{y}})$ is

$$\hat{V}(\bar{\bar{y}}) = \left(\frac{1}{n} - \frac{1}{N}\right)\frac{1}{n-1}\sum_{i=1}^{n}(\bar{y}_i - \bar{\bar{y}})^2$$
$$+ \frac{1}{Nn}\left(\frac{1}{m} - \frac{1}{M}\right)\frac{1}{m-1}\sum_{i=1}^{n}\sum_{j=1}^{m}(y_{ij} - \bar{y}_i)^2 \quad (8.29)$$

The cluster sampling given in this section considers only clusters f equal size. One may extend to the case of unequal cluster size. 'urther, instead of simple random sampling, one may use systematic ampling or unequal probability sampling. These extensions together ith the problem of allocating sample sizes are beyond the scope of his book.

EXERCISES

8.1 Let T_i be defined as in Section 8.2. Show that $E(T_i^2) = n/N$ and $E(T_i T_j) = n(n - 1)/N(N - 1)$, $i \neq j$.

8.2 A simple random sample of size n is taken from a population of size N. The estimator for the population mean is $\bar{y}$. Show that an unbiased estimator of $V(\bar{y})$ is $(1 - f)\ \Sigma_{i=1}^{n}\ (y_i - \bar{y})^2/[n(n - 1)]$.

8.3 It is desired to estimate the total number of automobiles in households in a city. There are 11,048 households in the city. A simple random sample of size 40 is taken and the following data obtained:

Number of automobiles	0	1	2	3	4
Number of households	6	21	10	2	1

Estimate the total number of automobiles and estimate the variance of the estimator.

8.4 A finite population consists of N units of which N_1 units possess a rare attribute. It is desired to estimate the parameter N_1. The following sampling design is used. The sample units are drawn one by one with equal probability until m units possessing the rare attribute are selected. Let n denote the sample size which is a random variable. The sample can be drawn either (a) with replacement or (b) without replacement. Show that an unbiased estimator of N_1 is $N(m - 1)/(n - 1)$ for each of these two cases.

8.5 The N individuals in a population are classified into one of two classes, say C and $\bar{C}$ (the complement of C). Let

$$y_i = \begin{cases} 1 & \text{if the } i\text{th unit is in C} \\ 0 & \text{otherwise} \end{cases}$$

The population mean is the population proportion p of individuals in C. A simple random sample of size n is taken and the sample mean is the sample proportion $\hat{p} = \Sigma_{i=1}^{n}\ y_i/n$. Show that $\hat{p}$ is an unbiased estimator of p and

$$V(\hat{p}) = \left(\frac{N - n}{N - 1}\right)\frac{pq}{n}$$

where $q = 1 - p$.

8.6 In Exercise 8.5 show that an unbiased estimator of $V(\hat{p})$ is

$$\left(\frac{N - n}{N}\right)\frac{\hat{p}\hat{q}}{n - 1}$$

where $\hat{q} = 1 - \hat{p}$. *Hint:* Use the result in Exercise 8.2.

8.7 The variance formula for the systematic sample mean may be written in terms of the *intraclass correlation coefficient*, i.e., the correlation coefficient between pairs of units that are in the same systematic sample. The intraclass correlation coefficient is defined as

$$\rho = \frac{E(y_{ij} - \mu)(y_{iu} - \mu)}{E(y_{ij} - \mu)^2}$$

where the numerator is the average over all $kn(n - 1)/2$ distinct pairs and the denominator over all kn values of y_{ij}. Show that

$$V(\bar{y}_{sy}) = \frac{kn - 1}{kn}\frac{S^2}{n}[1 + (n - 1)\rho]$$

8.8 Suppose that the values of the N units labeled from 1 to N follow a periodic variation (e.g., daily temperature, daily traffic flow). Discuss the effect of periodic variation on the systematic mean in terms of the selection of the value of k in relation to the length of the period.

8.9 For a population that consists of two strata, the primary purpose of a survey is to estimate the difference of the two stratum means, $\mu_1 - \mu_2$. Independent simple random samples of sizes n_1 and n_2 are taken from the two strata. (a) Show that $\bar{y}_1 - \bar{y}_2$ is an unbiased estimator. (b) Find the variance of $\bar{y}_1 - \bar{y}_2$. (c) Obtain the optimum allocation of sample sizes with a cost function $C = c_0 + c_1n_1 + c_2n_2$.

8.10 A finite population consists of k strata. Let $\bar{y}_{st}$ be the estimator of the population mean from a stratified random sample. Assume that the cost function is of the form $C = c_0 + \Sigma_{i=1}^{k} c_i n_i$. Obtain the optimum allocation of the sample sizes which minimizes the cost function subject to the constraint $V(\bar{y}_{st}) = v$.

8.11 Consider the estimation of the proportion p for a finite population in single-stage cluster sampling with equal cluster size. A simple random sample of n clusters is taken. An estimator is $\hat{p} = (1/n) \Sigma_{i=1}^{n} p_i$, where p_i is the proportion for the ith cluster. Show that $\hat{p}$ is unbiased and find the variance of $\hat{p}$.

8.12 Suppose $Z = g(X,Y)$ and $U = h(X,Y)$ are functions of the two random variables X and Y. Show that

$$\mathrm{Cov}(Z,U) = E(\mathrm{Cov}(g,h)|y) + \mathrm{Cov}(E(g|y),E(h|y))$$

8.13 In two-stage sampling of clusters of equal size, the estimator of the population mean is $\bar{\bar{y}}$. Show that an unbiased estimator of $V(\bar{\bar{y}})$ is

$$\left(\frac{1}{n} - \frac{1}{N}\right) \frac{1}{n-1} \sum_{i=1}^{n} (\bar{y}_i - \bar{\bar{y}})^2 + \frac{1}{Nn}\left(\frac{1}{m} - \frac{1}{M}\right) \frac{1}{m-1} \sum_{i=1}^{n} \sum_{j=1}^{m} (y_{ij} - \bar{y}_i)^2$$

8.14 In two-stage sampling of clusters of equal size, let a simple random sample of n clusters be taken at the first stage and a systematic sample of size m at the second stage, where $M = km$. Define the estimator of the population mean by

$$\bar{\bar{y}}_{sy} = \frac{1}{n} \sum_{i=1}^{n} \bar{y}_{ij}$$

where $\bar{y}_{ij}$ is the systematic mean for the ith cluster. Show that $\bar{\bar{y}}_{sy}$ is unbiased and find its variance.

REFERENCES

Cochran, W. G. (1977). *Sampling Techniques*, 3rd ed. Wiley, New York.

Neyman, J. (1934). On the two different aspects of the representative method: The method of stratified sampling and the method of purposive selection. J. Roy. Stat. Soc. 97:558-606.

Sukhatme, P. V., and Sukhatme, B. V. (1970). *Sampling Theory of Surveys with Applications*, 2nd ed. Iowa State University Press, Ames.

Tschuprow, A. A. (1923). On the mathematical expectation of the moments of frequency distributions in the case of correlated observations. Metron 2:461-493, 646-683.

9

INTERVAL ESTIMATION

9.1 INTRODUCTION: In Chapters 7 and 8 we introduced point estimation for estimating the value of a parameter. A method which uses an interval rather than a point estimator is the method of *interval estimation*. This method derives limits C_1 and C_2 which are functions of the sample values $\{X_i\}$ or functions of the sample values and known population parameters. The interval (C_1,C_2) is called the *confidence interval*, which is determined so that, in repeated sampling from the same population, the interval will contain the parameter θ a certain percentage of the time. Symbolically C_1 and C_2 are determined so that

$$P(C_1 < \theta < C_2) \geq 1 - \alpha$$

that is, the probability of the random interval (C_1,C_2) containing the population parameter θ is greater than or equal to $1 - \alpha$. The value α is a small positive number less than 1. The probability holds only for a large number of similarly drawn samples, where C_1 and C_2 are recalculated for each sample. The value $1 - \alpha$ is called the *confidence probability* or *confidence coefficient*.

These concepts were introduced by Neyman (1935). R. A. Fisher uses the terms *fiducial interval* and *fiducial probability* to indicate substantially the same concepts, though he restricts his results to sufficient statistics.

Another method to construct an interval estimate of a parameter is by the use of the Bayesian principle. Just like the Bayes estimate given in Section 7.14, a Bayes interval for θ is obtained through the posterior distribution of θ. The interpretation of the Bayes interval is different from that of the confidence interval. The confidence interval approach treats the interval as a *random interval* and the probability that this random interval will include the *fixed* parameter θ with a confidence probability $1 - \alpha$. On the contrary, the Bayesian approach treats the parameter θ as a random variable and the observed data is fixed. Inference is made by considering the posterior distribution.

9.2 CONFIDENCE INTERVALS: We first consider the construction of a confidence interval. If $\hat{\theta}$ is a sufficient estimator such that

$$\prod_{i=1}^{n} f(x_i;\theta) = g(x_1,x_2,\ldots,x_n|\hat{\theta})h(\hat{\theta};\theta)$$

then the problem of estimating the confidence limits becomes one of finding the limits $\nu_1(\theta)$ and $\nu_2(\theta)$ such that

$$\int_{\nu_1(\theta)}^{\nu_2(\theta)} h(\hat{\theta};\theta)\, d\hat{\theta} = 1 - \alpha$$

It follows that

$$P(\nu_1(\theta) < \hat{\theta} < \nu_2(\theta)) = 1 - \alpha$$

We then solve the equations $\nu_1(\theta) = \hat{\theta}$ and $\nu_2(\theta) = \hat{\theta}$ for θ and obtain the solutions C_2 and C_1, respectively.

Let us consider the problem graphically. In Fig. 9.1 the curves $\nu_1(\theta)$ and $\nu_2(\theta)$ are drawn so that $P(\nu_1(\theta) < \hat{\theta} < \nu_2(\theta)) = 1 - \alpha$ if $f(x_i;\theta)$ is continuous, and $P \geq 1 - \alpha$ if $f(x_i;\theta)$ is discrete. Let C_1 and C_2 be the intersections of the straight line $\hat{\theta} = \hat{\theta}_0$ with the curves $\nu_2(\theta)$ and $\nu_1(\theta)$, respectively (note the interchange of subscripts). $\hat{\theta}_0$ is the particular value of $\hat{\theta}$ obtained from the sample. The line segment (C_1,C_2) will intersect the line $\theta = \theta_0$ (the true value of the parameter) only if $\hat{\theta}$ falls between ν_{10} and ν_{20}. But the probability of the latter event is

$$P(\nu_{10} < \hat{\theta} < \nu_{20}) = 1 - \alpha$$

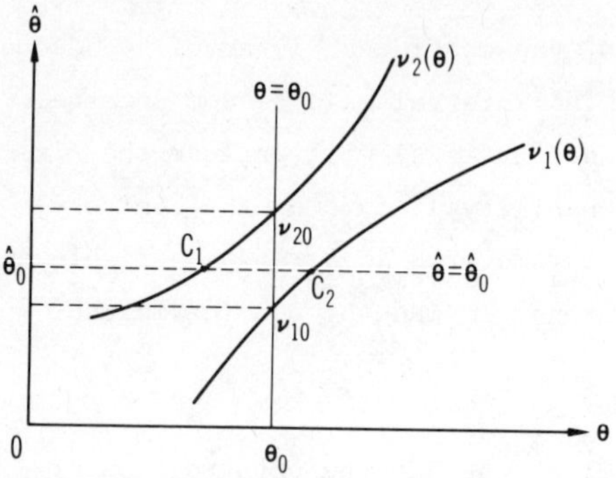

FIG. 9.1 Graphical representation of confidence interval.

Hence, $1 - \alpha$ is also the probability that the random interval (C_1, C_2) includes θ_0.

We may summarize as follows:

$$1 - \alpha = P(\nu_{10} < \hat{\theta} < \nu_{20} \mid \theta = \theta_0) = P(C_1C_2 \text{ intersects } \theta = \theta_0)$$
$$= P(C_1 < \theta_0 < C_2) \tag{9.1}$$

This does *not* imply that θ has a distribution or that on a given trial there is a probability that the true value θ_0 lies between C_1 and C_2. What *is* meant is that if a series of trials are made, in about $100(1 - \alpha)\%$ of these trials the *random interval* (C_1, C_2) will include θ_0, the true value of θ. In Section 9.5 the Bayesian approach is used; then θ has a distribution. This is the primary difference between Bayesian and non-Bayesian statistics.

9.3 SHORTEST CONFIDENCE INTERVAL: It is clear that there is an infinity of possible limits (ν_1, ν_2) such that $P(\nu_1 < \hat{\theta} < \nu_2) = 1 - \alpha$. For example, we might take $\nu_1 = -\infty$ so that $P(\hat{\theta} > \nu_2) = \alpha$, or we might take $P(\hat{\theta} > \nu_2) = P(\hat{\theta} < \nu_1) = \alpha/2$.* In determining which of the infinity of possible limits (ν_1, ν_2) to use, we shall usually wish to make the confidence interval as small as possible. If we consider

---

*It will be convenient, in general, to let $P(\hat{\theta} > \nu_2) = \alpha_2$ and $P(\hat{\theta} < \nu_1) = \alpha_1$, where $\alpha_1 + \alpha_2 = \alpha$.

only unbiased estimates which are asymptotically normally distributed, this interval can be made as small as possible by choosing the $\hat{\theta}$ with the smallest variance and selecting the limits (ν_1, ν_2) such that

$$\alpha_1 = \alpha_2 = \frac{\alpha}{2}$$

The maximum-likelihood (ML) estimator possesses this desired property. For detailed discussions of this topic, see Neyman (1935) and Wilks (1938).

For small samples, a confidence interval may be obtained by finding a function of X_1, X_2, ..., X_n and θ, say $Q(X_1,X_2,\ldots,X_n;\theta)$, such that the distribution of Q does not depend on θ. This function is called a *pivotal quantity*. Since the distribution of Q does not depend on θ, we have

$$P(q_1 < Q(X_1,X_2,\ldots,X_n;\theta) < q_2) = 1 - \alpha$$

where q_1 and q_2 are obtained from the distribution of Q and do not include θ. By reversing the inequalities, we may find C_1 and C_2 such that

$$P(C_1 < \theta < C_2) = 1 - \alpha$$

The limits C_1 and C_2, or equivalently q_1 and q_2, are determined such that the confidence interval is the shortest. A pivotal quantity can usually be found by considering the ML estimator and then making a transformation.

9.4 MORE THAN ONE UNKNOWN PARAMETER: If the distribution depends on several unknown parameters $\{\theta_i\}$, i = 1, 2, ..., h, there is a confidence interval for one of these parameters, say θ_1, if a function of the sample $(X_1,X_2,\ldots,X_n)$ and θ_1, $\phi(X_1,X_2,\ldots,X_n;\theta_1)$, can be found such that

$$\int_{\phi_1}^{\phi_2} f(\phi)\, d\phi = 1 - \alpha$$

where ϕ_1 and ϕ_2 are numerical values of $\phi = \phi(X_1, X_2, \ldots, X_n; \theta_1)$ and ϕ is independent of the other θ_i. In this case ϕ can serve as a pivotal quantity, and

$$P(\phi_1 < \phi\ (X_1, X_2, \ldots, X_n; \theta_1) < \phi_2) = 1 - \alpha$$

By reversing the inequalities $\phi < \phi_2$ and $\phi_1 < \phi$ we can find values of C_1 and C_2 such that

$$P(C_1 < \theta_1 < C_2) = 1 - \alpha$$

where C_1 and C_2 are functions of $(X_1, X_2, \ldots, X_n)$ and also ϕ_2 and ϕ_1, respectively.

The problem of finding a function ϕ which is independent of the other parameters is often quite difficult. The use of the usual t, used in t tests, to solve the Behrens-Fisher problem is an example of this (see Chapter 10). Hotelling introduced the term *nuisance parameters* to apply to these other parameters which appear in the distribution of the statistic but which we wish to eliminate when making statements concerning confidence limits for one of the parameters.

EXAMPLE 9.1 It is desired to construct the 95% confidence interval ($\alpha = .05$) for the mean μ of a $N(\mu, \sigma^2)$ population by use of a random sample of size n. We consider the following cases:

(a) σ^2 *Known*. The ML estimation of μ is $\bar{X} = \Sigma_{i=1}^{n} X_i/n$. Since $\bar{X}$ is $N(\mu, \sigma^2/n)$, it is evident that the shortest interval will be obtained by letting $\alpha_1 = \alpha_2 = .025$, that is,

$$\int_{-\infty}^{\nu_1} n\left(\mu, \frac{\sigma^2}{n}\right) d\bar{x} = \int_{\nu_2}^{\infty} n\left(\mu, \frac{\sigma^2}{n}\right) d\bar{x} = .025$$

where $n(\mu, \sigma^2/n)$ denotes the density function of the normal distribution with mean μ and variance σ^2/n. This is true because of the concentration of the probability, or area under the curve, about μ for the symmetrical normal distribution.

Now, the integral on the right above becomes

$$\sqrt{\frac{n}{2\pi\sigma^2}} \int_{\nu_2}^{\infty} e^{-n(\bar{x}-\mu)^2/2\sigma^2} \, d\bar{x} = .025$$

Let $z = \sqrt{n}(\bar{x} - \mu)/\sigma$, and the integral becomes

$$\int_{\sqrt{n}(\nu_2 - \mu)/\sigma}^{\infty} n(0,1) \, dz = .025$$

The random variable $Z = \sqrt{n}\,(\bar{X} - \mu)/\sigma$ has a standard normal distribution; hence, Z is a pivotal quantity. The value of the lower limit of this integral may be obtained from a table of areas for the standard normal curve. It will be found that

$$\frac{\sqrt{n}(\nu_2 - \mu)}{\sigma} = 1.96$$

and hence

$$\nu_2 = \mu + \frac{1.96\sigma}{\sqrt{n}}$$

Similarly,

$$\nu_1 = \mu - \frac{1.96\sigma}{\sqrt{n}}$$

It follows that

$$P\left(\mu - \frac{1.96\sigma}{\sqrt{n}} < \bar{X} < \mu + \frac{1.96\sigma}{\sqrt{n}}\right) = .95$$

Reversing the positions of μ and $\bar{X}$, we have

$$P\left(\bar{X} - \frac{1.96\sigma}{\sqrt{n}} < \mu < \bar{X} + \frac{1.96\sigma}{\sqrt{n}}\right) = .95 \tag{9.2}$$

Note that $C_1 = \bar{X} - (1.96\sigma/\sqrt{n})$ corresponds to $\nu_2 = \mu + (1.96\sigma/\sqrt{n})$, and similarly for C_2 and ν_1; that is, if

$$\bar{X} < \nu_2 = \mu + \frac{1.96\sigma}{\sqrt{n}}$$

then

$$\mu > \bar{X} - \frac{1.96\sigma}{\sqrt{n}} = C_1$$

Unless the nuisance parameter σ is known, these confidence limits are of little use.

(b) σ^2 *Unknown.* In this case we are concerned with two unknown parameters, μ and σ^2. It is known that $T = \sqrt{n}\ (\bar{X} - \mu)/S$ is distributed as Student's t with n - 1 degrees of freedom. Since the distribution of T does not depend on μ and σ^2, T is a pivotal quantity. Note that T is a function of only μ and the sample values n, $\bar{X}$, and S. It is possible then to find numerical values t_1 and t_2 such that

$$P(t_1 < T < t_2) = .95$$

Since T has a symmetrical distribution with its maximum density in the center, the shortest confidence interval with $\alpha = .05$ will result from setting

$$P(T > t_2) = P(T < t_1) = .025$$

In this case $t_1 = -t_2$. Since the values of t_1 and t_2 depend on n, we cannot find unique confidence limits as in (a) above. For n = 4 (3 degrees of freedom), $t_1 = -t_2 = -3.182$, and for n = 20 (19 degrees of freedom), $t_1 = -t_2 = -2.093$.

The reverse limits are obtained as follows: For $T = \sqrt{n}\ (\bar{X} - \mu)/S < t_2$, $\mu > \bar{X} - t_2 S/\sqrt{n}$, and similarly for $T > t_1$ we obtain $\mu < \bar{X} - t_1 S/\sqrt{n}$. Hence

$$P\left(\bar{X} - t_2 \frac{S}{\sqrt{n}} < \mu < \bar{X} - t_1 \frac{S}{\sqrt{n}}\right) = .95 \qquad (9.3)$$

where .95 is the confidence probability. The confidence limits are now independent of the nuisance parameter σ since T was used instead of Z, as in Section 6.11.

EXAMPLE 9.2 In Example 9.1(b), determine a 95% confidence interval for σ^2. We know that $V = (n - 1)S/\sigma^2$ is distributed as chi-square with n - 1 degrees of freedom. Let v_1 and v_2 be values such that

$$\int_{v_1}^{v_2} f(v)\ dv = .95$$

Then

$$P(v_1 < V < v_2) = .95$$

and for $V = (n - 1)S^2/\sigma^2 < v_2$, $\sigma^2 > (n - 1)S^2/v_2$. Similarly, for $V > v_1$, $\sigma^2 < (n - 1)S^2/v_1$. Hence, the confidence interval for σ^2 is

$$\frac{(n - 1)S^2}{v_2} < \sigma^2 < \frac{(n - 1)S^2}{v_1} \qquad (9.4)$$

As for Example 9.1(b), the values of v_1 and v_2 depend on n.

In this case the problem of selecting values of v_1 and v_2 in order to obtain the shortest confidence interval is more complicated because of the skewness of the χ^2 distribution. It is clear that the interval is proportional to

$$\frac{1}{v_1} - \frac{1}{v_2}$$

Let us illustrate the difficulty in selecting values of v_1 and v_2 in order to obtain the shortest confidence interval for the case n = 3 (2 degrees of freedom). We have

$$\int_{v_1}^{v_2} \frac{1}{2}e^{-v/2}\, dv = .95$$

(a) If we select v_1 and v_2 so that $\alpha_1 = \alpha_2 = .025$, then $v_2 = 7.38$, and $v_1 = .05066$. Hence

$$\frac{1}{v_1} - \frac{1}{v_2} = 19.8 - 0.1 = 19.7$$

and the confidence interval is $0.2S^2 < \sigma^2 < 39.6S^2$.

(b) If we select v_1 and v_2 so that $\alpha_1 = .05$ and $\alpha_2 = 0$, then $v_2 = \infty$ and $v_1 = .1026$. Hence

$$\frac{1}{v_1} - \frac{1}{v_2} = 9.7$$

The confidence interval now becomes

$$0 < \sigma^2 < 19.4S^2$$

which is much shorter than that obtained in (a) above.

(c) By minimizing $1/v_1 - 1/v_2$ subject to the relation

$$\int_{v_1}^{v_2} f(v)\, dv = 1 - \alpha$$

it is possible to show that the values v_1 and v_2 should satisfy the following relationship in order to provide the shortest confidence interval for σ^2:

$$v_1^2 f(v_1) = v_2^2 f(v_2)$$

In application, tables were computed to obtain the shortest confidence interval for σ^2 by Tate and Klett (1959). Improved confidence intervals were given by Cohen (1972).

9.5 BAYES INTERVAL: A different approach than that of confidence interval is the Bayes interval. From a Bayesian point of view, the parameter θ is a random variable and hence has a distribution. Therefore, probability statements concerning θ may be made by using the distribution of θ. When data are collected, the posterior distribution of θ, given data fixed, can be obtained as given in Section 7.14. The posterior density function of θ is denoted by $h(\theta|x_1,\ldots,x_n)$ which is proportional to the product of the prior density function $g(\theta)$ and the likelihood $f(x_1,\ldots,x_n|\theta)$. We can determine two limits a and b which are functions of the data $x_1, \ldots, x_n$ such that

$$\Pr(a < \theta < b) = \int_a^b h(\theta|x_1,\ldots,x_n)\, d\theta = 1 - \alpha$$

The interval (a,b) is a Bayes interval for θ. Note that in the above probability statement, θ is a random variable and a and b are functions of the fixed data $x_1, \ldots, x_n$.

EXAMPLE 9.3 In Example 7.12 a Bayes interval for μ is obtained as follows: The posterior distribution of μ given $x_1, \ldots, x_n$ is normal with mean

$$\left(\frac{1}{\tau^2} + \frac{n}{\sigma^2}\right)^{-1} \left(\frac{1}{\tau^2}\nu + \frac{n}{\sigma^2}\bar{x}\right)$$

and variance

$$\left(\frac{1}{\tau^2} + \frac{n}{\sigma^2}\right)^{-1}$$

Therefore the conditional probability of the interval

$$\frac{(1/\tau^2)\nu + (n/\sigma^2)\bar{x}}{1/\tau^2 + n/\sigma^2} - \frac{1.96}{\sqrt{1/\tau^2 + n/\sigma^2}}$$

$$< \mu < \frac{(1/\tau^2)\nu + (n/\sigma^2)\bar{x}}{1/\tau^2 + n/\sigma^2} + \frac{1.96}{\sqrt{1/\tau^2 + n/\sigma^2}} \qquad (9.5)$$

is .95. This provides a Bayes interval for μ with respect to the normal prior $N(\nu,\tau^2)$. Note that a different prior distribution will give a different posterior distribution, and hence will yield a different Bayes interval.

We may rewrite the above interval as

$$\frac{\sigma^2\nu/n\tau^2 + \bar{x}}{\sigma^2/n\tau^2 + 1} - \frac{1.96}{\sqrt{\sigma^2/n\tau^2 + 1}}\sqrt{\frac{\sigma^2}{n}}$$

$$< \mu < \frac{\sigma^2\nu/n\tau^2 + \bar{x}}{\sigma^2/n\tau^2 + 1} + \frac{1.96}{\sqrt{\sigma^2/n\tau^2 + 1}}\sqrt{\frac{\sigma^2}{n}}$$

When n is large or τ^2 is large (which corresponds to the case that the prior density becomes "flat"), the interval tends to

$$\bar{x} - 1.96\sqrt{\frac{\sigma^2}{n}} < \mu < \bar{x} + 1.96\sqrt{\frac{\sigma^2}{n}}$$

This is the same form given in Example 9.1(a).

EXERCISES

9.1 A random sample of size n is taken from a Bernoulli population with constant probability p. If we let $\alpha_1 = \alpha_2 = .025$ and assume that the sample estimator of p, $\bar{X}$, is approximately $N(p, p(1 - p)/n)$, show that the 95% confidence interval is $p_1 < p < p_2$, where p_1 and p_2 are solutions of the quadratic equation

$$n(\bar{X} - p)^2 = (1.96)^2(p - p^2)$$

9.2 Find the shortest $100(1 - \alpha)\%$ confidence interval for the difference between the means of two normal populations with the same unknown variance σ^2 with a sample of n_1 from the first population and of n_2 from the second population.

9.3 (a) Show that the confidence interval for the ratio of the variance of two normal populations

$$\theta = \frac{\sigma_1^2}{\sigma_2^2}$$

is given by

$$\frac{F_0}{F_2} < \theta < \frac{F_0}{F_1}$$

where

$$\int_{F_1}^{F_2} f(F)\, dF = 1 - \alpha$$

and $F_0 = s_1^2/s_2^2$ with $n_1 - 1$ and $n_2 - 1$ degrees of freedom, respectively,

(b) Given $n_1 = 12$ and $n_2 = 25$ and $\alpha = .10$, determine values of F_1 and F_2 if $\alpha_1 = \alpha_2$. Suppose $s_1^2 = 20$ and $s_2^2 = 10$, what are the 90% confidence limits?

9.4 Use the data in Exercise 6.38 to set up 90% confidence limits for the following ratios, used in statistical genetics:

(a) $\dfrac{\sigma_2^2 + 5\sigma_1^2}{\sigma_2^2}$

(b) $\dfrac{\sigma_1^2}{\sigma_2^2}$

(c) $\dfrac{\sigma_1^2}{\sigma_1^2 + \sigma_2^2}$

9.5 (For students who have studied advanced calculus.) Derive condition (c) of Example 9.2.

9.6 Using condition (c) of Example 9.2, derive the shortest confidence interval for σ^2 with $s^2 = 24.81$ for the data in Exercise 6.10, assuming μ and σ^2 unknown, $\alpha = .05$.

9.7 (a) Derive the general confidence limits for the mean of a linear function of the independently and normally distributed random variables $Y_1, Y_2, \ldots, Y_k$, that is,

$$L = \sum_{i=1}^{k} a_i Y_i$$

where $Y_i \sim N(\mu_i, \sigma^2)$, σ^2 is unknown, and an independent estimator S^2 of σ^2 is available; $\nu S^2/\sigma^2 \sim \chi^2(\nu)$.

(b) Apply your result to L_1 and L_2 in Exercise 6.8, assuming $r = 4$, $k = 4$, $T_0 = 50$, $T_1 = 105$, $T_2 = 95$, and $T_3 = 70$, and $S^2 = 12$ with 12 degrees of freedom. Let $\alpha = 0.05$.

9.8 Give a random sample of size n from the uniform distribution $f(x) = 1/\theta$, $0 < x < \theta$, and 0 otherwise. Show that $X_{(n)}/\theta$ is a pivotal quantity and use it to find a $100(1 - \alpha)\%$ confidence interval for θ.

9.9 A random sample of size n is taken from $f(x) = \theta e^{-\theta x}$, $x > 0$, and 0 otherwise. Find a $100(1 - \alpha)\%$ confidence interval for θ. *Hint:* Find the maximum-likelihood estimator of θ and its distribution.

9.10 Find a 95% confidence interval for the difference between the two means in Exercise 6.29.

9.11 Given a random sample of size n from a bivariate normal distribution, the maximum-likelihood estimator of $\mu_1 - \mu_2$ is given in Exercise 7.15. Use the ML estimator to obtain a $100(1 - \alpha)\%$ confidence interval for $\mu_1 - \mu_2$ when σ_1^2, σ_2^2, and σ_{12} are (a) known; (b) unknown.

9.12 Compute the Bayes interval for μ in Example 9.3 with $\nu = 0$, $\bar{x} = 0.5$, $\sigma^2 = 1$, and (a) $n = 10$, $\tau^2 = 1, 2, 3, 5, 20, 50$. (b) Compare the Bayes intervals with the confidence interval $\bar{x} - 1.96\sqrt{\sigma^2/n} < \mu < \bar{x} + 1.96\sqrt{\sigma^2/n}$.

9.13 Explain how to obtain a Bayes interval for p in Exercise 7.16.

9.14 Explain how to obtain a Bayes interval for m in Exercise 7.17.

REFERENCES

Cohen, A. (1972). Improved confidence intervals for the variance of a normal distribution. J. Amer. Stat. Assoc. 67:382-387.

Neyman, J. (1935). On the problem of confidence intervals. Ann. Math. Stat. 6:111-116.

Tate, R. F., and Klett, G. W. (1959). Optimal confidence intervals for the variance of a normal distribution. J. Amer. Stat. Assoc. 54:674-682.

Wilks, S. S. (1938). Shortest average confidence intervals for large samples. Ann. Math. Stat. 9:166-175.

10

TESTS OF HYPOTHESES

10.1 INTRODUCTION: Statistical inference concerns itself in general with two types of problems: *estimation of population parameters* and *tests of hypotheses*. In the preceding three chapters, we have considered problems of estimation. Desirable properties of a "good" estimator were discussed; and a principle of estimation, the maximum-likelihood method, was presented as a technique which in many cases may be easily used to obtain estimators possessing many of these desirable properties.

In this chapter, we propose to discuss the general problem of tests of hypotheses and present a principle, the *likelihood-ratio criterion*, which in many cases will provide a "good" *test criterion* to be used in testing hypotheses concerning population parameters. In discussing derived sampling distributions in Chapter 6, it will be recalled that the distributions of χ^2, t, and F were obtained and their uses in applied statistics in testing hypotheses were pointed out. In this discussion we wish to investigate the theoretical justification for selecting a particular test criterion for a particular problem in hand.

10.2 THE GENERAL PROBLEM: In order to test whether a given hypothesis (H_0, the *null hypothesis*) is supported by a given set of data, we must devise a rule of procedure, depending on the outcome of calculations obtained from the sample, to decide whether to reject or not to reject H_0. For example, in testing whether or not a given sample supports the hypothesis that the observations were randomly selected from N(0,1) we calculate $Z = \sqrt{n}\bar{X}$ and consider it a normal deviate with unit variance. After a choice of an allowable error or probability of rejecting H_0 when it is true, say α, we find two regions such that, if $\sqrt{n}\bar{X}$ is in one region, we reject H_0 and, if in the other region, we do not reject H_0. The first region will be called the *region of rejection*, R, and is defined so that the probability of the sample falling in R is α, if H_0 is true. Designate Z the *test criterion* and α the *significance level*.

As indicated earlier, there are many different criteria for judging the truth of a given hypothesis. For example, if we wish to test whether or not a given random sample of size n could have been drawn from N(0,1), we could use any one of the following tests (and probably many more):

1. $Z = \sqrt{n}\bar{X}$, a normal deviate with unit variance.
2. $T = \sqrt{n}\bar{X}/S$, Student's test.
3. $\chi^2 = (n - 1)S^2/1$, the χ^2 test for the agreement of the sample estimate of σ^2 and the population variance σ^2.
4. Tests for skewness or kurtosis in the distribution (evidence of nonnormality).
5. Tests for serial correlation in the observations.

The likelihood-ratio criterion, mentioned earlier, may be used to indicate which of the possible test criteria to use. Actually, the likelihood ratio defines a region of rejection R, which involves computing some test criterion such as one of those mentioned above.

The observations in the sample $\underline{X} = (X_1,X_2,\ldots,X_n)'$ may be thought of as representing the coordinates of a point in n-dimensional space. The space is divided into two regions--the region of rejection R and the region of nonrejection. If $\underline{X}$ falls in R, we shall reject H_0; otherwise accept it. The region R corresponds to the region outside the confidence interval discussed in the previous chapter and is defined so that the probability of rejecting a true hypothesis (the probability of $\underline{X}$ falling in R when H_0 is true) is the significance level α (for example, $\alpha = .05$ or $.01$). This will be indicated symbolically as

$$P(\underline{X} \in R|H_0) = \alpha$$

where $\underline{X} \in R$ means that the sample point $\underline{X}$ is contained in R.

As with confidence intervals, there may be a large number of possible regions R which satisfy this probability statement. For purposes of making tests of significance, it seems reasonable to select that R for which $P(\underline{X} \in R)$ is maximized if the true hypothesis is not H_0. That is, we want to reject the null hypothesis as often as possible when it is not true. Hence, we are led to consider the possible alternatives to H_0. Designate all alternative hypotheses as H_1. Symbolically we wish to maximize $P(\underline{X} \in R|H_1)$ for a fixed $P(\underline{X} \in R|H_0) = \alpha$. In future discussions we shall let $P(X \in R|H_1) = 1 - \beta$. The quantity $1 - \beta$ is called the *power* of the test, since it measures how powerful the test is in indicating a true difference from H_0 when such a difference actually exists. In most cases there will be an infinity of possible alternatives H_1, and $1 - \beta$ will be different for each H_1. Hence, in general, it will not be possible to maximize the power for all alternative hypotheses. However, in certain cases, such a test can be found and is called a *uniformly most powerful* test. It can be shown that the likelihood ratio criterion will produce a uniformly most powerful test if one exists.

An *unbiased* test is one for which the power is a minimum for $H = H_0$; that is, we reject H_0 the least number of times when H_0 is the true hypothesis. In case there is no overall uniformly most powerful test, it would appear that we should at least choose an unbiased test and, if possible, select the uniformly most powerful test from among the unbiased ones.

The theory of tests of hypotheses under discussion was introduced by Neyman and Pearson (1928, 1937, 1938). Tests of significance are considered from the point of view of errors of the first and second kind. In making a test of H_0, two types of errors may be committed: (I) we may reject H_0 when it is true; (II) we may accept H_0 when it is false. It follows that

$$P(I) = \alpha \qquad P(II) = \beta$$

Maximizing the power corresponds to minimizing the probability of committing a Type II error for a fixed probability of committing a Type I error.

The importance of taking into account the Type II error or the power of a test may be illustrated as follows: It is desired to determine whether or not a new variety of corn yields more than some standard variety. Suppose account is taken of a Type I error only and that the significance probability is fixed at $\alpha = .05$, which guarantees that if there is no real difference between the new variety and the standard, significance shall be indicated only 5% of the time. In this case it would not be necessary to perform an experiment at all. It would suffice to draw a bead at random from a bowl whose composition is 19 white and 1 red. If it is white, we accept H_0 (no difference between the varieties); if red, we reject H_0. Using the procedure, we shall always reject H_0 5% of the time, and this will be done whether H_0 is true or not. The defect in this procedure is now apparent. If there is a real difference, we shall recognize this only 5% of the time, which also, in this case, is the power of the test. It is now clear that we need a different testing procedure in order that the probability of detecting true differences when they exist may be large. The t test may be shown to maximize the power of the test under certain conditions. This will be discussed later.

Cramer (1946) has summarized the problem of selecting a "good" test criterion as follows: "In order that a test of the hypothesis H_0 should be judged to be good, we should accordingly require that the test has a small probability of rejecting H_0 when this hypothesis is true, but a large probability of rejecting H_0 when it is false. Of two tests corresponding to a probability α of rejecting H_0 when it is true, we should thus prefer the one that gives the largest probability of rejecting H_0 when it is false." Note that the two tests are comparable at the same α level; this α value is also called the *size* of the test [the size of a test is defined to be the supremum of P(I) over the null hypothesis space].

10.3 NEYMAN-PEARSON LEMMA: Before we discuss the Neyman-Pearson lemma, let us define two types of hypotheses. A hypothesis H_0 which specifies the values of all parameters in the population distribution is called a *simple hypothesis*; in other words, a simple hypothesis specifies that the distribution is one specific member of a family of distributions. If H_0 does not specify the values of all population parameters, it is called a *composite hypothesis*. The hypothesis H_0 must be taken from a set of admissible hypotheses, Ω, which usually depend upon the form of the distribution. For the normal parent distribution $N(\mu,\sigma^2)$, the parameter space Ω is $-\infty < \mu < \infty$, $0 < \sigma^2 < \infty$. A composite hypothesis then states that a distribution belongs to some subspace of the parameter space. We now state the Neyman-Pearson lemma:

NEYMAN-PEARSON LEMMA In testing a simple null hypothesis H_0 against a simple alternative hypothesis H_1, a sample $X_1, X_2, \ldots, X_n$ is obtained. Let $f_0(x_1,x_2,\ldots,x_n)$ and $f_1(x_1,x_2,\ldots,x_n)$ be the joint density functions under H_0 and H_1 respectively; then among all tests of size α $(0 < \alpha < 1)$, the test defined by the region of rejection R is most powerful where R is the region consisting of all sample points such that

$$\frac{f_0(x_1,x_2,\ldots,x_n)}{f_1(x_1,x_2,\ldots,x_n)} \leq k \tag{10.1}$$

the constant k being determined so that the size of the test is α.

To prove this lemma, let R* denote any region of rejection having size α. Assume that the random variables are continuous and for simplicity denote the multiple integral

$$\int_R \cdots \int f_0(x_1,x_2,\ldots,x_n)\, dx_1\, dx_2 \ldots dx_n$$

by $\int_R f_0(\underline{x})\, d\underline{x}$. Since both tests have the same size,

$$\int_R f_0(\underline{x})\, d\underline{x} = \int_{R^*} f_0(\underline{x})\, d\underline{x} = \alpha$$

Denoting the powers of the two tests with regions of rejection R and R* by $1 - \beta(R)$ and $1 - \beta(R^*)$, respectively, we wish to compare the two powers. The difference of the two powers is

$$[1 - \beta(R)] - [1 - \beta(R^*)] = \int_R f_1(\underline{x})\, d\underline{x} - \int_{R^*} f_1(\underline{x})\, d\underline{x} \qquad (10.2)$$

From Fig. 10.1, we see that $R = RR^* \cup R\bar{R}^*$ and $R^* = RR^* \cup \bar{R}R^*$. Therefore, Eq. (10.2) becomes

$$\int_R f_1(\underline{x})\, d\underline{x} - \int_{R^*} f_1(\underline{x})\, d\underline{x} = \int_{RR^*} f_1(\underline{x})\, d\underline{x} + \int_{R\bar{R}^*} f_1(\underline{x})\, d\underline{x}$$

$$- \int_{RR^*} f_1(\underline{x})\, dx - \int_{\bar{R}R^*} f_1(\underline{x})\, d\underline{x}$$

$$= \int_{R\bar{R}^*} f_1(\underline{x})\, d\underline{x} - \int_{\bar{R}R^*} f_1(\underline{x})\, d\underline{x}$$

Equation (10.1) defines the region R; hence in the region $R\bar{R}^*$ we have $f_1(\underline{x}) \geq f_0(\underline{x})/k$ and in the region $\bar{R}R^*$, $f_1(\underline{x}) \leq f_0(\underline{x})/k$. So

$$\int_{R\bar{R}^*} f_1(\underline{x})\, d\underline{x} - \int_{\bar{R}R^*} f_1(\underline{x})\, d\underline{x} \geq \frac{1}{k} \int_{R\bar{R}^*} f_0(\underline{x})$$

$$- \frac{1}{k} \int_{\bar{R}R^*} f_0(\underline{x})\, d\underline{x}$$

$$= \frac{1}{k} \left[\int_{R\bar{R}^*} f_0(\underline{x})\, d\underline{x} + \int_{RR^*} f_0(\underline{x})\, d\underline{x} \right.$$

$$\left. - \int_{\bar{R}R^*} f_0(\underline{x})\, d\underline{x} - \int_{RR^*} f_0(\underline{x})\, d\underline{x} \right]$$

$$= \frac{1}{k} \left[\int_R f_0(\underline{x})\, d\underline{x} - \int_{R^*} f_0(\underline{x})\, d\underline{x} \right] = \frac{1}{k} (\alpha - \alpha) = 0$$

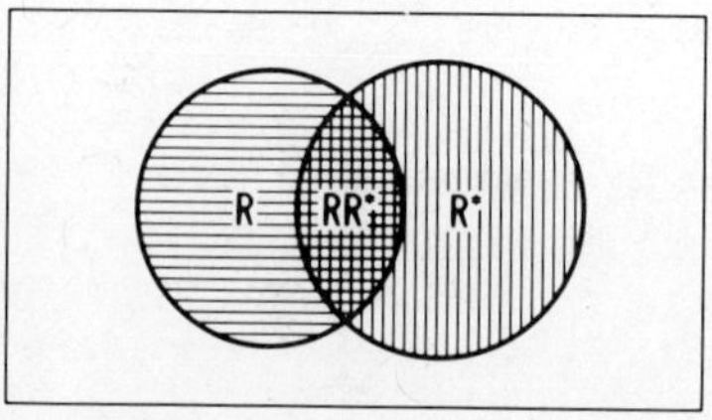

FIG. 10.1

Therefore, we have shown that the power of the test with region of rejection R is at least as large as any other test of the same size and the Neyman-Pearson lemma is proved.

The Neyman-Pearson lemma provides us a procedure to determine a most powerful test for testing simple H_0 against simple H_1. In fact, the test is constructed by taking the ratio of the likelihoods under H_0 and H_1. This leads us to consider the likelihood ratio criterion to be discussed later.

EXAMPLE 10.1 Let $X_1, X_2, \ldots, X_n$ be a random sample from $N(\mu,1)$. The simple null hypothesis is $H_0: \mu = \mu_0$ and the simple alternative hypothesis is $H_1: \mu = \mu_1$, where μ_0 and μ_1 are specified values. We assume that $\mu_0 < \mu_1$. By the Neyman-Pearson lemma, the most powerful test has the region of rejection R:

$$\frac{f_0(x_1,x_2,\ldots,x_n)}{f_1(x_1,x_2,\ldots,x_n)} = \frac{(2\pi)^{-n/2}\exp[-\Sigma(x_i - \mu_0)^2/2]}{(2\pi)^{-n/2}\exp[-\Sigma(x_i - \mu_1)^2/2]}$$

$$= \exp[-\Sigma(x_i - \mu_0)^2/2 + \Sigma(x_i - \mu_1)^2/2] \le k$$

Taking the logarithm, we have

$$-\frac{1}{2}\,\Sigma(x_i^2 - 2\mu_0 x_i + \mu_0^2) + \frac{1}{2}\,\Sigma(x_i^2 - 2\mu_1 x_i + \mu_1^2) \le \log k$$

$$-n(\mu_1 - \mu_0)\bar{x} + \frac{n}{2}\,(\mu_1^2 - \mu_0^2) \le \log k$$

where $\bar{x} = \Sigma x_i/n$. Since $\mu_1 > \mu_0$, the region of rejection is equivalent to $\bar{x} > k'$, where k' is determined such that the size of the test is α. Under H_0 the test statistic $\bar{X}$ is distributed as $N(\mu_0, 1/n)$, so k' satisfies the following equation:

$$P(\bar{X} \ge k') = \int_{k'}^{\infty} \frac{\sqrt{n}}{\sqrt{2\pi}}\, e^{-n(\bar{x}-\mu_0)^2/2}\, d\bar{x} = \alpha$$

We find

$$k' = \mu_0 + \frac{z_\alpha}{\sqrt{n}}$$

where z_α is the $100(1 - \alpha)$ percentage point of the standard normal distribution.

10.4 POWER FUNCTION FOR ONE-PARAMETER DISTRIBUTION: In order to illustrate the method of determining the power curve for a test, consider the problem in Example 10.1 of testing some hypothesis concerning the value of the mean μ of a normal population with unit variance $N(\mu,1)$. The sample observations may be n differences obtained from n pairs of observations such as in the corn variety problem mentioned earlier. In this case the population of differences would be assumed to have unit variance. Set up the null hypothesis $H_0: \mu = \mu_0$. The set of admissible hypotheses is $-\infty < \mu < \infty$, $\sigma^2 = 1$. The sample mean of the differences, $\bar{X}$, is to be used to test H_0 against some alternative H_1.

From the discussion on confidence intervals, we know that the probability of committing a Type I error is given by

$$P(I) = 1 - \sqrt{\frac{n}{2\pi}} \int_{\nu_1}^{\nu_{22}} e^{-n(\bar{x}-\mu_0)^2/2} \, d\bar{x} = \alpha$$

where $\nu_2 = \mu_0 + z_2/\sqrt{n}$ and $\nu_1 = \mu_0 + z_1/\sqrt{n}$. It will be shown later that for this particular problem, $\bar{X}$ is the "best" test criterion. If $\alpha = .05$, the shortest confidence interval was shown to result when $z_2 = -z_1 = 1.96$.

The integration is implified by setting $z = \sqrt{n}(\bar{x} - \mu_0)$ so that

$$\alpha = 1 - \frac{1}{\sqrt{2\pi}} \int_{z_1}^{z_2} e^{-z^2/2} \, dz$$

Hence, if the calculated $\bar{x}$ is greater than ν_2 or less than ν_1 or, what amounts to the same thing, if the value of z for a particular problem is greater than z_2 or less than z_1, then we say that we have evidence that H_0 is false.

In order to evaluate the power of this test to detect real differences, that is, $\mu = \mu_1 \neq \mu_0$, we obtain the power function. The power function will give the probability of rejecting H_0 when $\mu = \mu_1 \neq \mu_0$. The function should increase as the difference between μ_1 and μ_0 increases, since in such cases it would be increasingly desirable to reject H_0. The problem then is to determine the probability $1 - \beta$ of obtaining $\bar{X} \geq \nu_2$ or $\bar{X} \leq \nu_1$ from $N(\mu_1, 1)$ for a fixed α. Note that the power, $1 - \beta$, equals α for $\mu_1 = \mu_0$. Now,

$$1-\beta = 1 - \sqrt{\frac{n}{2\pi}} \int_{\nu_1}^{\nu_2} e^{-n(\bar{x}-\mu_1)^2/2} \, d\bar{x}$$

$$= 1 - \sqrt{\frac{1}{2\pi}} \int_{z_1}^{z_2} e^{-(1/2)[z-\sqrt{n}(\mu_1-\mu_0)]^2} \, dz$$

$$= \frac{1}{\sqrt{2\pi}} \int_{-\infty}^{b} e^{-u^2/2} \, du + \frac{1}{\sqrt{2\pi}} \int_{a}^{\infty} e^{-u^2/2} \, du$$

where

$$u = \sqrt{n}(\bar{x} - \mu_1) = z - \sqrt{n}(\mu_1 - \mu_0)$$

$$a = z_2 - \sqrt{n}(\mu_1 - \mu_0)$$

$$b = z_1 - \sqrt{n}(\mu_1 - \mu_0)$$

To illustrate the computations, let us compute $1 - \beta$ for $n = 25$, $\alpha = .05$, and (i) $z_1 = -\infty$, $z_2 = 1.645$ and (ii) $z_2 = -z_1 = 1.96$. We see that the value of the sum of the two integrals giving $1 - \beta$ may be obtained from Table C.1.b (Appendix C) for given $(\mu_1 - \mu_0)$ as set out in Table 10.1.

It might be helpful if these results were illustrated graphically as in Fig. 10.2. Let us consider the difference between the power for $5(\mu_1 - \mu_0) = d = 0$ and the power for $d = 1$. The small shaded area is the rejection region for $d = 0$, with $\alpha = .05$. The upper shaded area is the added amount to the rejection area for $d = 1$. The sum of the two shaded areas gives the *power* for rejecting the hypothesis that $\mu_1 = \mu_0$ when μ_1 is actually $\mu_0 + .2$, since $1 = 5(\mu_1 - \mu_0)$. In this case $1 - \beta = .2594$.

TABLE 10.1

$5(\mu_1 - \mu_0)$	(i)		(ii)		
	a	$1 - \beta$	a	b	$1 - \beta$
-3.0	4.645	.0000	4.960	1.040	.8508
-2.0	3.645	.0001	3.960	.040	.5160
-1.0	2.645	.0041	2.960	- .960	.1700
.0	1.645	.0500	1.960	-1.960	.0500
1.0	.645	.2594	.960	-2.960	.1700
2.0	- .355	.6387	- .040	-3.960	.5160
3.0	-1.355	.9123	-1.040	-4.960	.8508
4.0	-2.355	.9907	-2.040	-5.960	.9793

Graphs of various power functions for the above example, depending on the values of z_1 and z_2, with fixed $P(I) = \alpha$, and on the position of the minimum point are set out in Fig. 10.3:

i. $z_1 = -\infty$, $z_2 = 1.645$

ii. Minimum at $\mu_1 = \mu_0$, $z_2 = 1.96$, $z_1 = -1.96$

iii. $z_2 = \infty$, $z_1 = -1.645$

iv. Minimum at $\mu_1 < \mu_0$, for example, $z_2 = 1.75$, $z_1 = -2.33$

v. Minimum at $\mu_1 > \mu_0$, for example, $z_2 = 2.33$, $z_1 = -1.75$

From the definition of $1 - \beta$, we see that the minimum point is reached when

$$z_2 - \sqrt{n}(\mu_1 - \mu_0) \doteq -[z_1 - \sqrt{n}(\mu_1 - \mu_0)]$$

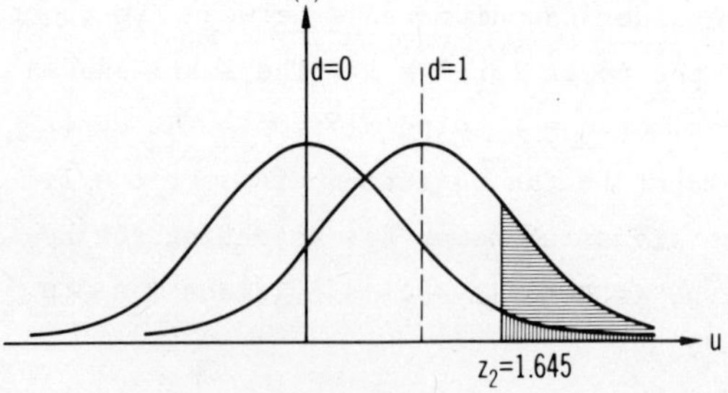

FIG. 10.2 Rejection region.

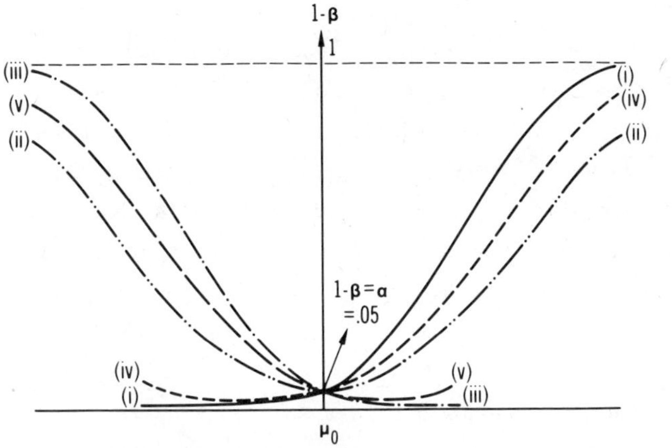

FIG. 10.3 Power functions.

or

$$z_1 + z_2 = 2\sqrt{n}(\mu_1 - \mu_0)$$

This relation is found by differentiating $1 - \beta$ with respect to μ and setting the derivative equal to zero. It follows that if $\mu_1 = \mu_0$, the minimum point is at $z_1 = -z_2$.

It is instructive to examine these *power curves* from the standpoint of the following types of alternative hypotheses:

1. If H_1 asserts $\mu > \mu_0$, power curve i is *uniformly most powerful*, since its power is greater than that of any other curve for all such H_1. In this case we are willing to accept H_0: $\mu = \mu_0$ even though the true hypothesis is $\mu < \mu_0$. Hence, the region of rejection is $Z > 1.645$ for $\alpha = .05$.

2. Similarly, power curve iii is uniformly most powerful for testing the hypothesis $\mu = \mu_0$ against the alternatives $\mu < \mu_0$. The region of rejection is $Z < -1.645$ for $\alpha = .05$.

3. There is no uniformly-most-powerful test for testing H_0 against the alternative H_1: $\mu \neq \mu_0$. This result is evident from a study of curves i and iii. Each of these is uniformly most powerful on opposite sides of $\mu = \mu_0$ but has practically no power on the other

side. No other single test can be as powerful as curve i on the right or curve iii on the left. For $H_1: \mu \neq \mu_0$ we must adopt some compromise rejection region. Neyman and Pearson have suggested using an *unbiased* test, which in this case would lead to the use of power curve ii. For this curve, $z_1 = -z_2$. It should be noted that curves iv and v are more powerful than ii on one tail but give a power less than α for some alternatives. It should be emphasized that the Type I error (probability of rejecting H_0 when $\mu = \mu_0$) is constant for all of these power curves.

10.5 COMPOSITE HYPOTHESES: We now consider some methods for testing composite hypotheses. These will be illustrated by use of the single-tiled t test. Given a random sample of size n from $N(\mu,\sigma^2)$, it is desired to test the null hypothesis $H_0: \mu = 0,\ 0 < \sigma^2 < \infty$, against the alternative hypothesis $H_1: \mu > 0,\ 0 < \sigma^2 < \infty$. The admissible hypotheses are $\Omega: -\infty < \mu < \infty,\ 0 < \sigma^2 < \infty$. The null hypothesis is composite since it does not specify the value of σ^2.

From our study of confidence intervals, it seems reasonable to use $-\infty < \bar{X} < t_2 S/\sqrt{n}$ as the acceptance region, where

$$\int_{t_2}^{\infty} f(t)\, dt = \alpha$$

The rejection region is $T = \bar{X}\sqrt{n}/S > t_2$. (If H_1 was $\mu < 0$, we should use the acceptance region $T > t_1 = -t_2$ or rejection region $T < t_1$. Both of these yield uniformly most powerful tests. However, if H_1 was $\mu \neq 0$, no uniformly most powerful test is available. In this case we might make use of the unbiased test with acceptance region $t_1' < T < t_2'$, where $t_1' = -t_2'$ and $\alpha_1 = \alpha_2 = \alpha/2$. A more rigorous treatment of this problem is beyond the scope of this text.)

The determination of the power for a composite test is, in general, quite complicated owing to the nonspecificity of certain of the parameters by the null hypothesis. In using the t test, H_0 does not specify the value of σ^2; hence, we must make use of S^2, the estimator of σ^2, from the sample. We must determine the probability that

$T = \bar{X}\sqrt{n}/S > t_2$ if the sample has been drawn from a population with mean $\mu \neq 0$. Now, when $T > t_2$, we reject H_0; and if μ is actually greater than zero, a correct decision has been made. The probability of making this correct decision is called the power of the test for a given μ ($\neq 0$) and α. It will be recalled that α is the probability of stating $\mu \neq 0$ when it actually is zero.

To evaluate the power of the t test, we recall the joint distribution of $\bar{X}$ and S^2 from Chapter 6:

$$f(\bar{x}, s^2) = f_1(\bar{x})\, f_2(s^2)$$

where

$$f_1(\bar{x}) = \sqrt{\frac{n}{2\pi}} \cdot \frac{1}{\sigma}\, e^{-n(\bar{x}-\mu)^2/2\sigma^2}$$

and

$$f_2(s^2) = \frac{[(n-1)/2]^{(n-1)/2}}{\sigma^2\Gamma[(n-1)/2]} \left(\frac{s^2}{\sigma^2}\right)^{(n-3)/2} e^{-(n-1)s^2/2\sigma^2}$$

Now, the power of the t test to detect a true mean $\mu = \mu_1$ is given by $P[T = \bar{X}\sqrt{n}/S > t_2 \mid \mu = \mu_1, P(I) = \alpha] = 1 - \beta$, where

$$P(I) = P(T > t_2 | \mu = 0) = \alpha$$

It will be recalled that the t distribution was obtained from that of $\bar{X}$ and S^2 in Chapter 6. Since $P(T > t_2) = P(\bar{X} > St_2/\sqrt{n})$, we may find the power of the test from

$$1 - \beta = \int_0^\infty f_2(s^2)\, d(s^2) \int_{st_2/\sqrt{n}}^\infty \sqrt{\frac{n}{2\pi}} \cdot \frac{1}{\sigma}\, e^{-n(\bar{x}-\mu_1)^2/2\sigma^2}\, d\bar{x}$$

where it is understood that we first determine t_2 so that $P(I) = \alpha$ and then $1 - \beta$.

Let $\mu_1 \sqrt{n}/\sigma = t_a$ (known value) and $\bar{x}\sqrt{n}/\sigma = t_a + y$; then

$$1 - \beta = \int_0^\infty f_2(s^2)\, d(s^2) \int_{y_2=(st_2/\sigma)-t_a}^\infty n(0,1)\, dy$$

But

$$\frac{s^2}{\sigma^2} = \frac{\chi^2}{n - 1}$$

and hence

$$1 - \beta = \int_0^\infty f(\chi^2)\, d(\chi^2) \int_{y_2}^\infty n(0,1)\, dy$$

where $y_2 = \chi t_2/\sqrt{n - 1} - t_a$.

The evaluation of $1 - \beta$ must be accomplished by some form of numerical integration over the region $y > \chi t_2/\sqrt{n - 1} - t_a$ and $\chi^2 > 0$. If we let

$$p = \int_0^{\chi^2} f(u)\, du$$

then

$$\frac{dp}{d\chi^2} = f(\chi^2)$$

Since $p = 0$ when $\chi^2 = 0$ and $p = 1$ when $\chi^2 = \infty$, we may change to a (p,y) coordinate system where $0 < p < 1$. We may now compute the power of the test to detect a given value of $t_a \neq 0$ for a fixed t_2 and n. In order to compute the power for $t = t_a$, we proceed as follows:

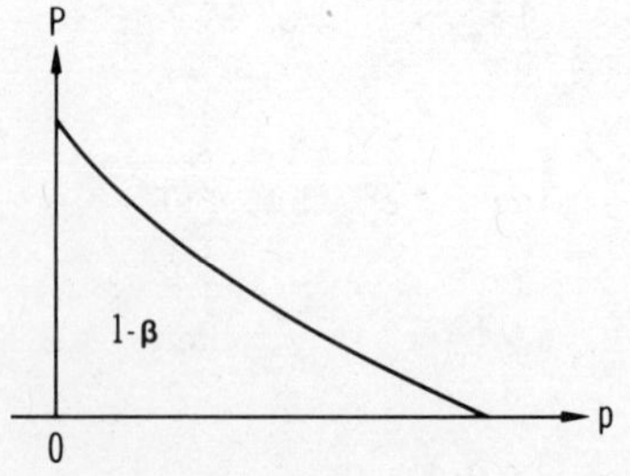

FIG. 10.4 Computation of power.

TABLE 10.2

p	χ	y_2	P
.90	2.146	3.282	.00052
.80	1.794	2.555	.00531
.70	1.552	2.054	.01996
.60	1.354	11.645	.04994
.50	1.177	1.281	.10004
.40	1.011	.9373	.17430
.30	.8446	.5941	.27622
.25	.7585	.4163	.33860
.20	.6680	.2295	.40924
.15	.5701	.0273	.48911
.10	.4590	- .2021	.58008
.075	.3949	- .3346	.63104
.05	.3203	- .4886	.68744
.025	.2250	- .6853	.75342
.020	.2010	- .7349	.76880
.015	.1739	- .7910	.78553
.010	.1418	- .8572	.80433
.005	.1001	- .9433	.82724
.0025	.07075	-1.0039	.84229
.00	.00	-1.1500	.87493

1. Set down successive values of p.
2. Ascertain the values of χ corresponding to each p.
3. Compute the value of y_2 for each χ.
4. Determine the area P under the normal curve between y_2 and ∞.

If we plot the values of P as ordinates with the corresponding p values as the abscissas, then $1 - \beta$ is given by the area under this curve as illustrated in Fig. 10.4. This area may be computed by some method of numerical integration, such as the trapezoidal rule or Simpson's rule.

Neyman and Tokarska (1936) have published values of t_a for $1 - \beta = .99, .95, .90(.10).10$.* Using the procedure outlined above, let us calculate $1 - \beta$ for $t_a = 1.15$, which is the value of t given in the Neyman and Tokarska tables, corresponding to $\alpha = .05$, $1 - \beta = .20$, and $n = 3$.† If $\alpha = .05$, then $t_2 = 2.920$ and $y_2 = 2.065\chi - 1.15$. We now obtain the entries in Table 10.2.

Using the trapezoidal rule, we obtain $1 - \beta = .2017$ as compared with the actual value of .20 mentioned earlier.

10.6 USE OF POWER FUNCTION TABLES IN PLANNING EXPERIMENTS: In case the experimenter has some knowledge in advance of the size of the *coefficient of variation*, that is, the standard deviation of any observation expressed as a percentage of the general mean, it will then be possible to make use of the tables of Neyman and Tokarska in the planning of experiments.

EXAMPLE 10.2 (Due to Neyman and Tokarska) A plant breeder wishes to compare a new variety V_1 with an established standard V_0. Let μ_1 and μ_0 denote the true mean yields of V_1 and V_0 per some unit of area, respectively. The hypothesis to be tested is H_0: $\mu_1 \leq \mu_0$, and the alternative hypothesis is H_1: $\mu_1 > \mu_0$. In other words, the plant breeder will consider his problem of producing a better variety as successfully accomplished whenever he obtains evidence that H_0 is not true and therefore that $\mu_1 > \mu_0$. It is desired to reduce the probability of an unjust rejection of H_0 to $\alpha = .01$. In a completely randomized experiment each variety is repeated $n' = 8$ times, and hence the pooled experimental error degrees of freedom is 14. According to previous experience the standard deviation of any single yield is expected to be $\sigma_0 = 6\%$ of the general mean yield. The experimenter now wishes to know the size of differences between the mean yields of varieties V_0 and V_1 (in favor of the new variety V_1) which he is likely to detect in his experiment in case they in fact exist.

---

*$1 - \beta$ may be found from the tables of Neyman and Tokarska by the relationship $\beta = P_{11}$.

†The n of the tables is the degrees of freedom.

Now in order to use Table II from Neyman and Tokarska (1936), we find the standard deviation of the difference of the two means as

$$\sigma = \sigma_0 \sqrt{\frac{2}{n'}} = 6\sqrt{\frac{2}{8}} = 3\% \text{ of the general mean}$$

But $\Delta = \rho\sigma = 3\rho\%$ of the general mean, where $\Delta = \mu_1 - \mu_0$ and $\rho = (\mu_1 - \mu_0)/\sigma$ by definition. Then, entering Table II opposite $n = 14$ degrees of freedom, we multiply the tabled values of ρ by 3 to obtain the entries in Table 10.3. The first pair of entries means that if the true difference in mean yield of V_1 over V_0 is as large as 15.54% of the general mean yield, then the experiment described will detect this difference in 99% of the cases. From the table it may be seen that a reasonable probability of detection such as .90 or .80 corresponds to true differences in yields exceeding 10% of the general mean and that differences under 5% have a probability of only .20 of being detected. The experimenter now possesses information enabling him to judge the adequacy of the proposed experiment. The experiment would be judged satisfactory if it is desired to discover differences over 10%. On the other hand, if the process of improving the particular varieties is well advanced, a difference as large as 5% may be as large as could be expected. In the latter case the proposed experiment is not satisfactory, and some modification is in order. Increased precision may be obtained by (1) increasing the number of repetitions and thereby increasing the degrees of freedom; (2) improving the experimental techniques and thereby decreasing the standard deviation of any single plot yield; or (3) increasing the size of α.

TABLE 10.3

(Level of Significance = .01)

Size of real differences	15.54	13.26	12.03	10.53	9.48	6.87	5.97	4.92	3.45
Probability of detection	.99	.95	.90	.80	.70	.40	.30	.20	.10

EXAMPLE 10.3 Tang (1938) has obtained the functional form of the power function of the analysis-of-variance tests and provides tables with illustrations of their uses. While the derivations are beyond the scope of this text, it is instructive to consider one of Tang's examples illustrating the use of his tables in planning experiments. A randomized-blocks experiment is planned to compare four treatments (k = 4) replicated five (n = 5) times. Let δ_i be the difference between the true ith-treatment effect and the true general mean, so that $\Sigma_{i=1}^k \delta_i = 0$. Suppose that for the experiment the δ_i have values -5, -4, 3, 6, expressed as percentages of the mean yield per plot. Further, suppose from past experience that the true standard deviation per plot, σ, is 10% of the general mean. In order to enter Tang's tables we calculate

$$\phi = \frac{\sqrt{(1/k)\ \Sigma_i\ \delta_i^2}}{\sigma/\sqrt{n}} = \frac{\sqrt{(25 + 16 + 9 + 36)/4}}{10/\sqrt{5}} = 1.04$$

Entering Table II from Tang, with degrees of freedom $f_1 = 3$, $f_2 = 12$, and $\phi = 1.04$, we find $P_{11} = .7$ roughly. This means that true treatment differences, such as those given above, would be significant at the 5% level in about 3 experiments out of 10 only.

In practice the true treatment differences are not known, but use may be made of the fact that if ϕ were as large as some specified value ϕ_0, say, the probability P_{11} of failing to detect the existence of treatment differences may be obtained from Tang's tables.

In a second example Tang considers a randomized-blocks experiment with k = 6 treatments and n = 7 blocks. Then $f_1 = 5$, $f_2 = 30$, and Table II, appropriate when using the 5% significance level, shows $P_{11} = .262$ for $\phi = 1.5$. In this case we would fail to detect the presence of treatment differences in about 1 in 4 times when ϕ is as large as 1.5 or when

$$\sqrt{\frac{1}{6}\ \Sigma_i\ \delta_i^2} = \frac{\sigma}{\sqrt{7}} \times 1.5 = 0.567\sigma$$

Assuming the standard deviation of a plot to be about 10% of the mean yield, then $\sqrt{(1/6)\Sigma_i \delta_i^2} = 5.67\%$ of the mean yield per plot. Now, there will be an unlimited number of sets of 6 values of δ_i, whose sum will be zero and having 5.67 as standard deviation. In order to obtain upper and lower bounds for at least one value of the δ_i, we consider the two extreme sets

$$\text{(a)} \quad \delta_1 = \delta_2 = \delta_3 = \delta_4 = \delta_5 = -\frac{\delta_6}{5}$$

$$\text{(b)} \quad \delta_1 = \delta_2 = \delta_3 = -\delta_4 = -\delta_5 = -\delta_6$$

For (a) we find

$$\delta_6 = \sqrt{k-1}\sqrt{\frac{1}{k}\Sigma_i \delta_i^2} = 12.68$$

and for (b)

$$\delta_6 = \sqrt{\frac{1}{k}\Sigma_i \delta_i^2} = 5.67$$

It may be proved, for this example, that there must be at least one δ, say δ_6, whose value lies between 12.68 and 5.67.

10.7 THE LIKELIHOOD-RATIO CRITERION: In Section 7.11 the method of maximum likelihood was presented as a general method, involving routine mathematical procedures, for obtaining an estimator of a population parameter possessing many desirable properties. In an analogous manner the *likelihood-ratio criterion* will now be presented as a general method, involving routine mathematical procedures, for obtaining a "good" test criterion.

The procedure for obtaining a likelihood-ratio criterion to be used in testing the hypothesis that $(\theta_1, \theta_2, \ldots, \theta_k)$ belongs to the subspace ω of the entire parameter space Ω on the basis of the random sample $X_1, X_2, \ldots, X_n$, drawn from the population with density function $f(x; \theta_1, \theta_2, \ldots, \theta_k)$, is set out below.

Let the likelihood be

$$L = \prod_{i=1}^{n} f(X_i;\theta_1,\theta_2,\ldots,\theta_k)$$

This likelihood will usually have a maximum as the parameters vary over the entire parameter space Ω. Denote this maximum value by

$$L(\hat{\theta}_1,\hat{\theta}_2,\ldots,\hat{\theta}_k)$$

or briefly as $L(\hat{\Omega})$. Similarly, L will usually have a maximum value in ω which shall be denoted as $L(\hat{\omega})$. Then the likelihood-ratio criterion for the hypothesis to be tested is

$$\lambda = \frac{L(\hat{\omega})}{L(\hat{\Omega})} \tag{10.3}$$

The estimators $\hat{\theta}_i$ of the population parameters θ_i, which are obtained as quantities to be substituted in L determining $L(\hat{\omega})$ and $L(\hat{\Omega})$, are derived by the method of maximum likelihood. It follows that λ is a function of the sample observations only, that is, it does not involve any population parameters.

Since L is positive as a result of being the product of density functions and $L(\hat{\omega})$ is less than or at most equal to $L(\hat{\Omega})$ because we are more restricted in maximizing L in ω than in Ω, λ will be a positive fraction. Its range will be from 0 to 1.

In order to use λ as a test criterion in applied statistics, it is necessary that we obtain the sampling distribution of λ on the assumption that the hypothesis being tested is true. We note that λ will be small if $L(\hat{\omega})$ is smaller than $L(\hat{\Omega})$. We shall wish to reject the hypothesis to be tested in case λ is small. When the exact sampling distribution of λ is complicated, we may use an asymptotic distribution. In large samples, the distribution of $-2 \log \lambda$ is approximately χ^2 with degrees of freedom equal to the difference of the number of parameters estimated under Ω and ω.

We now find a λ_α such that $P(\lambda \leq \lambda_\alpha) = \alpha$ on the assumption that the hypothesis to be tested is true. If the calculated value λ_0 is less than or equal to λ_α, that is, if $\lambda_0 \leq \lambda_\alpha$, we reject the hypothesis; otherwise we accept it. It should be noted that any monotonic function of λ may be used in place of λ as the test criterion.

As indicated earlier, a "good" test criterion is one which determines a region which maximizes the power of detecting true deviations from the null hypothesis for a given probability of committing a Type I error. In general, it will not be possible to find a region of rejection which will maximize the power for all alternatives to the null hypothesis. However for the simple case of only one alternative, H_1, and when both H_0 and H_1 are simple hypotheses, it was shown by the Neyman-Pearson lemma that the likelihood-ratio test defines a *best critical region*. In this case the whole parameter space contains only two points. If the alternative hypotheses H_1 specify the entire parameter space Ω other than ω to be a range of values on a line, then it is possible to choose a best critical region for each H_1. If this region is the same for each H_1, then the test is said to be uniformly most powerful.

EXAMPLE 10.4 Given a random sample of n from $N(\mu,1)$, the null hypothesis to be tested is H_0: $\mu = \mu_0$, which states that ω is a point while Ω is the whole μ axis. The likelihood of the sample is

$$L = \left(\frac{1}{\sqrt{2\pi}}\right)^n e^{-(1/2)\Sigma(X_i-\mu)^2}$$

or

$$L = \left(\frac{1}{\sqrt{2\pi}}\right)^n e^{-(1/2)\Sigma(X_i-\bar{X})^2-(n/2)(\bar{X}-\mu)^2}$$

Since the maximum-likelihood (ML) estimator for μ is $\bar{X}$, we find the maximum value of L in Ω to be

$$L(\hat{\Omega}) = \left(\frac{1}{\sqrt{2\pi}}\right)^n e^{-(1/2)\Sigma(X_i-\bar{X})^2}$$

Also,

$$L(\hat{\omega}) = \left(\frac{1}{\sqrt{2\pi}}\right)^n e^{-(1/2)\Sigma(X_i-\bar{X})^2-(n/2)(\bar{X}-\mu_0)^2}$$

The likelihood ratio becomes

$$\lambda = e^{-(n/2)(\bar{X}-\mu_0)^2}$$

If $\bar{X}$ is close to μ_0 in value, then the sample is reasonably consistent with the null hypothesis H_0 and λ will be close to 1 in value. Conversely, the sample will not be reasonably consistent with H_0, and λ will ordinarily be close to zero.

Now, suppose for the above example, or in general, the distribution of λ when H_0 is true is $g(\lambda)$ and $P(I) = \alpha$; then λ_α is determined so that

$$\int_0^{\lambda_\alpha} g(\lambda)\, d\lambda = \alpha$$

If the calculated λ, say λ_0, be less than λ_α, we would reject H_0, and vice versa. Note that $\lambda \leq \lambda_\alpha$ is equivalent to $\bar{X} \geq \nu_2$ or $\bar{X} \leq \nu_1$, which is the test criterion discussed in Section 10.4.

From the discussion in the above paragraph it follows that the likelihood-ratio method as described may not always lead to a unique test. If H_0 is a simple hypothesis, a unique distribution of λ may be obtained. On the other hand, if H_0 is a composite hypothesis, it will not in general be possible to obtain a unique distribution for λ and hence no unique test.

EXAMPLE 10.5 Given a random sample of size n from $N(\mu,\sigma^2)$, the null hypothesis to be tested is H_0: $\mu = 0$, σ^2 is unspecified, and the alternative hypothesis is H_1: $\mu \neq 0$. The entire parameter space is the half-plane of Fig. 10.5. The subspace specified by H_0 is the vertical line $\mu = 0$.

The likelihood of the sample is

$$L = \left(\frac{1}{\sigma\sqrt{2\pi}}\right)^n e^{-(1/2)\Sigma(X_i-\mu)^2/\sigma^2}$$

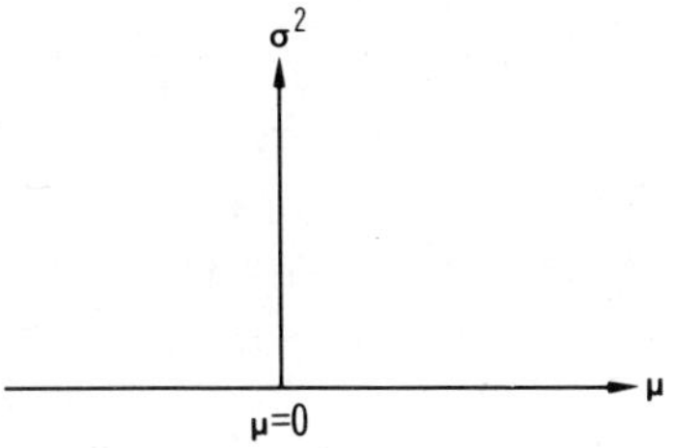

FIG. 10.5

The values of μ and σ^2 which maximize L in Ω have already been found to be

$$\hat{\mu} = \frac{1}{n}\Sigma X_i = \bar{X}$$

$$\hat{\sigma}^2 = \frac{1}{n}\Sigma(X_i - \bar{X})^2$$

Hence,

$$L(\hat{\Omega}) = \left[\frac{2\pi}{n}\Sigma(X_i - \bar{X})^2\right]^{-n/2} e^{-n/2}$$

Also,

$$L(\hat{\omega}) = \left[\frac{2\pi}{n}\Sigma X_i^2\right]^{-n/2} e^{-n/2}$$

Hence

$$\lambda = \left[\frac{\Sigma(X_i - \bar{X})^2}{\Sigma X_i^2}\right]^{n/2}$$

Now, we know that if H_0 is true

$$T = \frac{\sqrt{n}\bar{X}}{S} = \frac{\sqrt{n(n-1)}\bar{X}}{\sqrt{\Sigma(X_i - \bar{X})^2}}$$

has a t distribution with n - 1 degrees of freedom. Then since $\Sigma X_i^2 = \Sigma(X_i - \bar{X})^2 + n\bar{X}^2$,

$$\lambda^{2/n} = \frac{\Sigma(X_i - \bar{X})^2}{\Sigma(X_i - \bar{X})^2 + n\bar{X}^2}$$

$$= \frac{1}{1 + n\bar{X}^2/\Sigma(X_i - \bar{X})^2} = \frac{1}{1 + T^2/(n - 1)}$$

In this case the likelihood-ratio test becomes the t test since T^2 is a monotonic function of λ. Then

$$P(\lambda \leq \lambda_\alpha) = P(|T| \geq t_{\alpha/2}) = \alpha \tag{10.4}$$

The region for rejection for T is $|T| \geq t_{\alpha/2}$ and we see that large absolute values of T correspond to small values of λ.

10.8 TESTING THE EQUALITY OF TWO MEANS: In applied statistics the experimenter often wishes to compare two treatments. For example, it may be of interest to compare two different fertilizers, or two rations for hogs, or two teaching methods. We shall assume that the population distributions are normal with the same unknown variance; but the two population means may be different. The null hypothesis is H_0: $\mu_1 = \mu_2$ and the alternative hypothesis is H_1: $\mu_1 \neq \mu_2$.

Let two independent samples be taken from the two populations, that is, $\{X_{11},X_{12},\ldots,X_{1n_1}\}$ is a random sample from $N(\mu_1,\sigma^2)$ and $\{X_{21},X_{22},\ldots,X_{2n_2}\}$ is a random sample from $N(\mu_2,\sigma^2)$. We consider the likelihood-ratio test for testing H_0 against H_1. The entire parameter space is Ω: $\{-\infty < \mu_1 < \infty, -\infty < \mu_2 < \infty, 0 < \sigma^2 < \infty\}$ and the likelihood function is

$$L = (2\pi\sigma^2)^{-(n_1+n_2)/2} \exp\left[-\frac{1}{2\sigma^2}\sum_{i=1}^{2}\sum_{j=1}^{n_i}(X_{ij} - \mu_i)^2\right]$$

The maximum likelihood estimators are

$$\hat{\mu}_i = \bar{X}_i = \frac{1}{n_i}\sum_{j=1}^{n_i} X_{ij} \qquad i = 1, 2$$

$$\hat{\sigma}^2_\Omega = \frac{1}{n_1 + n_2} \sum_{i=1}^{2} \sum_{j=1}^{n_i} (X_{ij} - \bar{X}_i)^2$$

Therefore, the maximum of the likelihood is

$$L(\hat{\Omega}) = (2\pi\hat{\sigma}^2_\Omega)^{-(n_1+n_2)/2} e^{-(n_1+n_2)/2}$$

Under H_0: $\mu_1 = \mu_2 = \mu$ say, the parameter space is ω: $\{-\infty < \mu < \infty, 0 < \sigma^2 < \infty\}$ and the likelihood function becomes

$$L = (2\pi\sigma^2)^{-(n_1+n_2)/2} \exp\left[-\frac{1}{2\sigma^2} \sum_{i=1}^{2} \sum_{j=1}^{n_i} (X_{ij} - \mu)^2\right]$$

The maximum-likelihood estimators under H_0 are

$$\hat{\mu} = \frac{1}{n_1 + n_2} (n_1\bar{X}_1 + n_2\bar{X}_2)$$

$$\hat{\sigma}^2_\omega = \frac{1}{n_1 + n_2} \sum_{i=1}^{2} \sum_{j=1}^{n_i} (X_{ij} - \hat{\mu})^2$$

Hence the maximum is

$$L(\hat{\omega}) = (2\pi\hat{\sigma}^2_\omega)^{-(n_1+n_2)/2} e^{-(n_1+n_2)/2}$$

Taking the ratio of $L(\hat{\omega})$ and $L(\hat{\Omega})$, we obtain

$$\lambda = \frac{L(\hat{\omega})}{L(\hat{\Omega})} = \left(\frac{\hat{\sigma}^2_\Omega}{\hat{\sigma}^2_\omega}\right)^{(n_1+n_2)/2}$$

or

$$\lambda^{2/(n_1+n_2)} = \frac{\Sigma\Sigma(X_{ij} - \bar{X}_i)^2}{\Sigma\Sigma(X_{ij} - \hat{\mu})^2}$$

But

$$\Sigma\Sigma(X_{ij} - \hat{\mu})^2 = \Sigma\Sigma[(X_{ij} - \bar{X}_i) + (\bar{X}_i - \hat{\mu})]^2$$

$$= \Sigma\Sigma(X_{ij} - \bar{X}_i)^2 + \frac{n_1 n_2(\bar{X}_1 - \bar{X}_2)^2}{n_1 + n_2}$$

Hence

$$\lambda^{2/(n_1+n_2)} = \frac{1}{1 + n_1 n_2(\bar{X}_1 - \bar{X}_2)^2/(n_1 + n_2)\Sigma\Sigma(X_{ij} - \bar{X}_i)^2}$$

$$= \frac{1}{1 + T^2/(n_1 + n_2 - 2)}$$

The region of rejection for λ is $\lambda \leq \lambda_\alpha$, which is equivalent to the region of rejection $|T| \geq t_{\alpha/2}$ for T. The value $t_{\alpha/2}$ is determined from the distribution under H_0 of the statistic

$$T = \frac{\bar{X}_1 - \bar{X}_2}{\sqrt{S_p^2(1/n_1 + 1/n_2)}} \tag{10.5}$$

where

$$S_p^2 = \frac{1}{n_1 + n_2 - 2}\Sigma\Sigma(X_{ij} - \bar{X}_i)^2$$

is the pooled unbiased estimator of σ^2. But we know that under H_0, $\bar{X}_1 - \bar{X}_2$ is distributed as $N(0,\sigma^2(1/n_1 + 1/n_2))$ and $(n_1 + n_2 - 2)S_p^2/\sigma^2$ is distributed as $\chi^2(n_1 + n_2 - 2)$. The two random variables are independently distributed. Now T can be written as the ratio of a standard normal variate and the square root of $\chi^2/(n_1 + n_2 - 2)$ with $n_1 + n_2 - 2$ degrees of freedom. Therefore T has $t(n_1 + n_2 - 2)$ distribution under H_0, so the region of rejection is readily determined. It should be noted that the test criterion does not depend on the unknown variance σ^2.

10.9 BEHRENS-FISHER PROBLEM: In the previous section, it was assumed that the variances of the two populations were equal. When $\sigma_1^2 \neq \sigma_2^2$, the t test is no longer valid. Under H_0: $\mu_1 = \mu_2$, the distribution of $\bar{X}_1 - \bar{X}_2$ is $N(0, \sigma_1^2/n_1 + \sigma_2^2/n_2)$. An unbiased estimator of $\sigma_1^2/n_1 + \sigma_2^2/n_2$ is $S_1^2/n_1 + S_2^2/n_2$, where

$$S_i^2 = \frac{1}{n_i - 1} \sum_{j=1}^{n_i} (X_{ij} - \bar{X}_i)^2$$

which is the sample unbiased estimator of σ_i^2, i = 1, 2. One may wish to consider the statistic

$$T' = \frac{\bar{X}_1 - \bar{X}_2}{\sqrt{S_1^2/n_1 + S_2^2/n_2}} \qquad (10.6)$$

But T' does not have a t distribution since the square of the denominator is not distributed as χ^2/ν, with ν degrees of freedom.

We know that $(n_1 - 1)S_1^2/\sigma_1^2$ is distributed as chi-square with $n_1 - 1$ degrees of freedom and similarly for $(n_2 - 1)S_2^2/\sigma_2^2$. Hence a denominator analogous to that of T^2 is

$$\frac{(n_1 - 1)S_1^2/\sigma_1^2 + (n_2 - 1)S_2^2/\sigma_2^2}{n_1 + n_2 - 2}$$

and an analogous numerator is

$$\frac{(\bar{X}_1 - \bar{X}_2)^2}{\sigma_1^2/n_1 + \sigma_2^2/n_2}$$

If we set

$$T = \frac{\bar{X}_1 - \bar{X}_2}{\sqrt{\sigma_1^2/n_1 + \sigma_2^2/n_2}} \Bigg/ \sqrt{\frac{(n_1 - 1)S_1^2/\sigma_1^2 + (n_2 - 1)S_2^2/\sigma_2^2}{n_1 + n_2 - 2}}$$

and let $\theta = \sigma_2^2/\sigma_1^2$ and $\ell = S_2^2/S_1^2$, then

$$T' = \frac{T\sqrt{(n_1 - 1) + (n_2 - 1)\ell/\theta}}{\sqrt{n_1 + n_2 - 2}} \quad \frac{\sqrt{1/n_1 + \theta/n_2}}{\sqrt{1/n_1 + \ell/n_2}}$$

is an analogous expression to the ordinary T. The exact distribution of T' is not known, and even if known would be of little practical use since it would undoubtedly involve the population parameters σ_1^2 and σ_2^2. Two methods have been devised for using T' to make the desired test. They are described by Bartlett (1936, 1939) and Welch (1937).

Several other methods have been suggested for testing the difference between two means when the population variances are unequal. One of the oldest methods is the Behrens-Fisher test (Behrens, 1964) which is based on Fisher's (1935) concept of *fiducial probability*. Sukhatme (1938) has prepared tables to be used in connection with this test. Since there is considerable controversy regarding the validity of this test, it will not be presented here.

Two approximate tests have become quite popular. One by Cochran and Cox (1957) utilizes a weighted mean of the tabular t values for the two samples, weighted by the two sample variances. Compute $D = \bar{X}_1 - \bar{X}_2$ and

$$S_D = \sqrt{\frac{S_1^2}{n_1} + \frac{S_2^2}{n_2}}$$

The approximate tabular value for $T' = D/S_D$ is

$$t'_\alpha = \frac{W_1 t_1 + W_2 t_2}{W_1 + W_2} \tag{10.7}$$

where $W_i = S_i^2/n_i$ and t_i is $t_{\alpha/2}$ for $n_i - 1$ degrees of freedom. The significance level and power of this test were studied by Cochran (1964) and Lauer and Han (1974).

Another approximate test was suggested by Smith (1936) and further expanded by Satterthwaite (1946). The test criterion is also T', but the approximate tabular value is t_α with f degrees of freedom, where

$$f = \frac{s_D^4}{(s_1^2/n_1)^2/(n_1 - 1) + (s_2^2/n_2)^2/(n_2 - 1)} \tag{10.8}$$

For other solutions of this problem, see Scheffé (1970).

10.10 TESTING THE HOMOGENEITY OF VARIANCES: Suppose we have k independent samples, $\{X_{i1},\ldots,X_{in_i}\}$, i = 1, ..., k, each from populations which are $N(\mu_i,\sigma_i^2)$. It is desired to test

$$H_0: \sigma_1^2 = \sigma_2^2 = \cdots = \sigma_k^2$$

against the alternative hypothesis that at least one equality does not hold. Let us consider the likelihood-ratio test. The likelihood function is

$$L = \prod_{i=1}^{k} (2\pi\sigma_i^2)^{-n_i/2} \exp\left[-\sum_{j=1}^{n_i} \frac{(X_{ij} - \mu_i)^2}{2\sigma_i^2}\right]$$

The maximum-likelihood estimators under the entire parameter space are

$$\hat{\mu}_i = \bar{X}_i = \frac{1}{n_i}\sum_{j=1}^{n_i} X_{ij} \qquad i = 1, \ldots, k$$

$$\hat{\sigma}_i^2 = \frac{-1}{n_i}\sum_{j=1}^{n_i} (X_{ij} - \bar{X}_i)^2$$

Therefore, the maximum of the likelihood is

$$L(\hat{\Omega}) = \prod_{i=1}^{k} (2\pi\hat{\sigma}_i^2)^{-n_i/2} e^{-n_i/2}$$

Under H_0: $\sigma_i^2 = \sigma^2$ for all i, the maximum-likelihood estimators for μ_i and σ^2 are

$$\hat{\mu}_i = \bar{X}_i \qquad i = 1, \ldots, k$$

$$\hat{\sigma}_0^2 = \frac{1}{N} \Sigma\Sigma(X_{ij} - \bar{X}_i)^2$$

where $N = \Sigma n_i$. Hence the maximum of the likelihood is

$$L(\hat{\omega}) = (2\pi\hat{\sigma}_0^2)^{-N/2} e^{-N/2}$$

The likelihood-ratio criterion is

$$\lambda = \frac{L(\hat{\omega})}{L(\hat{\Omega})} = \frac{\Pi_{i=1}^k (\hat{\sigma}_i^2)^{n_i/2}}{(\hat{\sigma}_0^2)^{N/2}} \tag{10.9}$$

The null hypothesis H_0 is rejected if $\lambda \leq \lambda_\alpha$. This test is slightly biased (the power may be made smaller than the level of significance). Bartlett (1937) suggested replacing the sample size n_i by $n_i - 1$ in λ and made the test to be unbiased. He proposed the criterion

$$M = \frac{\nu \log S_p^2 - \Sigma_{i=1}^k \nu_i \log S_i^2}{1 + \frac{1}{3(k-1)}\left(\Sigma_{i=1}^k 1/\nu_i - 1/\nu\right)} \tag{10.10}$$

where

$$\nu_i = n_i - 1 \qquad \nu = \Sigma\nu_i = N - k$$

$$S_i^2 = \frac{1}{\nu_i} \sum_{i=1}^{n_i} (X_{ij} - \bar{X}_i)^2$$

$$S_p^2 = \frac{\Sigma\nu_i S_i^2}{\nu}$$

Under H_0, M is approximately distributed as a χ^2 distribution with $k - 1$ degrees of freedom when the sample sizes are large. Hence H_0 is rejected if $M > \chi_\alpha^2(k - 1)$. This test is usually referred to as *Bartlett's test*. The exact critical values for Bartlett's test when the sample sizes are equal are given by Glaser (1976). It should be noted that Box (1953) has shown that Bartlett's test is sensitive to nonnormality.

Another test statistic was suggested by Hartley (1950). We shall let $n_i = n$, i.e., all sample sizes are equal. Define $S^2_{max} = \max S^2_i$ and $S^2_{min} = \min S^2_i$; then

$$F_{max} = \frac{S^2_{max}}{S^2_{min}} \tag{10.11}$$

is a test statistic. When the population variances are all equal, it is expected that the sample variances, especially when the sample size n is large, will be about the same, because the sample variances are consistent. Under the alternative hypothesis, the sample variances will vary according to the σ^2 values and F_{max} will be large. Hence the null hypothesis of homogeneity of variances is rejected if a large value of F_{max} is observed. The upper 5% and 1% points of F_{max} were tabulated by David (1952).

Another relatively simple test statistic was developed by Cochran (1941). He suggested the statistic

$$C_{max} = \frac{S^2_{max}}{\Sigma S^2_i} \tag{10.12}$$

This is the ratio of the largest sample variance to the total. The upper 5% and 1% points of C_{max} were tabulated by Eisenhart and Solomon (1947).

For the special case k = 2, the likelihood-ratio test in (10.9) the F_{max} test in (10.11) and the Cochran's test in (10.12) are equivalent to the ratio of the two sample variances. Hence an F test is used. The proof is left as an exercise for the reader (see Exercise 10.10).

The above tests are given for independent samples. When the samples are correlated, several tests, including extensions of Bartlett's test and the F_{max} test, are given in Han (1968).

10.11 RELATIONSHIP BETWEEN TESTING HYPOTHESIS AND CONFIDENCE INTERVAL: In Chapter 9 a random interval (C_1,C_2) was constructed to estimate the population parameter θ. It was stated that if a series of trials are made, in about $100(1 - \alpha)\%$ of these trials the random interval (C_1,C_2) will include θ_0, the true value of θ. In this chapter, a test was constructed for testing H_0: $\theta = \theta_0$. The relationship between testing hypothesis and confidence interval is that if the confidence interval does not include the null hypothesis value θ_0, i.e., the interval estimator does not include θ_0, we have reason to reject the null hypothesis H_0. If the confidence probability is $1 - \alpha$, then the probability of the Type I error is α.

We can illustrate this relationship by considering the test for H_0: $\mu = \mu_0$ against H_1: $\mu \neq \mu_0$ in the normal distribution $N(\mu,\sigma^2)$ with σ^2 known. The test statistic based on a random sample of size n is

$$Z = \frac{\sqrt{n}(\bar{X} - \mu_0)}{\sigma}$$

The distribution of Z when $\mu = \mu_0$ is N(0,1). Hence

$$P\left(-Z_{\alpha/2} < \frac{\sqrt{n}(\bar{X} - \mu_0)}{\sigma} < Z_{\alpha/2}\right) = 1 - \alpha$$

where $Z_{\alpha/2}$ is the $100(1 - \alpha/2)$ percentage point of N(0,1). Inverting the inequalities, we have

$$\bar{X} - Z_{\alpha/2}\frac{\sigma}{\sqrt{n}} < \mu_0 < \bar{X} + Z_{\alpha/2}\frac{\sigma}{\sqrt{n}}$$

The limits are precisely the confidence limits for estimating μ. Hence if H_0 is true, the confidence interval would include the true value μ_0 with probability $1 - \alpha$. If the confidence interval does not include μ_0, we have evidence that H_0 is false and H_0 should be rejected.

10.12 SEQUENTIAL PROBABILITY RATIO TEST: In the previous sections, statistical test procedures are given for fixed sample sizes. In some practical situations, the observations may be obtained in a sequential order. For example, in clinical trials, patients usually arrive one after another; in a manufacturing process, items may be produced one by one. Hence a sequential statistical analysis may be more appropriate than a fixed sample size analysis in those situations. When sequential tests are used, it can terminate at an early stage with relatively small sample size. If the sampling cost is high, it would be advantageous to adopt a sequential plan.

It was stated earlier that the likelihood-ratio criterion is a procedure for obtaining a "good" test. When observations are obtained one by one in a sequential manner, we can use the likelihood ratio criterion at each stage of the sampling process. The procedure of the sequential probability ratio test for testing a simple hypothesis H_0: $\theta = \theta_0$ against a simple alternative hypothesis H_1: $\theta = \theta_1$ for $f(x;\theta)$ is given as follows:

1. At the nth stage, n = 1, 2, ..., a random sample $\{X_1, X_2, \ldots, X_n\}$ is obtained. The joint probability density function under H_0 is

$$L_{0n} = \prod_{i=1}^{n} f(x_i;\theta_0)$$

and the joint probability density function under H_1 is

$$L_{1n} = \prod_{i=1}^{n} f(x_i;\theta_1)$$

2. The ratio is computed:

$$\lambda_n = \frac{L_{0n}}{L_{1n}} \tag{10.13}$$

3. Two constants a and b are determined such that a < b. Then we accept H_0 if $\lambda_n \geq b$, accept H_1 if $\lambda_n \leq a$, and continue sampling if $a < \lambda_n < b$.

If the test procedure at the nth stage results in either accepting H_0 or accepting H_1, the test is completed. If it results in continuing sampling, we take an additional sample, go back to step (1) and consider the (n + 1)th stage. The test procedure continues in this way until it accepts either H_0 or H_1. One may wonder whether the sampling would continue forever. However, Wald (1947), the inventor of the sequential probability ratio test, showed that the sampling will terminate with probability 1.

The constants a and b are chosen by the experimenter. Their values determine the probabilities of Type I error α and Type II error β. In practice, we can select α and β first, then determine a and b to satisfy the specified values of α and β. Although exact solutions for a and b to satisfy the required α and β are difficult, approximations to the solutions are possible. This is done as follows.

Let E_{0n} be the set of all sample points $\{X_1, X_2, \ldots, X_n\}$ which lead to accept H_0 at the nth stage with continuing sampling in the previous n - 1 stages, and, similarly, E_{1n} be the set of all sample points which lead to accept H_1. Then

$$P(\text{Accept } H_1|H_1) = \int_{E_{1n}} \cdots \int L_{1n}\, dx_1 \cdots dx_n$$

$$P(\text{Accept } H_0|H_1) = \int_{E_{0n}} \cdots \int L_{1n}\, dx_1 \cdots dx_n$$

In the region E_{1n}, we have $L_{1n} \geq L_{0n}/a$. Therefore,

$$P(\text{Accept } H_1|H_1) \geq \int_{E_{1n}} \cdots \int \frac{L_{0n}}{a}\, dx_1 \cdots dx_n$$

or

$$1 - \beta \geq \frac{\alpha}{a}$$

$$a \geq \frac{\alpha}{1 - \beta} \tag{10.14}$$

where $0 < \beta < 1$. Similarly, in the region E_{0n}, we have $L_{1n} \leq L_{0n}/b$. So

$$P(\text{Accept } H_0|H_1) \leq \int_{E_{0n}} \cdots \int \frac{L_{0n}}{b}\, dx_1 \cdots dx_n$$

or

$$\beta \leq \frac{1 - \alpha}{b}$$

$$b \leq \frac{1 - \alpha}{\beta} \tag{10.15}$$

Given α and β, we have shown that a is bounded below by $\alpha/(1 - \beta)$ and b is bounded above by $(1 - \alpha)/\beta$. In practice, we may set $a = \alpha/(1 - \beta)$ and $b = (1 - \alpha)/\beta$ and carry out the sequential probability ratio test as stated from steps 1 to 3. With these choices of a and b, the actual probabilities of Type I and Type II errors will not be α and β but their values will be close. If we let α' and β' be the actual probabilities of the two types of errors, then from the above inequalities,

$$a = \frac{\alpha}{1 - \beta} \geq \frac{\alpha'}{1 - \beta'}$$

and

$$\frac{1 - \alpha'}{\beta'} \geq \frac{1 - \alpha}{\beta} = b$$

Multiplying these two inequalities by $(1 - \beta)(1 - \beta')$ and $\beta'\beta$, respectively, and adding the results, we have

$$\alpha + \beta \geq \alpha' + \beta' \tag{10.16}$$

This shows that the sum of the actual probabilities of errors, $\alpha' + \beta'$, is bounded by $\alpha + \beta$. Further,

$$\alpha' \leq \frac{\alpha'}{1 - \beta'} \leq \frac{\alpha}{1 - \beta}$$

$$\beta' \leq \frac{\beta'}{1 - \alpha'} \leq \frac{\beta}{1 - \alpha} \tag{10.17}$$

So not only the sum is bounded but also the individual probability is bounded. When α and β are small, which is usually the case, α' and β' will not be too far away from α and β, respectively.

EXAMPLE 10.6 Consider testing the mean of a normal distribution $N(\mu,1)$. The null hypothesis is H_0: $\mu = \mu_0$ and the alternative hypothesis is H_1: $\mu = \mu_1$, where $\mu_0 < \mu_1$ are known constants. The probability density functions under H_0 and H_1 are, respectively,

$$L_{0n} = \left(\frac{1}{2\pi}\right)^{n/2} e^{-(1/2)\Sigma(x_i-\mu_0)^2}$$

and

$$L_{1n} = \left(\frac{1}{2\pi}\right)^{n/2} e^{-(1/2)\Sigma(x_i-\mu_1)^2}$$

The ratio is

$$\lambda_n = \frac{L_{0n}}{L_{1n}} = \exp\{-\tfrac{1}{2}[\Sigma(x_i - \mu_0)^2 - \Sigma(x_i - \mu_1)^2]\}$$

$$= \exp[-(\mu_1 - \mu_0)\Sigma x_i + \tfrac{n}{2}(\mu_1^2 - \mu_0^2)]$$

So the sequential probability ratio test is

1. To accept H_0 if $\lambda_n \geq \frac{1 - \alpha}{\beta}$
2. To accept H_1 if $\lambda_n \leq \frac{\alpha}{1 - \beta}$
3. To continue sampling if $\frac{\alpha}{1 - \beta} < \lambda_n < \frac{1 - \alpha}{\beta}$

Equivalently we may take the logarithm and express the sequential probability ratio test in terms of $\log \lambda_n$. After simplification, the sequential testing procedure becomes

1. To accept H_0 if $\sum_{i=1}^{n} x_i \leq \frac{1}{\mu_1 - \mu_0} \log \frac{\beta}{1 - \alpha} + \frac{n}{2}(\mu_1 + \mu_0)$
2. To accept H_1 if $\sum_{i=1}^{n} x_i \geq \frac{1}{\mu_1 - \mu_0} \log \frac{1 - \beta}{\alpha} + \frac{n}{2}(\mu_1 + \mu_0)$
3. To continue sampling if

$$\frac{1}{\mu_1 - \mu_0} \log \frac{1 - \beta}{\alpha} + \frac{n}{2}(\mu_1 + \mu_0) < \sum_{i=1}^{n} x_i < \frac{1}{\mu_1 - \mu_0} \log \frac{\beta}{1 - \alpha} + \frac{n}{2}(\mu_1 + \mu_0)$$

This testing procedure is depicted in Fig. 10.6. The two parallel lines correspond to

(i) $\Sigma x_i = \frac{1}{\mu_1 - \mu_0} \log \frac{\beta}{1 - \alpha} + \frac{n}{2}(\mu_1 + \mu_0)$

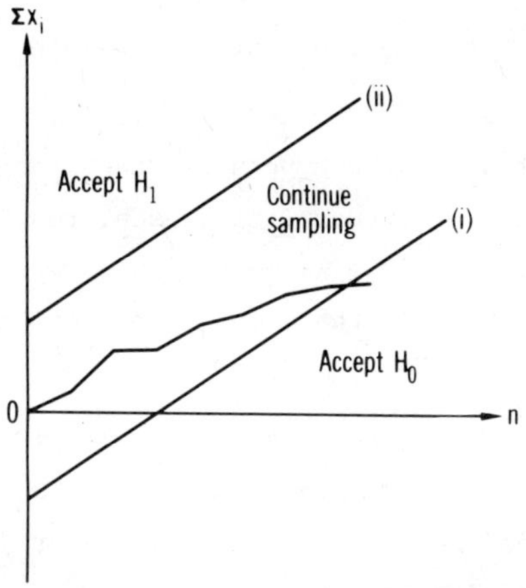

FIG. 10.6 Sequential probability ratio test.

$$\text{(ii)} \quad \Sigma x_i = \frac{1}{\mu_1 - \mu_0} \log \frac{1 - \beta}{\alpha} + \frac{n}{2}(\mu_1 + \mu_0)$$

When the sample sum Σx_i crosses either one of the two lines, we stop sampling and accept either H_0 or H_1.

EXERCISES

10.1 Suppose a single observation is taken from $f(x) = \theta x^{\theta-1}$, $0 < x < 1$, and 0 otherwise. The null hypothesis H_0: $\theta = 2$ is rejected if $x > 0.9$. Find (a) the probability of Type I error; (b) the probability of Type II error if $\theta = 4$.

10.2 Let X have the exponential distribution $f(x) = \theta e^{-\theta x}$, $x > 0$, $\theta > 0$, and 0 otherwise. Use the Neyman-Pearson lemma to find the most powerful test for testing H_0: $\theta = \theta_0$ against H_1: $\theta = \theta_1$ ($\theta_1 < \theta_0$) based on a random sample of size n.

10.3 Let $X_1, X_2, \ldots, X_n$ be independent Bernoulli trials with $P(X_i = 1) = p$ and $P(X_i = 0) = 1 - p$. Consider testing H_0: $p = 1/2$ against H_1: $p = 1/4$; show that the most powerful test has the region of rejection $\Sigma X_i \leq k$. Use normal approximation to find n and k if $P(I) = P(II) = .05$.

10.4 Construct the power curve for $H_1: \mu_1 < \mu_0$. Consider values of $5(\mu_1 - \mu_0) = \pm 4, \pm 3, \pm 2, \pm 1, 0$ given the conditions of the example of Section 10.4.

10.5 Discuss rejection regions for testing the hypothesis that the difference between the population means of two variates, each, respectively, from $N(\mu_i, \sigma^2)$, $i = 1, 2$, is zero. Take $\Omega: -\infty < \mu_1, \mu_2 < \infty$, $0 < \sigma^2 < \infty$, and $H_0: \mu_1 = \mu_2$. *Hint:* Consider the example of Section 10.5.

10.6 Set up the admissible hypotheses Ω and the region of rejection R for testing the hypothesis that the population variance of $N(0, \sigma^2)$ is σ_0^2, based on a random sample of size n. What can be said regarding the power of the test for various alternatives?

10.7 Determine Ω and the region of rejection R for testing the hypothesis that the variances of two normal populations are equal ($\sigma_1^2 = \sigma_2^2$) against the alternative hypothesis that $\sigma_1^2 = a\sigma_2^2$ (a > 1). Since the test criterion is $F = s_1^2/s_2^2$, the null hypothesis is rejected if $F > F_0$. Show that

$$1 - \beta = I_x\left(\frac{n_2}{2}, \frac{n_1}{2}\right)$$

where I_x indicates the incomplete beta function, and

$$x = \left(1 + \frac{n_1}{n_2}F\right)^{-1} \qquad n_1 = (\text{df})_1 \qquad n_2 = (\text{df})_2$$

Complete the following power table for $n_1 = 2$, $n_2 = 10$, and $F_a = F_0/a$:

a	F_a	x	$1 - \beta$
		.3466	.005
		.3981	.010
		.4782	.025
1	4.1025	.5493	.050
		.6310	.100
		.7579	.250
		.8706	.500

10.8 Derive the λ criterion for testing the null hypothesis H_0: $m = m_0$, given a random sample of size n from the Poisson distribution

$$f(x) = e^{-m}\frac{m^x}{x!} \qquad x = 0, 1, \ldots, m \geq 0, \text{ and 0 otherwise}$$

The parameter space Ω is the half-line $m \geq 0$.

10.9 Given two independent random samples of sizes n_1 and n_2 from each of two populations $N(0,\sigma_1^2)$ and $N(0,\sigma_2^2)$, derive the λ criterion for testing the null hypothesis H_0: $\sigma_1^2 = \sigma_2^2 = \sigma^2$ (unspecified). The entire parameter space Ω is the quarter-plane determined by $\sigma_1^2 > 0$, $\sigma_2^2 > 0$. The subspace ω is the line $\sigma_1^2 = \sigma_2^2 = \sigma^2$ (unspecified).
Hint: First determine the joint distribution of $\Sigma X_{1i}^2/n_1$ and $\Sigma X_{2i}^2/n_2$, the sample estimates of σ_1^2 and σ_2^2, respectively. Show that the criterion reduces to Snedecor's F.

10.10 Repeat Exercise 10.9 when the samples are from the populations $N(\mu_1,\sigma_1^2)$ and $N(\mu_2,\sigma_2^2)$.

10.11 Given a random sample of size n from $N(\mu,\sigma^2)$, show that the λ criterion for testing the null hypothesis H_0: $\sigma^2 = \sigma_0^2$, μ unspecified, reduces to χ^2. The entire parameter space is determined so that both μ and σ^2 are unspecified.

10.12 Repeat Example 10.4 when the random sample of size n is from $N(\mu,\sigma^2)$ and σ^2 is unspecified by H_0.

10.13 Let $X_1, X_2, \ldots, X_k$ have the multinomial distribution

$$f(x) = \frac{n!}{\prod_{i=1}^{k} x_i!}\prod_{i=1}^{k} p_i^{x_i} \qquad \sum_{i=1}^{k} p_i = 1 \qquad \sum_{i=1}^{k} x_i = n$$

Derive the likelihood ratio criterion for H_0: $p_i = p_{0i}$, $i = 1, \ldots, k$.

10.14 Let $\{X_1,\ldots,X_{n_1}\}$ and $\{Y_1,\ldots,Y_{n_2}\}$ be two independent random samples from Poisson distributions with parameters m_1 and m_2, respectively. Find the likelihood-ratio criterion for testing H_0: $m_1 = m_2$ against H_1: $m_1 \neq m_2$.

10.15 Let $\{X_1, \ldots, X_n\}$ be a random sample from

$$f(x) = \begin{cases} \theta e^{-\theta x} & x > 0,\ \theta > 0 \\ 0 & \text{otherwise} \end{cases}$$

Find the likelihood ratio test for testing H_0: $\theta = \theta_0$ against H_1: $\theta \neq \theta_0$.

10.16 Given a random sample of size n from a bivariate normal distribution, the maximum-likelihood estimator of $\mu_1 - \mu_2$ was derived in Exercise 7.15. Derive the likelihood-ratio test for testing H_0: $\mu_1 - \mu_2 = \mu_0$ against H_1: $\mu_1 - \mu_2 \neq \mu_0$ when σ_1^2, σ_2^2, and σ_{12} are (a) known, (b) unknown. This test is usually referred to as the comparison of means in the paired samples.

10.17 Let Y_i, i = 1, ..., n be a random sample from $N(\alpha + \beta(x_i - \bar{x}), \sigma^2)$ where x_i are given constants and $\bar{x} = \Sigma x_i/n$. Derive the likelihood-ratio test for testing H_0: $\beta = \beta_0$ against H_1: $\beta \neq \beta_0$; assume that σ^2 is unknown.

10.18 Let (X,Y) have the joint density function

$$f(x,y) = \begin{cases} 4\dfrac{xy}{\theta_1\theta_2} e^{-x^2/\theta_1 - y^2/\theta_2} & x > 0,\ y > 0,\ \theta_1 > 0,\ \theta_2 > 0 \\ 0 & \text{otherwise} \end{cases}$$

Find the likelihood-ratio criterion for testing H_0: $\theta_1 = \theta_2$ against H_1: $\theta_1 \neq \theta_2$ based on a random sample of size n.

10.19 Consider the Bernoulli variable X with P(X = 1) = p and P(X = 0) = 1 - p. Derive the sequential probability ratio test for testing H_0: $p = p_0$ against H_1: $p = p_1$ where $p_1 > p_0$; use $\alpha = .05$ and $\beta = .05$. Graph the continuing sampling, acceptance, and rejection regions.

10.20 Suppose X has the density function $f(x) = \theta e^{-\theta x}$, $x > 0$, $\theta > 0$, and 0 otherwise. Derive the sequential probability ratio test for testing H_0: $\theta = \theta_0$ against H_1: $\theta = \theta_1$ where $\theta_1 < \theta_0$; use $\alpha = .05$ and $\beta = .05$. Graph the continuing sampling, acceptance, and rejection regions.

REFERENCES

Bartlett, M. S. (1936). The information available in small samples. Proc. Cambridge Phil. Soc. 32: 560ff.

Bartlett, M. S. (1937). Properties of sufficiency and statistical tests. Proc. Roy. Soc., Ser. A 160: 268-282.

Bartlett, M. S. (1939). Complete simultaneous fiducial distribution. Ann. Math. Stat. 10: 129-138.

Behrens, W. V. (1964). The comparison of means of independent normal distribution with different variances. Biometrics 20: 16-27.

Box, G. E. P. (1953). Nonnormality and tests of variance. Biometrika 40: 318-335.

Cochran, W. G. (1941). The distribution of the largest of a set of estimated variances as a fraction of their total. Ann. Eugenics 2: 47-52.

Cochran, W. G. (1964). Approximate significance levels of the Behrens-Fisher test. Biometrics 20: 191-195.

Cochran, W. G., and Cox, G. M. (1957). *Experimental Design*, 2nd ed. Wiley, New York.

Cramer, H. (1946). *Mathematical Methods of Statistics*. Princeton University Press, Princeton, N.J.

David, H. A. (1952). Upper 5 and 1 percent points of the maximum F-ratio. Biometrika 39: 422-424.

Eisenhart, C., and Solomon, H. (1947). Significance of the largest of a set of sample estimates of variance. In *Techniques of Statistical Analysis*, ed. C. Eisenhart, M. W. Hastay, and W. A. Wallis. McGraw-Hill, New York.

Fisher, R. A. (1935). The fiducial argument in statistical inference. Ann. Eugenics 6: 391-398.

Glaser, R. E. (1976). Exact critical values for Bartlett's test for homogeneity of variances. J. Amer. Stat. Assoc. 71: 488-490.

Han, C. P. (1968). Testing homogeneity of a set of correlated variances. Biometrika 55: 317-326.

Hartley, H. O. (1950). The maximum F-ratio as a short-cut test for heterogeneity of variance. Biometrika 37: 308-312.

Lauer, G. N., and Han, C. P. (1974). Power of Cochran's test in the Behrens-Fisher problem. Technometrics 16: 545-549.

Neyman, J., and Pearson, E. S. (1928). On the use and interpretation of certain test criteria for purposes of statistical inference. Biometrika 20A: 175-240, 263-294.

Neyman, J., and Pearson, E. S. (1937). Statistical estimation. Phil. Trans. Roy. Soc. A236: 333-380.

Neyman, J., and Pearson, E. S. (1938). The problem of the most efficient tests of statistical hypotheses. Phil. Trans. Roy. Soc. A237: 289-337.

Neyman, J., and Tokarska, B. (1936). Errors of the second kind in testing Student's hypothesis. J. Amer. Stat. Assoc. 31: 318-326.

Satterthwaite, F. E. (1946). An approximate distribution of estimates of variance components. Biometrics Bull. 2: 110-114.

Scheffé, H. (1970). Practical solutions of the Behrens-Fisher problem. J. Amer. Stat. Assoc. 65: 1501-1508.

Smith, H. F. (1936). The problem of comparing the results of two experiments with unequal errors. J. Sci. Ind. Research (India) 9: 211-212.

Sukhatme, P. V. (1938). On the Fisher-Behrens test of significance for the difference in means of two normal samples. Sankhyā 4, 39-48.

Tang, P. C. (1938). The power function of the analysis of variance tests with tables and illustrations of their use. Stat. Res. Mem. 2: 126-149.

Wald, A. (1947). *Sequentical Analysis*. Wiley, New York.

Welch, B. L. (1937). Significance of the difference between two means when the population variances are unequal. Biometrika 29: 350-362.

11

NONPARAMETRIC TESTING PROCEDURES

11.1 INTRODUCTION: In Chapter 10, testing procedures were obtained for parameters of a specified distribution. We now consider testing procedures for which the distribution is not given or the experimenter is unwilling to assume a specific form of the underlying parent population distribution. When the parent distribution is not known, it is natural to consider hypotheses concerning the distribution function. For example, the null hypothesis may be that the cumulative distribution function of a continuous random variable at a given value is p, or two distribution functions are equal, or the distribution takes a particular form. The test for the hypothesis that the cumulative distribution function is a specific function is usually referred to as goodness-of-fit test. The chi-square test is widely used as a goodness-of-fit test. It is also used in the analysis of categorical data in contingency tables. This chapter discusses some of the nonparametric testing procedures. The sign test and the Mann-Whitney-Wilcoxon test will be studied in detail. For other nonparametric testing procedures, the readers may consult books on nonparametric statistics, e.g., Conover (1971) and Gibbons (1971).

11.2 SIGN TEST: We first consider a test for quantiles of a distribution. Suppose F(x) is the cumulative distribution function of the continuous random variable X. The *quantile* of the distribution of order p, denoted by x_p, is defined by $P(X \leq x_p) = F(x_p) = p$. In particular, $x_{.25}$, $x_{.50}$, and $x_{.75}$ are called the *first quartile*, *median*, and the *third quartile*, respectively. A hypothesis of interest is that $x_p = a$ where a is a given value, i.e., H_0: $F(a) = p$. This hypothesis simply states that 100p% of the distribution lies below a.

Supposing a random sample of size n, $\{X_1,\ldots,X_n\}$, is taken from F(x), let us count the number of observations whose values are less than a. Let Y be the number of observations such that $a - X_i$, i = 1, 2, ..., n, have positive signs, then, under H_0, Y has a binomial distribution. This can be seen if we let $Y_i = 1$ if $a - X_i > 0$ and $Y_i = 0$ if $a - X_i < 0$ (note that the probability is zero for $a - X_i = 0$ because X is a continuous random variable). Under H_0, $P(Y_i = 1) = P(a - X_i > 0) = P(X_i < a) = p$. So the Y_i terms are independent Bernoulli random variables and $Y = \Sigma_{i=1}^{n} Y_i$ has a binomial distribution.

The total number of positive signs, Y, can be used as the test criterion for H_0. This test based on the number of positive signs is called the *sign test*. The region of rejection depends on the alternative hypothesis whether it is one-sided or two-sided. If the alternative hypothesis is H_1: $F(a) > p$, then H_0 is rejected for large values of Y. When H_1 is $F(a) < p$, H_0 is rejected for small values of Y. For H_1: $F(a) \neq p$, H_0 is rejected when Y is either too large or too small.

EXAMPLE 11.1 Let the number of hours that patients stay in a hospital from the time of admission to release have the cumulative distribution function F(x). It is desired to test the median number of hours to be 48. (This may be of interest because certain insurance companies write health policies which require patients to pay the hospital cost incurred in the first two days.) Hence H_0: $x_{.5} = 48$, or $F(48) = .5$. Let the alternative be H_1: $F(48) < .5$. Suppose 16 patients are observed and H_0 is rejected if there are 4 or less positive signs. The α level of this test is

$$\alpha = \sum_{x=0}^{4} \binom{16}{x} \left(\frac{1}{2}\right)^{16} = .038$$

Suppose the data are

77, 46, 108, 68, 210, 92, 72, 59,
24, 64, 72, 260, 54, 168, 94, 148

There are 2 positive signs; so H_0 is rejected at the level .038. The power of the test is

$$1 - \beta = P\{Y \leq 4\} = \sum_{y=0}^{4} \binom{16}{y} p^y (1 - p)^{16-y}$$

When p = 1/2, the binomial distribution is symmetrical and normal approximation can be used for large n. The mean and variance of Y are (1/2)n and (1/4)n, respectively. Therefore, from Example 6.7, the random variable

$$Z = \frac{Y - (1/2)n}{(1/2)\sqrt{n}} \tag{11.1}$$

is approximately distributed as N(0,1). The region of rejection for an α level test and the power function can be obtained from this normal approximation. The power function for the above one-sided test is

$$1 - \beta = P\{Y \leq 4\} \doteq P\left(Z < \frac{4.5 - 16p}{\sqrt{16p(1 - p)}}\right)$$

where .5 added to the integer 4 is the continuity correction.

11.3 MANN-WHITNEY-WILCOXON TEST: Suppose there are two independent random samples $\{X_1, X_2, \ldots, X_n\}$ and $\{Y_1, Y_2, \ldots, Y_m\}$. We wish to test the null hypothesis that these two samples are taken from identical distributions against the alternative hypothesis that the distribution of the Y's is shifted in location. So the null hypothesis is H_0: $F(x) = F(y)$ and the alternative hypothesis is H_1: $F(x) = F(y + \theta)$, $\theta \neq \theta$. (Note that the null hypothesis is equivalent to $\theta = 0$. Although we have expressed the hypotheses in terms of θ, the distributional form is not specified. The test derived below is distribution-free, so it is a nonparametric test.)

The *Mann-Whitney-Wilcoxon test* is based on pooling the two samples together and arranging the pooled sample from the smallest observation to the largest observation. If the two samples are from identical distributions, we would expect that the X observations and the Y observations will mix very well. On the other hand, if there is a location shift for Y, the Y observations would cluster either to the right or to the left of the pooled sample depending on whether $\theta > 0$ or $\theta < 0$. A test statistic can be constructed by using these phenomena.

Let U be the total number of X observations preceeding each of the Y observations in the pooled sample. Then U may be used as a test statistic for testing H_0. When H_0 is false, we would expect that U is either large or small. The value of U can be computed by letting

$$Z_{ij} = \begin{cases} 1 & \text{if } X_i < Y_j, \quad i = 1, \ldots, n \quad j = 1, \ldots, m \\ 0 & \text{if } X_i > Y_j \end{cases} \tag{11.2}$$

Then

$$U = \sum_{j=1}^{m} \sum_{i=1}^{n} Z_{ij} \tag{11.3}$$

If we let

$$Z_{.j} = \sum_{i=1}^{n} Z_{ij} \qquad j = 1, \ldots, m$$

$Z_{.j}$ is the number of X observations preceeding Y_j. So $U = \Sigma_{j=1}^{m} Z_{.j}$ is the total number.

Under the null hypothesis that X and Y have identical distributions, $P(Z_{ij} = 1) = P(X_i < Y_j) = P(X_i > Y_j) = P(Z_{ij} = 0) = 1/2$. Hence

$$E(Z_{ij}) = P(X_i < Y_j) = \frac{1}{2}$$

$$E(Z_{ij}^2) = P(X_i < Y_j) = \frac{1}{2}$$

$$E(Z_{ij}Z_{i\ell}) = P(X_i < Y_j \text{ and } X_i < Y_\ell) = \frac{1}{3} \qquad j \neq \ell$$

$$E(Z_{ij}Z_{kj}) = P(X_i < Y_j \text{ and } X_k < Y_j) = \frac{1}{3} \qquad i \neq k$$

$$E(Z_{ij}Z_{k\ell}) = P(X_i < Y_j \text{ and } X_k < Y_\ell) = \frac{1}{4} \qquad i \neq k \qquad j \neq \ell$$

The mean and variance of U can now be obtained from the Z's:

$$E(U) = \sum_{j=1}^{m} \sum_{i=1}^{n} E(Z_{ij}) = \frac{mn}{2} \tag{11.4}$$

$$\sigma_U^2 = \sum_{ijk\ell}\sum\sum\sum \mathrm{Cov}(Z_{ij}, Z_{k\ell})$$

$$= \sum_{ijk\ell}\sum\sum\sum \left[E(Z_{ij}Z_{k\ell}) - E(Z_{ij})E(Z_{k\ell})\right]$$

To evaluate the sum, we consider four cases: (1) $i = k$, $j = \ell$; (2) $i = k$, $j \neq \ell$; (3) $i \neq k$, $j = \ell$; (4) $i \neq k$, $j \neq \ell$. Case 4 need not be considered because $E(Z_{ij}Z_{k\ell}) = E(Z_{ij})E(Z_{k\ell})$ and the terms are equal to zero; hence we need only consider the first three cases. The numbers of terms for the first three cases in the summation are (1) mn, (2) $m^2n - mn$, and (3) $mn^2 - mn$; hence we have

$$\sigma_U^2 = mn\left(\frac{1}{2} - \frac{1}{4}\right) + (m^2n - mn)\left(\frac{1}{3} - \frac{1}{4}\right) + (mn^2 - mn)\left(\frac{1}{3} - \frac{1}{4}\right)$$

$$= \frac{1}{12} mn(m + n + 1) \tag{11.5}$$

When m and n are small, the region of rejection was tabulated by Mann and Whitney (1947). When both m and n are large, it can be shown that

$$\frac{U - mn/2}{\sqrt{(1/12)mn(m + n + 1)}} \tag{11.6}$$

is distributed approximately as $N(0,1)$. Using this approximation, we can determine the rejection region. If the alternative hypothesis is two-sided, i.e., H_1: $F(x) = F(y + \theta)$, $\theta \neq 0$, the rejection region is

$$\left[\frac{U + 1/2 - mn/2}{\sqrt{(1/12)mn(m + n + 1)}} < -z_{\alpha/2}\right] \cup \left[\frac{U - 1/2 - mn/2}{\sqrt{(1/12)mn(m + n + 1)}} > z_{\alpha/2}\right]$$

where 1/2 is the continuity correction, α is the probability of Type I error, and $z_{\alpha/2}$ is the $100(1 - \alpha/2)$ percentage point of the standard normal distribution.

If the alternative hypothesis is one-sided, e.g., H_1: $F(x) = F(y + \theta)$, $\theta > 0$, the rejection region is

$$\frac{U - 1/2 - mn/2}{\sqrt{(1/12)mn(m + n + 1)}} > z_\alpha$$

For H_1: $F(x) = F(y + \theta)$, $\theta < 0$, the rejection region becomes

$$\frac{U + 1/2 - mn/2}{\sqrt{(1/12)mn(m + n + 1)}} < -z_\alpha$$

11.4 GOODNESS OF FIT: When the distribution of a random variable is unknown and the experimenter suspects that the random variable Y follows a particular distribution $F_0(y)$, he or she may test the hypothesis H_0: $F(y) = F_0(y)$. Such a test is called a *goodness-of-fit test*. Goodness-of-fit tests are also used in simulation studies (also known as Monte Carlo studies). In such studies the researchers often generate random variables on a computer from a specified distribution such as a uniform or normal distribution. It is desirable to test whether the generated random variables actually follow the specified distribution.

If the null distribution $F_0(y)$ is completely specified, H_0 is a simple hypothesis; otherwise H_0 is a composite hypothesis. Consider now the simple hypothesis. Suppose a random sample of size n is taken from F(y). Let the range of y be partitioned into k mutually exclusive and exhaustive classes. If Y has a discrete distribution, the k classes may be naturally defined. For example, in the experiment of tossing a die, there are 6 possible outcomes which define the 6 classes. Let p_i be the probability that an observation belongs to the ith class, i = 1, ..., k. Then the experimenter may wish to test the hypothesis that the die is a balanced die, i.e., H_0: $p_i = 1/6$,

$i = 1, \ldots, 6$. If Y has a continuous distribution, the k classes may be constructed arbitrarily. Let y_i be the boundary of the ith and $(i + 1)$th classes. Then

$$p_1 = \int_{-\infty}^{y_1} f(y)\, dy$$

$$p_{i+1} = \int_{y_i}^{y_{i+1}} f(y)\, dy \qquad i = 1, \ldots, k - 2$$

$$p_k = \int_{y_{k-1}}^{\infty} f(y)\, dy$$

where f(y) is the density function of Y. We shall denote the probability for the ith class under H_0 by p_{0i}.

Suppose in a random sample of size n, $X_1, X_2, \ldots, X_k$ are the numbers of observations in the respective classes, $\Sigma_{i=1}^{k} X_i = n$. Then the X's have a multinomial distribution

$$f(x_1, x_2, \ldots, x_k) = \frac{n!}{\Pi_{i=1}^{k} x_i!} \Pi_{i=1}^{k} p_i^{x_i} \qquad \Sigma_{i=1}^{k} x_i = n$$

From Section 5.10, the means, variances, and covariances of the X's under H_0 are

$$\mu_i = np_{0i}$$

$$\sigma_i^2 = np_{0i}(1 - p_{0i})$$

$$\sigma_{ij} = -np_{0i}p_{0j}$$

When H_0 is true, we expect that the number of observations in the ith class is $np_{0i} = E_i$, say. The classical test statistic for H_0 is

$$\chi^2 = \sum_{i=1}^{k} \frac{(X_i - np_{0i})^2}{np_{0i}} = \sum_{i=1}^{k} \frac{(O_i - E_i)^2}{E_i} \qquad (11.7)$$

where O_i and E_i indicate the observed and expected frequencies, respectively. The value of χ^2 is expected to be small when H_0 is true and H_0 should be rejected when χ^2 is large. The rejection region is determined from the distribution of χ^2 under H_0. The exact distribution of χ^2 is not easy to find. We shall obtain an asymptotic distribution for χ^2 when n is large.

In Example 6.7, it was shown that the binomial distribution can be approximated by the normal distribution when n is large. The multinomial distribution is an extension of the binomial distribution. It can be approximated by the multivariate normal distribution. Although there are k classes, the number of random variables is k - 1. Hence the multivariate normal distribution is also of dimension k - 1. So the distribution of $\underline{X} = (X_1, X_2, \ldots, X_{k-1})'$ for large n is approximated by the (k - 1)-dimensional normal distribution with mean vector $\underline{\mu}$ and covariance matrix $\underline{\Sigma}$, where the components of $\underline{\mu}$ and $\underline{\Sigma}$ are μ_i and σ_{ij}, respectively.

Now we show that (11.7) can be written as the quadratic form

$$(\underline{X} - \underline{\mu})'\underline{\Sigma}^{-1}(\underline{X} - \underline{\mu})$$

and that it has a χ^2 distribution with k - 1 degrees of freedom. The inverse of $\underline{\Sigma}$ has $(1/n)(\delta_{ij}/p_{0i} + 1/p_{0k})$ as its (i,j)th element, where $\delta_{ij} = 1$ if i = j and $\delta_{ij} = 0$ if $i \neq j$. Hence

$$\begin{aligned}(\underline{X} - \underline{\mu})'\underline{\Sigma}^{-1}(\underline{X} - \underline{\mu}) &= \sum_{i=1}^{k-1}\sum_{j=1}^{k-1} \frac{1}{n}\left(\frac{\delta_{ij}}{p_{0i}} + \frac{1}{p_{0k}}\right)(X_i - np_{0i})(X_j - np_{0j}) \\ &= \sum_{i=1}^{k-1} \frac{1}{n}\left(\frac{1}{p_{0i}} + \frac{1}{p_{0k}}\right)(X_i - np_{0i})^2 \\ &\quad + \sum_{i=1}^{k-1} \frac{1}{np_{0i}}(X_i - np_{0i}) \sum_{j\neq i}^{k-1} (X_j - np_{0j})\end{aligned}$$

Utilizing $\Sigma_{i=1}^{k} p_{0i} = 1$, it can be shown easily that the last equation is equal to (11.7). The distribution of $(\underline{X} - \underline{\mu})'\underline{\Sigma}^{-1}(\underline{X} - \underline{\mu})$ is found as follows. Since $\underline{\Sigma}$ is a symmetric and positive definite matrix, there exists a nonsingular matrix $\underline{B}$ such that $\underline{\Sigma} = \underline{BB}'$. Then

$$(\underline{X} - \underline{\mu})'\underline{\Sigma}^{-1}(\underline{X} - \underline{\mu}) = [\underline{B}^{-1}(\underline{X} - \underline{\mu})]'[\underline{B}^{-1}(\underline{X} - \underline{\mu})]$$

If we let $\underline{Z} = \underline{B}^{-1}(\underline{X} - \underline{\mu})$, then the components of $\underline{Z}$ are distributed independently as N(0,1), which can be seen from Section 5.11. The random variable $\underline{Z}$ has dimension k - 1, so $\underline{Z}'\underline{Z}$ is distributed as $\chi^2(k - 1)$. This is a large sample distribution. Usually n should be large enough so that $np_{0i} \geq 5$ for all i. The chi-square test rejects H_0: $F(y) = F_0(y)$ if

$$\chi^2 = \sum_{i=1}^{k} \frac{(X_i - np_{0i})^2}{np_{0i}} > \chi^2_\alpha(k - 1) \tag{11.8}$$

where $\chi^2_\alpha(k - 1)$ is the $100(1 - \alpha)$ percentage point of $\chi^2(k - 1)$.

EXAMPLE 11.2 Genetic data provide examples of the use of χ^2 in goodness-of-fit tests. Some possible theoretical genetic ratios are 3:1, 1:1, 9:7, 15:1, and 63:1.

In a study of chlorophyll inheritance in corn, Lindstrom (1918) found 98 green and 24 yellow seedlings in one progeny of 122 young corn plants. Presumably green is dominant to yellow and is segregated in a ratio 3:1, so that we should expect $E_1 = 91.5$ green seedlings and $E_2 = 30.5$ yellow seedlings if this ratio is correct. To test this hypothesis, we calculate

$$\chi^2_0 = \frac{(98 - 91.5)^2}{91.5} + \frac{(24 - 30.5)^2}{30.5} = 1.85$$

By interpolation, the probability of obtaining a chi-square value of this size or larger on the assumption that the genetic ratio is 3:1 is .18. Hence, the 3:1 ratio is not rejected as a possible fit to the data.

EXAMPLE 11.3 Federer (1946) fitted a normal curve to the frequency distribution below of rubber content (percentage) in 378 guayule plants:

Class center	1.5	2.5	3.5	4.5	5.5	6.5	7.5	8.5
Frequency	1	1	2	33	139	155	42	5

The mean and standard deviation of the 378 observations were found to be 6.07 and .892, respectively. Using the tables for the normal curve, Federer calculated the expected values for the respective classes set out below:

Class	<4.0	4.0-5.0	5.0-6.0	6.0-7.0	7.0-8.0	>8.0	Total
Observed frequency	4	33	139	155	42	5	378
Expected frequency	3.8	39.7	133.4	144.7	50.6	5.8	378.0

The frequencies of the first three classes in the first table have been pooled in the second table in order that the number in any class be not less than 5, approximately as suggested by Fisher (1970). Using the formula developed in this section Federer found $\chi_0^2 = 3.302$. The proof of the rule given by Fisher for determining the degrees of freedom to be assigned χ_0^2 for this example is beyond the scope of this text. This rule states that the correct number of degrees of freedom may be found by subtracting the total number of restrictions imposed on the data from the total number of classes. In the example this would be 6 - 3, since the sum, mean, and standard deviation of the sample and hypothetical curve have been equated. The probability of obtaining a chi-square value as large as or larger than 3.30 with 3 degrees of freedom lies between .5 and .3. Hence, we have no evidence to reject the hypothesis that the rubber contents in the 378 guayule plants are normally distributed.

11.5 CONTINGENCY TABLES: When the observations are classified according to two attributes and arranged into a two-way table, it is called a 2 × 2 contingency table. Consider the ordinary 2 × 2 contingency table used in applied statistics:

Expectations

	B_1	B_2	
A_1	$np_{1.}p_{.1}$	$np_{1.}p_{.2}$	$np_{1.}$
A_2	$np_{2.}p_{.1}$	$np_{2.}p_{.2}$	$np_{2.}$
	$np_{.1}$	$np_{.2}$	n

Observations

	B_1	B_2	
A_1	n_{11}	n_{12}	$n_{1.}$
A_2	n_{21}	n_{22}	$n_{2.}$
	$n_{.1}$	$n_{.2}$	$n_{..}$

(11.9)

In the tables, A_1 and A_2 are the two categories for attribute A and B_1 and B_2 are the two categories for attribute B; the probability that an observation belongs to A_1 is $p_{1.}$; similarly, it belongs to B_1 with probability $p_{.1}$, $p_{1.} + p_{2.} = 1$, $p_{.1} + p_{.2} = 1$. The observed numbers in the (i,j)th cell is n_{ij}; also,

$$n_{i.} = \sum_{j=1}^{2} n_{ij} \qquad n_{.j} = \sum_{i=1}^{2} n_{ij} \qquad \text{and} \qquad n_{..} = n$$

A hypothesis of interest is that the two classifications are independent. The null hypothesis can be written as H_0: $p_{ij} = p_{i.}p_{.j}$ for all i and j, where p_{ij} is the probability that an observation belongs to the (i,j)th cell. We may use the chi-square test given in Section 11.4. In this case k = 4. If there is no a priori knowledge of the values of $p_{i.}$ and $p_{.j}$, we usually use their maximum-likelihood estimates, which are

$$\hat{p}_{i.} = \frac{n_{i.}}{n} \qquad \hat{p}_{.j} = \frac{n_{.j}}{n} \qquad i, j = 1, 2 \tag{11.10}$$

The test statistic is

$$\chi^2 = \sum_{i=1}^{k} \frac{(O_i - E_i)^2}{E_i} = \sum_{i=1}^{2} \sum_{j=1}^{2} \frac{(n_{ij} - n\hat{p}_{i.}\hat{p}_{.j})^2}{n\hat{p}_{i.}\hat{p}_{.j}}$$

The approximate distribution of this test statistic under H_0 is $\chi^2(1)$ when n is large. If the calculated χ^2 is greater than the tabular $\chi^2_\alpha(1)$, we say that we have evidence from this sample that the two classifications are not independent. In other words, there would be evidence of an interaction between the two classifications.

EXAMPLE 11.4 Robertson (1937) segregated the F_2 progeny of a barley cross in the following manner:

	F	f	Total
V	1178	273	1451
v	291	156	447
Total	1469	429	1898

We calculate

$$\chi_0^2 = \frac{(54.97)^2}{1123.03} + \frac{(-54.97)^2}{327.97} + \frac{(-54.97)^2}{345.97} + \frac{(54.97)^2}{101.03} = 50.547$$

The probability of obtaining a chi-square this large or larger on the assumption of independence of the classification is extremely small. Hence, there is considerable evidence in favor of association of these two attributes.

An extension of the 2 × 2 contingency table is the r × c contingency table when attribute A has r categories and attribute B has c categories. The null hypothesis of independence is H_0: $p_{ij} = p_{i.}p_{.j}$, $i = 1, \ldots, r$, $j = 1, \ldots, c$. The chi-square test statistic is

$$\chi^2 = \sum_{i=1}^{r} \sum_{j=1}^{c} \frac{(n_{ij} - n\hat{p}_{i.}\hat{p}_{.j})^2}{n\hat{p}_{i.}\hat{p}_{.j}} \tag{11.12}$$

where $\hat{p}_{i.} = n_{i.}/n$ and $\hat{p}_{.j} = n_{.j}/n$. In this case $k = rc$ and the number of parameters estimated is $(r - 1) + (c - 1)$. So the number of degrees of freedom associated with the χ^2 is $rc - 1 - (r - 1) - (c - 1) = (r - 1)(c - 1)$. The approximate distribution of χ^2 is $\chi^2(r - 1)(c - 1)$ provided n is large. The null hypothesis is rejected if the calculated χ^2 is greater than the tabular $\chi^2_\alpha(r - 1)(c - 1)$.

In the chi-square test, we are evaluating the probability that an observed set of frequencies $X_1, X_2, \ldots, X_k$ could have resulted from a multinomial distribution with probabilities $p_1, p_2, \ldots, p_k$. The requirements that must be met in order that the χ^2 approximation may be used to evaluate this probability are (1) the frequencies follow the multinomial distribution; (2) the expectations are large enough so that the normal approximation is satisfactory; and (3) any estimation of the p's should be efficient. For discussion related to this problem consult Cochran (1952, 1954).

For the 2 × 2 contingency table, Yates (1934) suggested a correction for continuity that .5 be subtracted from the absolute value of each deviation in computing χ^2 in order to correct for a slight bias in determining the true probability levels. This slight bias arises from the fact that the χ^2 distribution is continuous, whereas the frequencies are discrete. Let us rewrite the 2 × 2 contingency table as

	B_1	B_2	
A_1	a	b	a + b
A_2	c	d	c + d
	a + c	b + d	n

(11.13)

and n = a + b + c + d. Using the correction for continuity, we increase b and c by .5 and decrease a and d by .5 when ad > bc (or decrease b and c by .5 and increase a and d by .5 when ad < bc). If this is done, the chi-square test statistic becomes

$$\chi^2 = \frac{n(|ad - bc| - .5n)^2}{(a + b)(c + d)(a + c)(b + d)} \tag{11.14}$$

A discussion of the use of continuity correction on 2 × 2 contingency table was given by Conover (1974). The correction for continuity is not very important for χ^2 with more than a single degree of freedom and is never to be used when adding χ^2 values.

EXAMPLE 11.5 The question of whether a school bond should be issued in a city was voted on. A random sample of 80 voters produced the following table:

	Yes	No	
Democrats	22	9	31
Republicans	19	16	35
Independents	11	3	14
	52	28	80

The null hypothesis to be tested is that the vote for the school bond is independent of the political parties. The calculated χ^2 value is

$$\chi_0^2 = \frac{(22 - 20.15)^2}{20.15} + \frac{(19 - 22.75)^2}{22.75} + \frac{(11 - 9.10)^2}{9.10} + \frac{(9 - 10.85)^2}{10.85} + \frac{(16 - 12.25)^2}{12.25} + \frac{(3 - 4.9)^2}{4.9} = 3.38$$

The upper 5% point of $\chi^2(2)$ is 5.99. Hence we do not reject the null hypothesis.

11.6 FISHER'S EXACT TEST: The chi-square test for the 2 × 2 contingency table is an approximate test and the approximation is satisfactory when the expected numbers are large. When the frequencies are small, it is customary to consider the exact test given by Fisher (1970). An illustration of the use of Fisher's exact test was given in Section 1.2.

Let us use the table in (11.13) and assume that the marginal frequencies are fixed in repeated sampling. The number of ways to obtain the frequency a out of a + b is $\binom{a+b}{a}$, c out of c + d is $\binom{c+d}{c}$, and a + c out of n is $\binom{n}{a+c}$. Hence the exact probability of observing the set of entries in (11.13) is

$$P = \frac{\binom{a+b}{a}\binom{c+d}{c}}{\binom{n}{a+c}} = \frac{(a+b)!(c+d)!(a+c)!(b+d)!}{n!a!b!c!d!} \tag{11.15}$$

This is actually the probability for the hypergeometric distribution. To test the null hypothesis that the two attributes are independent, we compute the above probability and probabilities associated with more extreme tables. That is, if a is the smallest frequency among the 4 entries and we denote the probability in (11.15) by P(a), we then compute P(a), P(a - 1), ..., P(0). The sum P(a) + P(a - 1) + ··· + P(0) gives the observed significance level. We reject the null hypothesis if this sum is smaller than the nominal α level, for the data indicates that it is unlikely that the observed entries have come from the population with independent attributes.

An extension of the Fisher's exact test to more than two samples is given by Carr (1980).

EXAMPLE 11.6 A random sample of 22 individuals is taken and they are asked if they believe that there is a gasoline shortage in the country. The following results are obtained:

	Yes	No	
Male	10	2	12
Female	6	4	10
	16	6	22

We wish to investigate whether the answer to the question is associated with the respondent's sex. The smallest frequency is 2. We compute

$$P(2) = \frac{\binom{12}{2}\binom{10}{4}}{\binom{22}{6}} = .186$$

$$P(1) = \frac{\binom{12}{1}\binom{10}{5}}{\binom{22}{6}} = .040$$

$$P(0) = \frac{\binom{12}{0}\binom{10}{6}}{\binom{22}{6}} = .003$$

The sum of the three terms is .229. Hence we do not reject the null hypothesis and conclude that the answer to the question is independent of respondent's sex.

EXERCISES

11.1 In Example 11.1, use normal approximation to compute the power when p = .2, .3, .4.

11.2 Two types of rope, A and B, are tested for strength. Both type A and type B are tied on a testing machine. If type A breaks down first, a negative sign is given; otherwise a positive sign is given. Twenty-five sample pairs are tested and 5 positive signs are observed. Test the hypothesis that the two types of rope are equal in strength.

11.3 In Section 11.3, let us assign rank 1 to the smallest observation in the pooled sample, rank 2 to the second smallest observation and so on, and rank n + m to the largest observation. Let T be the sum of the ranks for the y's in the pooled sample. Show that

$$T = U + \frac{1}{2} m(m + 1)$$

The statistic T is the Wilcoxon rank sum test statistic.

11.4 A class of 20 students are divided at random into two groups. Two teaching methods--method A using a videotape equipment and method B without such equipment (otherwise the teaching materials are the same)--are assigned at random to the two groups. After the course is completed, identical tests are given to the two groups of students. Test scores for the two groups are

Method A:	53,	54,	58,	62,	63,	75,	79,	81,	83,	84
Method B:	26,	52,	59,	60,	65,	69,	74,	80,	86,	97

Use the Mann-Whitney-Wilcoxon test to test the effectiveness of using the videotape equipment.

11.5 (Some of Beall's data given by Neyman, 1939.) Fit a Poisson curve to the following distribution of European corn borers in 120 groups of 8 hills each. Use the method illustrated in Example 11.3 to calculate a goodness-of-fit chi-square.

No. of borers	0	1	2	3	4	5	6	7	8	9	10	11	12
Observed frequency	24	16	16	18	15	9	6	5	3	4	3	0	1

11.6 The following table gives the observed frequencies of hospital lengths of stay for 1310 patients at St. Joseph's Hospital as reported by Welch et al. (1977):

Time (Days)	Frequency	Time (Days)	Frequency
<3.5	244	39.5-42.5	16
3.5-6.5	181	42.5-45.5	17
6.5-9.5	163	45.5-48.5	12
9.5-12.5	119	48.5-51.5	6
12.5-15.5	108	51.5-54.5	8
15.5-18.5	82	54.5-57.5	8
18.5-21.5	82	57.5-60.5	6
21.5-24.5	76	60.5-63.5	3
24.5-27.5	35	63.5-66.5	3
27.5-30.5	42	66.5-69.5	4
30.5-33.5	33	69.5-72.5	4
33.5-36.5	16	>72.5	27
36.5-39.5	15		

The sample mean of the patients was computed to be $\bar{x} = 17.044$ days. Fit the data with an exponential distribution.

11.7 Calculate chi-square corrected for continuity for Example 11.4.

11.8 A random sample of 30 individuals are asked if they like a certain brand of coffee. The following table is observed:

	Like	Dislike
Male	16	2
Female	6	6

Use Fisher's exact test for testing the association between sex and taste.

11.9 Blankenship (1967) reported the percentage of institutional budget allotted to library and the numbers of male and female heads of libraries as follows:

Percentage of budget	Number of male head librarians	Number of female head librarians
Under 2%	9	6
2-4%	78	60
4-6%	92	109
Over 6%	19	18

Test if sex is associated with the attraction of the allotment of budget.

11.10 The observed frequencies of hair color of three groups of people, chosen from different geographical regions, are as follows:

	Light hair	Red hair	Dark hair
Region A	9	2	9
Region B	6	3	21
Region C	15	15	20

Do these data indicate that hair color is associated with geographical region?

11.11 A city is divided into four districts; Northeast (NE), Northwest (NW), Southeast (SE), and Southwest (SW). A random sample of 400 residents are asked if they like the idea of building a new city hall on the east side of the downtown area. The following data are obtained:

	NE	NW	SE	SW
Yes	20	40	30	60
No	70	40	60	30
No opinion	10	20	10	10

Test the hypothesis that opinion is independent of district.

REFERENCES

Blankenship, W. C. (1967). Head librarians: How many men? How many women? College and Research Libraries 28: 41-48.

Carr, W. E. (1980). Fisher's exact test extended to more than two samples of equal size. Technometrics 22: 269-270.

Cochran, W. G. (1952). The χ^2 test of goodness of fit. Ann. Math. Stat. 23: 315-345.

Cochran, W. G. (1954). Some methods of strengthening the common χ^2 tests. Biometrics 10: 417-451.

Conover, W. J. (1971). *Practical Nonparametric Statistics*. Wiley, New York.

Conover, W. J. (1974). Some reasons for not using the Yates continuity correction on 2 × 2 contingency tables. J. Amer. Stat. Assoc. 69: 374-382.

Federer, W. T. (1946). Variability of certain seed, seedling, and young plant characters of guayule. U. S. Department of Agriculture Technical Bulletin 919.

Fisher, R. A. (1970). *Statistical Methods for Research Workers*, 14th ed. Oliver Boyd, Edinburgh.

Gibbons, J. D. (1971). *Nonparametric Statistical Inference*. McGraw-Hill, New York.

Lindstrom, E. W. (1918). Cornell University Agriculture Experiment Station Memorandum 13.

Mann, H. B., and Whitney, D. R. (1947). On a test of whether one of two random variables is stochastically larger than the other. Ann. Math. Stat. 18: 50-60.

Neyman, J. (1939). On a new class of contagious distributions, applicable in entomology and bacteriology. Ann. Math. Stat. 10: 35-57.

Robertson, D. W. (1937). Material inheritance in barley. Genetics 22: 104-113.

Welch, R. L. W., Smith, T. L., and Denker, M. W. (1977). Estimating hospital mean length of stay from censored data. Amer. Stat. Assoc. Proc. Social Stat. Sec. 554-556.

Yates, F. (1934). Contingency tables involving small numbers and χ^2 test. J. Roy. Stat. Soc. Suppl. 1: 217-235.

12

INFERENCE BASED ON CONDITIONAL SPECIFICATION

12.1 INTRODUCTION: In nonparametric procedures, it is not necessary to specify the form of the parent population. When an inference procedure calls for a specification of the parent population, as in the case of parametric statistical inference, the problem of specification arises. It is well known to research workers, in substantive areas in which statistics is used as an important part of their research methodology, that it is extremely unlikely that any particular specification will represent exactly the natural or social phenomena under investigation. In other words, before selecting some particular specification to represent such phenomena; for example, the relationship among certain variables in a regression model (see Chapter 13) or certain side assumptions such as normality and independence, the research worker has already used his or her past experience and that of other workers with such or similar data, and possibly intuition, in making this initial specification choice. Should there be still some uncertainty as regards the inclusion or not of certain assumptions in the initial specification, the research worker may wish to perform objective tests using available data as an aid in dealing with such uncertainties as regards the choice of a final specification. Such tests used as a preliminary analysis of the specification of the parent population are called *preliminary tests.*

When the research worker, either acting as a specialist in a substantive field as well as his or her own statistician or working in conjunction with a consulting statistician, has considerable experience and knowledge about the investigation undertaken, he or she will be often able to choose an appropriate final specification for the investigation. Such a specification is referred to as an *unconditional specification*. However, should the research worker be uncertain regarding the initial specification, he or she may wish to verify the assumptions and determine the final specification based on available data, usually by using preliminary tests. In such a case, the research worker has a *conditional specification*.

With the above description of conditional specification, any attempt of checking assumptions in the initial specification by preliminary test and the effect of the test on subsequent statistical inference cannot be ignored and must be incorporated into the entire inference procedure. Although this has been pointed out in the literature by Bancroft and Han (1977), Moran (1973), and Bancroft (1968), it has not been systematically studied before in a basic statistical theory text such as this one. The treatment given here will be minimal because the formulas involved are usually very complicated. The research worker should consult the bibliography by Bancroft and Han (1977) for a particular conditional specification problem in hand. The simple case concerning normal populations is treated in this chapter. Inference based on conditional specification in regression and analysis of variance will be given in the next two chapters.

12.2 POOLING MEANS: In applied statistics, it often happens that the experimenter has two independent samples that are taken under similar conditions; for example, the two samples may be obtained from different but similar laboratories, or at different times. We know that if they are from the same population, it is advantageous to pool the two samples and to make statistical inference based on the pooled sample because of the increase in degrees of freedom. If the two samples come from different populations, then the two samples should not be pooled. However, in certain practical situations, the experimenter is uncertain whether the two populations are identical and wishes to leave the decision whether to pool the two samples to the outcome of a preliminary test.

Let us consider the estimation of the mean μ_1 of a normal population. A random sample of size n_1, $\{X_{11}, X_{12}, \ldots, X_{1n_1}\}$, is taken from $N(\mu_1, \sigma^2)$. The sample mean $\bar{X}_1 = \Sigma_{i=1}^{n_1} X_{1i}/n_1$ is an estimator of μ_1. However, a second sample of size n_2, $\{X_{21}, X_{22}, \ldots, X_{2n_2}\}$, is suspected to have come from the same population. It is assumed that the population variance of the second sample is the same as that of the first population, but the mean μ_2 may differ from μ_1. Suppose the second sample comes from $N(\mu_2, \sigma^2)$. If $\mu_2 = \mu_1$, we should pool the two samples and use

$$\frac{n_1\bar{X}_1 + n_2\bar{X}_2}{n_1 + n_2}$$

as an estimator for μ_1, where $\bar{X}_2 = \Sigma_{i=1}^{n_2} X_{2i}/n_2$. In order to resolve the uncertainty whether $\mu_2 = \mu_1$, a preliminary test is used to test: $H_0: \mu_1 = \mu_2$. Let σ^2 be known, then the test statistic is

$$\frac{|\bar{X}_1 - \bar{X}_2|}{\sigma\sqrt{1/n_1 + 1/n_2}}$$

If H_0 is accepted, the pooled estimator is used; otherwise only the first sample is used for estimating μ_1. The *preliminary test estimator* of μ_1 is defined as

$$\bar{X}' = \begin{cases} \bar{X}_1 & \text{if } |\bar{X}_1 - \bar{X}_2| \geq z_{\alpha/2}\sigma\sqrt{\dfrac{1}{n_1} + \dfrac{1}{n_2}} \\ \dfrac{n_1\bar{X}_1 + n_2\bar{X}_2}{n_1 + n_2} & \text{if } |\bar{X}_1 - \bar{X}_2| < z_{\alpha/2}\sigma\sqrt{\dfrac{1}{n_1} + \dfrac{1}{n_2}} \end{cases} \qquad (12.1)$$

where $z_{\alpha/2}$ is the $100(1 - \alpha/2)$ percentage point of $N(0,1)$. This estimator is also called the *sometimes-pool estimator*. If only the first sample is used for estimating μ_1, $\bar{X}_1$ is called the *never-pool estimator*. If we always pool the two samples and use $(n_1\bar{X}_1 + n_2\bar{X}_2)/(n_1 + n_2)$, it is called the *always-pool estimator*.

The properties of the estimator $\bar{X}'$ were studied by Bennett (1952), Kale and Bancroft (1967), and Mosteller (1948). The bias $E(\bar{X}') - \mu_1$

and the mean square error $MSE(\bar{X}') = E(\bar{X}' - \mu_1)^2$ depend on the two sample sizes, n_1, n_2, the significance level α, and the difference of the two population means $\mu_1 - \mu_2$. Discussions of these are given in the above-mentioned papers.

When the population variance σ^2 is unknown, the preliminary test for $H_0: \mu_1 = \mu_2$ is a t test. The preliminary test estimator of μ_1 becomes

$$\bar{X}^* = \begin{cases} \bar{X}_1 & \text{if } |\bar{X}_1 - \bar{X}_2| \geq t_{\alpha/2} S_p \sqrt{\dfrac{1}{n_1} + \dfrac{1}{n_2}} \\ \dfrac{n_1\bar{X}_1 + n_2\bar{X}_2}{n_1 + n_2} & \text{if } |\bar{X}_1 - \bar{X}_2| < t_{\alpha/2} S_p \sqrt{\dfrac{1}{n_1} + \dfrac{1}{n_2}} \end{cases} \tag{12.2}$$

where S_p^2 is the pooled unbiased estimator of σ^2 and $t_{\alpha/2}$ is the $100(1 - \alpha/2)$ percentage point of the t distribution with $n_1 + n_2 - 2$ degrees of freedom.

Define the relative efficiency of $\bar{X}^*$ as compared to $\bar{X}_1$ as

$$e = \frac{1/MSE(\bar{X}^*)}{1/MSE(\bar{X}_1)} = \frac{MSE(\bar{X}_1)}{MSE(\bar{X}^*)} \tag{12.3}$$

The relative efficiency is a function of α and $\delta = (\mu_2 - \mu_1)/\sigma$. Figure 12.1 gives an example of the function and shows the general behavior of the relative efficiency. It is seen that the relative efficiency has a maximum value (> 1) at $\delta = 0$. When δ increases, the value of e decreases crossing the line $e = 1$; it decreases to a minimum, then increases to unity as δ tends to infinity. Hence when μ_2 is very different from μ_1, the preliminary test would almost always reject H_0, and $\bar{X}_1$ is used as the estimator. Figure 12.1 gives only the relative efficiency for positive δ. Since it can be shown that the relative efficiency is symmetrical about $\delta = 0$, similar curves can be obtained for negative δ.

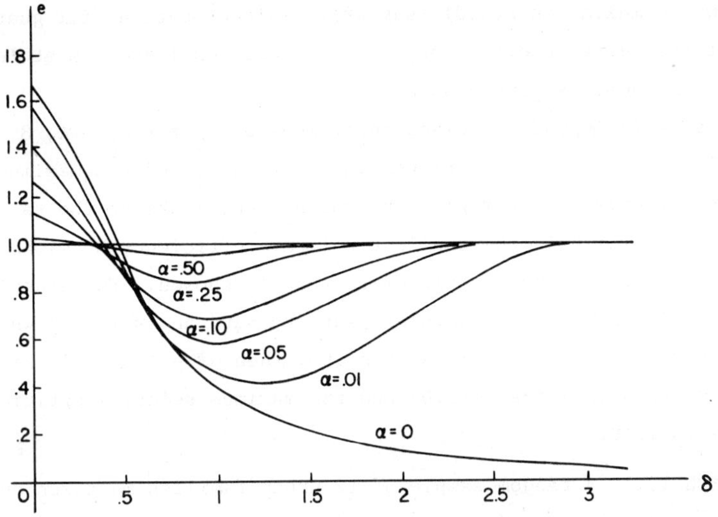

FIG. 12.1 Relative efficiency of $\overline{X}^*$ to $\overline{X}_1$ for $n_1 = 12$, $n_2 = 8$.

The experimenter usually wants to select an estimator with high relative efficiency. The relative efficiency e depends on n_1, n_2, α, and $\delta = (\mu_2 - \mu_1)/\sigma$. The sample sizes n_1 and n_2 are usually fixed and δ is unknown. So the only parameter at the disposal of the experimenter is α, the significance level of the preliminary test. In testing statistical hypothesis, the levels .01 and .05 are often used in practice. However, in a preliminary test situation, these choices may not be appropriate because we must consider their effect on subsequent analysis. Therefore, the experimenter must be careful in the selection of the level of the preliminary test. A criterion of selection is given in Han and Bancroft (1968) and described as follows:

Denote the relative efficiency as a function of α and δ, $e(\alpha,\delta)$. If the experimenter does not know the size of δ and is willing to accept an estimator which has a relative efficiency of no less than e_0, then among the set of estimators with $\alpha \in A$, where $A = \{\alpha: e(\alpha,\delta) \geq e_0$ for all $\delta\}$, the estimator is chosen to maximize $e(\alpha,\delta)$ over all α and δ. Since $\max_\alpha e(\alpha,\delta) = e(\alpha,0)$, he selects the $\alpha \in A$

(say α^*) which maximizes $e(\alpha,0)$ (say e^*). This criterion will guarantee that the relative efficiency of the chosen estimator is at least e_0 and it may become as large as e^*.

The table in Appendix B gives the values of e_0 for $n_i = 4, 8, 12, 16, 20$ ($i = 1, 2$), the corresponding α^* to use, and the maximum relative efficiency e^*. For given n_1 and n_2, one looks at the table and picks up e_0 which is the smallest relative efficiency he wishes to accept. The recommended α^* level is readily chosen. For example, if $n_1 = n_2 = 12$ and the experimenter wants to have an estimator which has a relative efficiency no less than .80, then $\alpha^* = .20$ would be used because this maximizes $e(\alpha,0)$ and the maximum relative efficiency obtainable is 1.19.

EXAMPLE 12.1 A random sample of 12 urban families is taken and their reductions in gasoline consumption in one month following the announcement of a conservation program are recorded. The sample mean of the reduction is $\bar{X}_1 = 5.4$ gallons and the sample variance is $S_1^2 = 10.6$. Similarly a random sample of 16 rural families yields $\bar{X}_2 = 3.7$ and $S_2^2 = 12.8$. It is suspected that the average reduction in gasoline consumption is the same for the respective urban and rural population. In such a case it is advantageous to pool the two samples to estimate the urban population mean, say μ_1. The t value is computed as 1.29. The significance level for the t test is determined by using Appendix B. If the investigator is willing to accept an estimator with relative efficiency no less than .79 but may be as high as 1.21, he or she selects $\alpha = 0.20$ for $n_1 = 12$, $n_2 = 16$. The critical value based on 26 degrees of freedom is 1.32 and the t test is not significant. Hence, the two sample means are pooled. The estimate of the urban population mean reduction is

$$\frac{1}{28}(12 \times 5.4 + 16 \times 3.7) = 4.43 \text{ gallons}$$

Note that this is also the estimate of the rural population mean reduction.

12.3 BAYESIAN POOLING: In order to use the Bayesian procedure, the experimenter must specify not only the parent population distribution but also the prior distribution. If the experimenter has a priori information about the distribution of δ--in particular, if δ is normally distributed with mean 0 and variance a^2--the estimator of μ_1 will be chosen to maximize the product of the likelihood and the a priori probability. Let σ^2 be known. The likelihood function in terms of the sufficient statistics is

$$f(\bar{x}_1,\bar{x}_2) = \frac{\sqrt{n_1 n_2}}{2\pi\sigma^2} \exp\left\{\frac{-1}{2\sigma^2}[n_1(\bar{x}_1 - \mu_1)^2 + n_2(\bar{x}_2 - \mu_2)^2]\right\}$$

Let the prior distribution of δ be $N(0,a^2)$ and that of μ_1 locally uniform. The prior density function is

$$g(\delta) = \frac{1}{\sqrt{2\pi}a} \exp\left\{-\frac{1}{2a^2\sigma^2}(\mu_1 - \mu_2)^2\right\} \tag{12.4}$$

Multiplying the prior density and the likelihood and normalizing the result (normalizing means multiplying by an appropriate constant to make the function as a density function), we have the joint posterior distribution of μ_1 and δ. Integrating out δ, we obtain the posterior distribution of μ_1, which is normal. The posterior mean is used as the estimate of μ_1, and

$$\hat{\mu}_1 = \frac{n_1(n_2a^2 + 1)\bar{x}_1 + n_2\bar{x}_2}{n_1(n_2a^2 + 1) + n_2} \tag{12.5}$$

We note that as $a^2 \to \infty$, $\hat{\mu}_1 \to \bar{x}_1$ and only the first sample is used for estimating μ_1.

The mean square error of μ_1 after averaging over the distribution of δ is

$$E(\hat{\mu}_1 - \mu_1)^2 = \frac{(n_2a^2 + 1)\sigma^2}{n_1(n_2a^2 + 1) + n_2} \tag{12.6}$$

which is smaller than the mean square error of the sometimes-pool estimator.

When σ^2 is unknown and the prior distribution of σ^2 is independent of μ_1 and δ, the Bayesian pooled estimator of μ_1 is the same as given in Eq. (12.5).

EXAMPLE 12.2 In Example 12.1, if the prior distribution of δ is N(0,0.5) and that of μ_1 is locally uniform and independent of the prior distribution of σ^2, the estimate of the average reduction, by Eq. (12.5), is

$$\hat{\mu}_1 = \frac{12(16 \times 0.5 + 1)(5.4) + 16(3.7)}{12(16 \times 0.5 + 1) + 16} = 5.18$$

We now have two estimators. Whether the experimenter uses the sometimes-pool estimator or the estimator $\hat{\mu}_1$ depends on the experimental situation. If the experimenter has the information that δ is normally distributed, $\hat{\mu}_1$ should be used. If there is no information available about the distribution of δ, the sometimes-pool estimator may be used according to the criterion given in Section 12.2.

12.4 BEHRENS-FISHER PROBLEM UNDER CONDITIONAL SPECIFICATION: In Section 10.9 the Behrens-Fisher problem was discussed in terms of unconditional specification. In such a case it was known to the experimenter that the two variances of the normal populations are unequal and the t test is not valid for testing the equality of means. In applied statistics, the experimenter may not be certain as to whether the two variances are equal. The experimenter often utilizes the data on hand to check the assumption of equality of variances. This is usually accomplished by using a preliminary test of hypothesis H_{00}: $\sigma_1^2 = \sigma_2^2$. If H_{00} is accepted, the t test is used to test the equality of the means; otherwise a procedure for the Behrens-Fisher problem is used. Since the experimenter's specification is made to depend on the outcome of the preliminary test, the testing procedure belongs to the area of inference based on conditional specification.

Suppose we have a random sample of size n_1, $\{X_{11}, X_{12}, \ldots, X_{1n_1}\}$, from $N(\mu_1, \sigma_1^2)$, and a second independent random sample of size n_2,

$\{X_{21}, X_{22}, \ldots, X_{2n_2}\}$, from $N(\mu_2, \sigma_2^2)$. We are interested in testing H_0: $\mu_1 = \mu_2$ against H_1: $\mu_1 \neq \mu_2$ but it is uncertain whether $\sigma_1^2 = \sigma_2^2$. The preliminary test for H_{00}: $\sigma_1^2 = \sigma_2^2$ versus H_{01}: $\sigma_1^2 \neq \sigma_2^2$ is a two-sided F test (a one-sided test may be used if it is appropriate). If H_{00} is accepted, a t test with pooled estimator of variance is used in the final test for H_0: $\mu_1 = \mu_2$. If H_{00} is rejected, we have the Behrens-Fisher problem. There are several solutions to the Behrens-Fisher problem; we shall only consider the solution suggested by Cochran and Cox (1957) and given in Section 10.9. Let $\bar{X}_1$, $\bar{X}_2$, and S_1^2, S_2^2 be the sample means and variances respectively. The pooled estimate of variance is

$$S_p^2 = \frac{(n_1 - 1)S_1^2 + (n_2 - 1)S_2^2}{n_1 + n_2 - 2}$$

Also let the preliminary test have significance level α_0, the t test have level α_1, and the test of $\mu_1 = \mu_2$ with $\sigma_1^2 \neq \sigma_2^2$ have level α_2. The test procedure has the rejection region given by

$$\left.\begin{array}{lll} \dfrac{|\bar{X}_1 - \bar{X}_2|}{S_p\sqrt{1/n_1 + 1/n_2}} > t_0 & & \text{when } d_1 \leq \dfrac{S_1^2}{S_2^2} \leq d_2 \\[3ex] \dfrac{|\bar{X}_1 - \bar{X}_2|}{\sqrt{W_1 + W_2}} > \dfrac{W_1 t_1 + W_2 t_2}{W_1 + W_2} & & \text{when } \dfrac{S_1^2}{S_2^2} < d_1 \\ & & \text{or } > d_2 \end{array}\right\} \tag{12.7}$$

where t_0 is the $100(1 - \alpha_1/2)$ percentage point of the t distribution with $n_1 + n_2 - 2$ degrees of freedom, $W_i = S_i^2/n_i$, t_i is the $100(1 - \alpha_2/2)$ percentage point of the t distribution with $n_i - 1$ degrees of freedom, and d_1 and d_2 are the appropriate critical points for a two-sided F test of level α_0.

The manner in which α_0, α_1, and α_2 are chosen gives the rise to different procedures. We suggest that one selects $\alpha_1 = \alpha_2 = \alpha$. Lauer and Han (1974) studied this test procedure and recommend that for $\alpha = .05$, $n_1 = n_2 = n$, and $n > 3$, one should use $\alpha_0 = 5/n^2$. For $n = 3$, $\alpha_0 = .47$ is optimal. The optimality refers to the situation which, with the above selection of α_0, P(Type I error) $\leq \alpha$, and the highest power is obtained.

EXERCISES

12.1 A random sample of 12 kernels of mature Iodent corn was tested by an experienced worker A for crushing resistance. The measurements in pounds of resistance were 41, 44, 29, 37, 41, 47, 40, 49, 44, 52, 38, 42. A second random sample of 12 kernels from the same bin of shelled corn when tested in the same manner by a less experienced worker yielded measurements in pounds of resistance of 36, 45, 42, 25, 33, 42, 43, 42, 39, 33, 50, 34. Using a proper preliminary test of H_1: $\mu_1 = \mu_2$, if indicated, combine the two samples to obtain a better estimate of the population mean crushing resistance μ, assuming that the population crushing resistance measurements of the two samples are from $N(\mu_1,\sigma_1^2)$ and $N(\mu_2,\sigma_2^2)$, respectively, with $\sigma_1^2 = \sigma_2^2 = \sigma^2$, but there is uncertainty as to whether $\mu_1 = \mu_2 = \mu$.

12.2 In Exercise 12.1, suppose that there was some uncertainty as to whether $\sigma_1^2 = \sigma_2^2 = \sigma^2$, the interest now is only in testing H_0: $\mu_1 = \mu_2 = \mu$. Using a proper preliminary test of H_1: $\sigma_1^2 = \sigma_2^2 = \sigma^2$, make the indicated test for H_0: $\mu_1 = \mu_2 = \mu$.

12.3 In Exercise 12.1, suppose the prior distribution of $\delta = (\mu_2 - \mu_1)/\sigma$ is $N(0,.5)$. Find the Bayesian pooled estimator of μ_1.

12.4 Considering the estimator $\bar{X}'$ in Eq. (12.1), show that

$$E(\bar{X}') = \mu_1 + \frac{n_1}{n_1 + n_2} E((\bar{X}_2 - \bar{X}_1)|A)P(A)$$

where A represents the event that $|\bar{X}_2 - \bar{X}_1| < z_{\alpha/2}\, \sigma\sqrt{1/n_1 + 1/n_2}$, and the expectation is the conditional expectation given that the event A has happened.

12.5 Let r_i, $i = 1, 2$, be the two sample correlation coefficients obtained from two bivariate normal populations with corresponding unknown population correlation coefficients ρ_i. We wish to estimate ρ_2, say, when it is suspected, though not known with certainty, that $\rho_1 = \rho_2$. A preliminary test for H_0: $\rho_1 = \rho_2$ is used to resolve the uncertainty. If H_0 is rejected, use r_2 to estimate ρ_2. Define $Z_i = (1/2) \log [(1 + r_i)/(1 - r_i)]$ which is referred to as the Fisher's Z transformation. The test statistic for H_0 is

$$\frac{|Z_1 - Z_2|}{\sqrt{1/(n_1 - 3) + 1/(n_2 - 3)}}$$

which is approximately distributed as N(0,1) when n_i, i = 1, 2, are large. When H_0 is accepted, we first pool Z_1 with Z_2, i.e.,

$$Z = \frac{(n_1 - 3)Z_1 + (n_2 - 3)Z_2}{n_1 + n_2 - 6}$$

then find its transformed r value as the estimate of ρ_2. In an experiment Lush (1931) had been interested in estimating the 1928 Brahmans' population coefficient between initial weight and gain of steers and was uncertain as to whether or not to use only the few pairs of observations ($n_2 = 11$) available for 1928 and obtain the sample $r_2 = .123$ as an estimate of ρ_2 or to combine the sample $r_1 = .570$ obtained from the 1927 data ($n_1 = 13$) with r_2 in a proper manner in estimating ρ_2. Use the pooling methodology described above to test H_0: $\rho_1 = \rho_2$ at the recommended $\alpha = .25$ significance level (see Srivastava and Bancroft, 1967) and obtain an estimate of ρ_2 based on the outcome of the preliminary test.

12.6 Bancroft (1944) considered the pooling of two independent sample variances for estimating a population variance. In Exercise 12.1, suppose that the experimenter is interested in estimating σ_1^2, but he is uncertain whether $\sigma_1^2 = \sigma_2^2$. He adopts the following pooling methodology. Test H_0: $\sigma_1^2 = \sigma_2^2$ by an F test. If H_0 is rejected, he uses S_1^2 to estimate σ_1^2; otherwise S_2^2 is pooled with S_1^2 and use $[(n_1 - 1)S_1^2 + (n_2 - 1)S_2^2]/(n_1 + n_2 - 2)$ as the estimator. Use the above procedure and the data in Exercise 12.1 to estimate σ_1^2.

REFERENCES

Bancroft, T. A. (1944). On biases in estimation due to the use of preliminary test of significance. Ann. Math. Stat. 15: 190-204.

Bancroft, T. A. (1968). *Topics in Intermediate Statistical Methods*, Vol. 1. Iowa State University Press, Ames.

Bancroft, T. A., and Han, C.-P. (1977). Inference based on conditional specification: A note and a bibliography. Int. Stat. Rev. 45: 117-127.

Bennett, B. M. (1952). Estimation of means on the basis of preliminary tests of significance. Ann. Inst. Stat. Math. 21: 539-556.

Cochran, W. G., and Cox, G. M. (1957). *Experimental Design*, 2nd ed. Wiley, New York.

Han, C.-P., and Bancroft, T. A. (1968). On pooling means when variance is unknown. J. Amer. Stat. Assoc. 63: 1333-1342.

Kale, B. K., and Bancroft, T. A. (1967). Inference for some incompletely specified models involving normal approximations to discrete data. Biometrics 23: 335-348.

Lauer, G. N., and Han, C.-P. (1974). Power of Cochran's test in the Behrens-Fisher problem. Technometrics 16: 545-549.

Lush, J. L. (1931). Predicting gains in feeder cattle and pigs. J. Agri. Res. 42: 853-881.

Moran, P. A. P. (1973). Problems and mistakes in statistical analysis. Commun. Stat. 2: 245-257.

Mosteller, F. (1948). On pooling data. J. Amer. Stat. Assoc. 43: 231-242.

Srivastava, S. R., and Bancroft, T. A. (1967). Inferences concerning a population correlation coefficient from one or possibly two samples subsequent to a preliminary test of significance. J. Roy. Stat. Soc. B29: 282-291.

13

REGRESSION ANALYSIS

13.1 INTRODUCTION: In previous chapters, we have presented some of the basic statistical concepts of estimation and tests of significance. We presented as our basic estimation procedure the method of maximum likelihood, because it has certain optimum properties such as giving a minimum-variance estimator and a sufficient estimator if the latter exists. Now we propose to consider the problem of predicting the value of some dependent random variable Y on the basis of information on one or more other fixed variables, x_1, x_2 .... A dependent variable will be understood to have a probability distribution, while a fixed variable does not have a probability distribution. Another way of saying this is that inferences are to be made regarding the variability of Y for this particular set of x values. The Y's are expected to fluctuate from sample to sample, while the x values remain fixed. For example, we might wish to estimate the yield of wheat, Y, for different amounts x of a standard fertilizer applied to the soil. Or, more exactly, we might estimate this same yield on the basis of the amount of nitrogen, x_1, phosphate, x_2, and potash, x_3, applied to the soil. There is a great demand for information on the effect of temperature and precipitation on yields. The economist tries to predict future employment and price relationships on the basis of past data on the same and other economic variables. The engineer has the problem of estimating the probable length of life of roads or other structures in terms of such things as probable use, type of construction, and weather conditions. The doctor must decide on the basis of certain measurements how much of a given drug can be safely administered to a patient.

As indicated in a previous chapter on bivariate distributions, the expected value of Y for a single fixed x gave the so-called regression curve of Y on x. A first approximation to this curve was indicated to be a straight line of the form

$$E(Y|X = x) \doteq \alpha + \beta x$$

In fact, if X and Y were distributed as a bivariate normal,

$$E(Y|X = x) \equiv \alpha + \beta x$$

Y is called the dependent random variable and x the fixed variable; x is also often called the independent variable. However, it should be understood that in general the straight line is only an approximation to the true relationship between Y and x. It is well known that we can approximate a short interval of most functions by a straight line; hence, if we collect data for a small range of x, it is possible that a straight line would fit the data quite well even though the true relationship were curvilinear.

Let us assume that the measured value of Y can be written as

$$Y = E(Y|X = x) + \varepsilon$$

where ε represents some residual or error, the amount of variation in Y not accounted for by the regression curve of Y on x. We postulate that the regression curve is selected so that the residuals are of a random nature, with the usual added assumptions that $E(\varepsilon) = 0$, $V(\varepsilon) = \sigma^2$, and the ε terms are uncorrelated. Further, the normality assumption is added if tests of hypotheses or confidence intervals are desired. If Y is a linear function of p fixed variables, we might write

$$Y = \alpha + \beta_1 x_1 + \beta_2 x_2 + \cdots + \beta_p x_p + \varepsilon$$

where $E(Y\ ) = \alpha + \beta_1 x_1 + \beta_2 x_2 + \cdots + \beta_p x_p$. This equation assumes that the only error involved is ε; in other words, there is no error in the x's. Hence we are considering only a one-variable normal distribution with the mean of Y being approximated by a simple linear function of the x's. If the x's are not measured without error (in other words, the X's have probability distributions of their own), we are led to consider the more complicated problem of multivariate analysis. Also, we are considering here only regression equations which are linear in the regression coefficients, α and β_i. Methods of handling multivariate problems and problems of nonlinearity of the regression coefficients are beyond the scope of this book. It should be emphasized that nonlinearity of the x's can be handled by the introduction of substitute terms in the regression equation. For example, x_2 might very well represent x_1^2.

In order to estimate the relationship between Y and x (or between Y and $x_1, x_2, \ldots, x_p$, for p fixed variables), assuming a random sample of n sets of observations on Y and the respective x's, n simultaneous observations will be obtained on Y and x. We can write each observed value Y_i in terms of estimates of $E(Y_i)$ and ε_i as

$$Y_i = \hat{Y}_i + e_i \qquad i = 1, 2, \ldots, n$$

where $\hat{Y}_i$ is the estimate of $E(Y_i)$ and e_i the estimate of ε_i. Hence if E(Y) is a linear relationship,

$$Y_i = \alpha + \beta x_i + \varepsilon_i = a + bx_i + e_i$$

where the estimators of α and β are a and b, respectively. We can represent these two equations graphically, as in Fig. 13.1. E(Y) is the true regression line and $\hat{Y}$ the observed regression line. For a

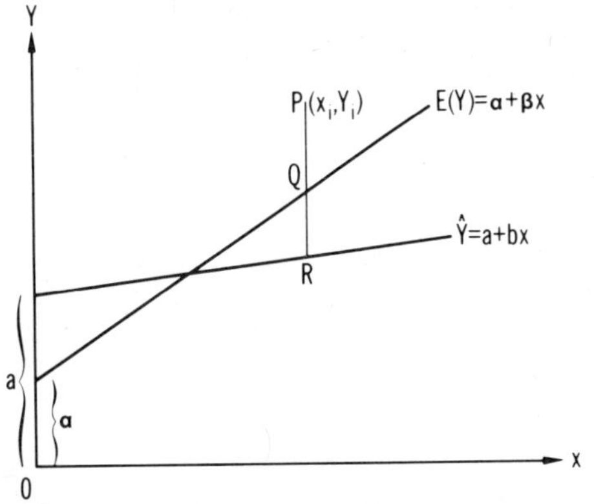

FIG. 13.1 True and estimated regression lines.

given observation (x_i, Y_i), the true error is given by ε_i(QP), the estimated residual by e_i(RP), and the difference between $\hat{Y}$ and E(Y) by $\delta_i (= \varepsilon_i - e_i = QR)$.

In order to obtain the values of a and b, at least two sets of observations (x,Y) are required. Of course, if only two sets are obtained, $\hat{Y}$ will pass through the two points, both values of e will be 0, and the values of a and b can be determined by a simultaneous solution of the two equations representing the two points. However, if more than two sets of observations are obtained, we shall have the situation pictured in Fig. 13.1 with a sample residual e corresponding to each set (x,Y). When there are more than two sets of points, some new method of determining a and b must be found. In the chapter on point estimation (Chapter 7), we discussed several principles of estimation; for example, a and b could be determined by the method of moments. There are many other methods of determining these estimates of the parameters α and β, in order to obtain the "best" linear fit to the data.

In any case, it seems reasonable to make the $\{e_i\}$ as small as possible. But what do we do to make these $\{e_i\}$ small? Many courses are suggested, among which are:

1. Minimize the sum of the absolute values of the e's.
2. Minimize the greatest of the absolute residuals.
3. Minimize the sum of the squares of the residuals.

Method 3, called the *method of least squares*, is probably the easiest to apply and has certain optimum properties. It shall be shown that for fixed x values this method produces a linear unbiased estimate of Y which has minimum variance. Also if the errors $\{e_{ij}\}$ are normally independently and identically distributed (NID), the method of least squares produces the same estimates as does the method of maximum likelihood. The method of moments will also give the same estimates for NID errors, provided the regression equation is linear in the parameters.

In the derivations which follow for estimating the parameters in the regression equation, for example, α and β, we need postulate only that the errors are uncorrelated and have the same variance. When the usual tests of hypotheses (such as t and F) and confidence limits are introduced, it is necessary to assume further that the errors are normally and independently distributed. Of course, uncorrelated normal errors are independent. In general we shall assume NID errors, with the understanding that the assumption of normality can be omitted if the investigator is interested only in point estimates and not in confidence limits or tests of hypotheses.

13.2 REGRESSION OF Y ON A SINGLE FIXED VARIABLE: Let us first consider the problem of estimating the best linear relationship between Y and a single fixed variable, x, so that

$$Y = \alpha + \beta x + \varepsilon = a + bx + e \tag{13.1}$$

where α and β are the parameters and a and b their respective estimates. ε is the true error and e the residual about the estimated regression line ($e = Y - \hat{Y}$, where $\hat{Y} = a + bx$). We assume that a sample of n x values are selected (without error) and the corresponding values of the Y's then measured. If it is further assumed that the true errors (ε) are distributed with zero means and the same variance σ^2 and they are uncorrelated, the method of least squres will produce unbiased and minimum-variance estimates of the parameters α and β. A proof will be given in the next section for the general case.

The error sum of squares (SS_e) is

$$SS_e = \sum_{i=1}^{n} e_i^2 = \Sigma(Y_i - a - bx_i)^2$$

a and b are to be determined so as to minimize SS_e. The estimating equation for a is

$$\frac{\partial SS_e}{\partial a} = 0 \qquad \Sigma Y_i = na + b\Sigma x_i$$

Hence

$$a = \frac{\Sigma Y_i}{n} - b\frac{\Sigma x_i}{n} = \bar{Y} - b\bar{x} \tag{13.2}$$

If we insert this value of a in SS_e we find that

$$SS_e = \Sigma[(Y_i - \bar{Y}) - b(x_i - \bar{x})]^2$$

Hence we might as well have written

$$\hat{Y}_i = \bar{Y} + b(x_i - \bar{x})$$

instead of

$$\hat{Y}_i = a + bx_i$$

and similarly

$$E(Y_i) = \mu + \beta(x_i - \bar{x})$$

where

$$\mu = \alpha + \beta\bar{x}$$

and $\bar{Y}$ is the least-squares estimator of μ. The least-squares equation for b is

$$\frac{\partial SS_e}{\partial b} = 0 \qquad b = \frac{\Sigma(x_i - \bar{x})(Y_i - \bar{Y})}{\Sigma(x_i - \bar{x})^2} \tag{13.3}$$

If we substitute this value of b in the equation for SS_e,

$$SS_e = \Sigma Y_i^2 - n\bar{Y}^2 - b^2\Sigma(x_i - \bar{x})^2$$

Hence we can say that the regression has resulted in a reduction of

$$b^2\Sigma(x_i - \bar{x})^2 = b\Sigma(x_i - \bar{x})(Y_i - \bar{Y})$$

in the sum of squares for Y, since $n\bar{Y}^2$ is the reduction attributable to the mean (the sum of squares of deviations from the mean is $\Sigma Y^2 - n\bar{Y}^2$). The proportional reduction in the sum of squares attributable to the regression on x is usually indicated as

$$\frac{b^2\Sigma(x_i - \bar{x})^2}{\Sigma(Y_i - \bar{Y})^2} = \frac{[\Sigma(x_i - \bar{x})(Y_i - \bar{Y})]^2}{\Sigma(x_i - \bar{x})^2\Sigma(Y_i - \bar{Y})^2} \equiv r^2 \tag{13.4}$$

where r is the correlation coefficient between x and Y. That is,

$$SS_e = (1 - r^2)\Sigma(Y_i - \bar{Y})^2$$

In terms of the parameters μ and β and the true errors $\{\varepsilon_i\}$,

$$\bar{Y} = \frac{\Sigma Y_i}{n} = \frac{\Sigma[(\mu + \beta(x_i - \bar{x}) + \varepsilon_i)]}{n} = \mu + \bar{\varepsilon} \tag{13.5}$$

$$Y_i - \bar{Y} = \mu + \beta(x_i - \bar{x}) + \varepsilon_i - \mu - \bar{\varepsilon} = \beta(x_i - \bar{x}) + (\varepsilon_i - \bar{\varepsilon})$$

$$b = \frac{\Sigma(x_i - \bar{x})(Y_i - \bar{Y})}{\Sigma(x_i - \bar{x})^2} = \frac{\Sigma(x_i - \bar{x})[\beta(x_i - \bar{x}) + (\varepsilon_i - \bar{\varepsilon})]}{\Sigma(x_i - \bar{x})^2}$$

$$= \beta + \frac{\Sigma(x_i - \bar{x})\varepsilon_i}{\Sigma(x_i - \bar{x})^2} \tag{13.6}$$

$$a = \bar{Y} - b\bar{x} = \mu + \bar{\varepsilon} - \left[\beta + \frac{\Sigma(x_i - \bar{x})\varepsilon_i}{\Sigma(x_i - \bar{x})^2}\right]\bar{x}$$

$$= \alpha + \bar{\varepsilon} - \frac{\Sigma(x_i - \bar{x})\varepsilon_i}{\Sigma(x_i - \bar{x})^2}\,\bar{x} \tag{13.7}$$

since $\Sigma(x_i - \bar{x}) = 0$.

Since $E(\varepsilon) = 0$, we see that $\bar{Y}$, b, and a are respectively unbiased estimators of μ, β, and α. And since the e's are independently distributed with the same variance,

$$V(\bar{Y}) = \frac{\sigma^2}{n} \qquad V(b) = \frac{\sigma^2}{\Sigma(x_i - \bar{x})^2} \tag{13.8}$$

and

$$V(a) = \left[\frac{1}{n} + \frac{\bar{x}^2}{\Sigma(x_i - \bar{x})^2}\right]\sigma^2 = \frac{\Sigma x_i^2}{n\Sigma(x_i - \bar{x})^2}\sigma^2 \qquad (13.9)$$

The predicted value of Y for a given x, say x_k, is

$$\hat{Y}_k = \bar{Y} + b(x_k - \bar{x}) \qquad (13.10)$$

EXAMPLE 13.1 Consider the problem of fitting a simple linear regression of Y (logarithm to the base 10 of the United States death rate for males per 1000 population) on x (age of those males 40 years or more). Ipsen and Feigl (1970) observed that "for age 40 and over there is a regular increase which can be well expressed by a linear regression of logarithmic death rate on age." Table 13.1 gives the United States death rates for males (per 1000 population) by age in 1970. (The age interval 85 and over is coded 9 as an approximation.) The fitted regression line for the data in Table 13.1 is

$$\hat{Y} = 1.5017 + 0.1732(x - 4.5)$$

For the age group 65-69, x = 5, the predicted Y value is

$$\hat{Y} = 1.5017 + 0.1732(5 - 4.5) = 1.5883$$

TABLE 13.1 Death Rates for Males (per 1000 Population) by Age--United States, 1970

Age	Coded Age, x	Male Death Rate	$\log_{10}$(Death Rate), y
40-44	0	4.85	.686
45-49	1	7.57	.879
50-54	2	11.79	1.072
55-59	3	18.57	1.269
60-64	4	27.86	1.445
65-69	5	41.18	1.614
70-74	6	58.93	1.770
75-79	7	86.77	1.939
80-84	8	123.87	2.093
≥85	9	178.22	2.250

Source: United States Department of Health, Education, and Welfare, Public Health Service, *Vital Statistics of the United States, 1970,* Vol 2, *Mortality,* Part A,(Washington, D.C., 1974).

13.3 REGRESSION OF Y ON p FIXED VARIABLES: We now consider the problem of estimating the best linear relationship between Y and p fixed variables,

$$\begin{aligned} Y &= \beta_1 x_1 + \beta_2 x_2 + \cdots + \beta_p x_p + \varepsilon \\ &= b_1 x_1 + b_2 x_2 + \cdots + b_p x_p + e \qquad (13.11) \end{aligned}$$

where the β's are the parameters, the b's are their respective estimators, and the x's are known constants. Note that the above model is in a general form. If we let $x_1 \equiv 1$, then b_1 becomes the intercept. Suppose a random sample of size n $(n > p)$ is taken, so we have

$$\begin{aligned} Y_1 &= \beta_1 x_{11} + \beta_2 x_{12} + \cdots + \beta_p x_{1p} + \varepsilon_1 \\ Y_2 &= \beta_1 x_{21} + \beta_2 x_{22} + \cdots + \beta_p x_{2p} + \varepsilon_2 \\ &\vdots \\ Y_n &= \beta_1 x_{n1} + \beta_2 x_{n2} + \cdots + \beta_p x_{np} + \varepsilon_n \end{aligned}$$

In matrix notation, the above system of equations can be written as

$$\underline{Y} = \underline{X}\underline{\beta} + \underline{\varepsilon} \qquad (13.12)$$

where

$$\underline{Y} = \begin{bmatrix} Y_1 \\ Y_2 \\ \vdots \\ Y_n \end{bmatrix} \qquad \underline{X} = \begin{bmatrix} x_{11} & x_{12} & \cdots & x_{1p} \\ x_{21} & x_{22} & \cdots & x_{2p} \\ \vdots & \vdots & & \vdots \\ x_{n1} & x_{n2} & \cdots & x_{np} \end{bmatrix}$$

$$\underline{\beta}' = [\beta_1, \beta_2, \ldots, \beta_p] \qquad \underline{\varepsilon}' = [\varepsilon_1, \varepsilon_2, \ldots, \varepsilon_n]$$

The model in (13.12) is known as a linear model since it is a linear function of the β's. The equations involving the b's can be written as

$$\underline{Y} = \underline{X}\underline{b} + \underline{e} \qquad (13.13)$$

where

$$\underline{b}' = [b_1, b_2, \ldots, b_p] \qquad \underline{e}' = [e_1, e_2, \ldots, e_n]$$

We assume that $E(\underline{\varepsilon}) = \underline{0}$ and $\underline{V}(\underline{\varepsilon}) = E(\underline{\varepsilon}\underline{\varepsilon}') = \sigma^2\underline{I}$, so $E(\underline{Y}) = \underline{X}\underline{\beta}$. The least-squares estimators of $\underline{\beta}$ are obtained by minimizing

$$\begin{aligned} SS_e &= \underline{e}'\underline{e} = (\underline{Y} - \underline{X}\underline{b})'(\underline{Y} - \underline{X}\underline{b}) \\ &= \underline{Y}'\underline{Y} - 2\underline{b}'\underline{X}'\underline{Y} + \underline{b}'\underline{X}'\underline{X}\underline{b} \end{aligned}$$

Using the matrix differentiation defined in Appendix A, we have

$$\frac{\partial SS_e}{\partial \underline{b}} = -2\underline{X}'\underline{Y} + 2\underline{X}'\underline{X}\underline{b} = \underline{0}$$

or

$$\underline{X}'\underline{X}\underline{b} = \underline{X}'\underline{Y} \tag{13.14}$$

There are p unknown and p equations. These p equations are called normal equations. The $p \times p$ matrix $\underline{X}'\underline{X}$ has elements

$$\underline{X}'\underline{X} = \begin{bmatrix} \Sigma x_{i1}^2 & \Sigma x_{i1}x_{i2} & \cdots & \Sigma x_{i1}x_{ip} \\ \Sigma x_{i1}x_{i2} & \Sigma x_{i2}^2 & \cdots & \Sigma x_{i2}x_{ip} \\ \vdots & \vdots & & \vdots \\ \Sigma x_{i1}x_{ip} & & & \Sigma x_{1p}^2 \end{bmatrix}$$

It is seen that $\underline{X}'\underline{X}$ is a symmetric matrix. If $(X'X)^{-1}$ exists, i.e., $(\underline{X}'\underline{X})$ is of rank p, then

$$\underline{b} = (\underline{X}'\underline{X})^{-1}\underline{X}'\underline{Y} = \underline{C}\ \underline{X}'\underline{Y} \tag{13.15}$$

where $\underline{C} = (\underline{X}'\underline{X})^{-1}$. The estimator $\underline{b}$ is the least-squares estimator of $\underline{\beta}$ which minimizes the sum of the squares of the residuals.

Let us study some properties of $\underline{b}$. The least-squares estimator $\underline{b}$ is an unbiased estimate of $\underline{\beta}$. Since

$$\underline{b} = \underline{C}\underline{X}'\underline{Y} = \underline{C}\underline{X}'(\underline{X}\underline{\beta} + \underline{\varepsilon}) = \underline{\beta} + \underline{C}\underline{X}'\underline{\varepsilon} \tag{13.16}$$

taking expectation and using $E(\underline{\varepsilon}) = 0$, we have

$$E(\underline{b}) = \underline{\beta} + \underline{C}\underline{X}'E(\underline{\varepsilon}) = \underline{\beta} \tag{13.17}$$

The covariance matrix of $\underline{b}$ is

$$E(\mathbf{b} - \boldsymbol{\beta})(\mathbf{b} - \boldsymbol{\beta})' = E(\mathbf{CX}\boldsymbol{\varepsilon}\boldsymbol{\varepsilon}'\mathbf{X}'\mathbf{C})$$
$$= \mathbf{CX}E(\boldsymbol{\varepsilon}\boldsymbol{\varepsilon}')\mathbf{X}'\mathbf{C} = \sigma^2\mathbf{C} \tag{13.18}$$

In terms of the elements of the matrix $\mathbf{C}$, the variances and the covariances of the elements of $\mathbf{b}$ can be expressed as $V(b_j) = c_{jj}\sigma^2$ and $\mathrm{Cov}(b_j, b_k) = c_{jk}\sigma^2$. We can also find the variance of any linear combination of the elements of $\mathbf{b}$. For example, the variance of $L = \Sigma_{j=1}^{p} \ell_j b_j$ is

$$V(L) = \sigma^2(\Sigma\ell_j^2 c_{jj} + 2\Sigma_{j<k}\Sigma\ell_j\ell_k c_{jk}) \tag{13.19}$$

where the ℓ's are constants.

We now show that the least-squares estimator is the best linear unbiased estimator (BLUE). Here "best" refers to minimum variance and "linear" means linear functions of the Y's. To show that $\mathbf{b}$ is the BLUE, let $\mathbf{b}^*$ be any other linear unbiased estimator of $\boldsymbol{\beta}$. We can write $\mathbf{b}^* = \mathbf{GY}$ where $\mathbf{G}$ is a $p \times n$ matrix. Since $\mathbf{b}^*$ is unbiased, $E(\mathbf{b}^*) = E(\mathbf{G}(\mathbf{X}\boldsymbol{\beta} + \boldsymbol{\varepsilon})) = \mathbf{GX}\boldsymbol{\beta} = \boldsymbol{\beta}$ for all $\boldsymbol{\beta}$. This implies that $\mathbf{GX} = \mathbf{I}$. Let $\mathbf{H} = \mathbf{G} - \mathbf{CX}'$; then $\mathbf{HX} = (\mathbf{G} - \mathbf{CX}')\mathbf{X} = \mathbf{0}$. Using $\mathbf{HX} = \mathbf{0}$ we can write

$$\mathbf{b}^* = (\mathbf{CX}' + \mathbf{H})(\mathbf{X}\boldsymbol{\beta} + \boldsymbol{\varepsilon}) = \boldsymbol{\beta} + (\mathbf{CX}' + \mathbf{H})\boldsymbol{\varepsilon}$$

Hence the covariance matrix of $\mathbf{b}^*$ is

$$E(\mathbf{b}^* - \boldsymbol{\beta})(\mathbf{b}^* - \boldsymbol{\beta})' = E((\mathbf{CX}' + \mathbf{H})\boldsymbol{\varepsilon}\boldsymbol{\varepsilon}'(\mathbf{CX}' + \mathbf{H})')$$
$$= (\mathbf{C} + \mathbf{HH}')\sigma^2$$

The diagonal elements of $(\mathbf{C} + \mathbf{HH}')\sigma^2$ are the variances of the elements of $\mathbf{b}^*$. So

$$V(b_j^*) = \left(c_{jj} + \sum_{i=1}^{n} h_{ji}^2\right)\sigma^2 \geq c_{jj}\sigma^2 = V(b_j)$$

Therefore, the least-squares estimator $\mathbf{b}$ has the smallest variance among all linear unbiased estimators. This proves the following important theorem.

GAUSS-MARKOFF THEOREM Suppose we have the model $\underline{Y} = \underline{X}\underline{\beta} + \underline{\varepsilon}$ with $E(\underline{\varepsilon}) = \underline{0}$, $E(\underline{\varepsilon}\underline{\varepsilon}') = \sigma^2\underline{I}$. Then among all linear unbiased estimators of $\underline{\beta}$, the least-squares estimator $\underline{b} = \underline{C}\underline{X}'\underline{Y}$ has the smallest variances, i.e., the least-squares estimator is the BLUE.

The above result may be generalized in two directions: (1) The BLUE of any linear function $\underline{L}\underline{\beta}$, where L is a $r \times p$ matrix, is $\underline{L}\underline{b}$. (2) The covariance matrix of $\underline{\varepsilon}$ may be generalized to $E(\underline{\varepsilon}\underline{\varepsilon}') = \sigma^2\underline{K}$, where $\underline{K}$ is a known $n \times n$ positive definite matrix. Since $\underline{K}$ is positive definite, there exists a nonsingular matrix $\underline{P}$ such that $\underline{P}\underline{K}\underline{P}' = I$. Now making the transformation $\underline{Y}^* = \underline{P}\underline{Y} = \underline{P}\underline{X}\underline{\beta} + \underline{P}\underline{\varepsilon} = \underline{X}^*\underline{\beta} + \underline{\varepsilon}^*$, where $\underline{X}^* = \underline{P}\underline{X}$ and $\underline{\varepsilon}^* = \underline{P}\underline{\varepsilon}$, we have $E(\underline{\varepsilon}^*) = \underline{0}$ and $E(\underline{\varepsilon}^*\underline{\varepsilon}^{*\prime}) = \sigma^2\underline{P}\underline{K}\underline{P}' = \sigma^2\underline{I}$. So the conditions of the Gauss-Markoff theorem are met and the BLUE of $\underline{\beta}$ is

$$\begin{aligned} \underline{b} &= (\underline{X}^{*\prime}\underline{X}^*)^{-1}\underline{X}^{*\prime}\underline{Y}^* \\ &= (\underline{X}'\underline{K}^{-1}\underline{X})^{-1}\underline{X}'\underline{K}^{-1}\underline{Y} \end{aligned} \tag{13.20}$$

This estimator is usually referred to as the generalized least-squares estimator.

13.4 DISTRIBUTION OF $\underline{b}$ UNDER NORMALITY ASSUMPTION:

In addition to the assumptions made about the ε's in the previous sections, we assume that the ε's are normally distributed. Then the joint density of the ε's or the likelihood function is

$$L = (2\pi\sigma^2)^{-n/2} e^{-(\underline{Y}-\underline{X}\underline{\beta})'(\underline{Y}-\underline{X}\underline{\beta})/2\sigma^2}$$

We can now obtain the maximum-likelihood estimators of $\underline{\beta}$ and σ^2. The logarithm of L is

$$\log L = -\frac{n}{2}\log 2\pi - \frac{n}{2}\log \sigma^2 - \frac{1}{2\sigma^2}(\underline{Y}-\underline{X}\underline{\beta})'(\underline{Y}-\underline{X}\underline{\beta})$$

Taking the derivation of log L with respect to $\underline{\beta}$ and σ^2, we obtain the maximum-likelihood estimators of $\underline{\beta}$ and σ^2:

$$\begin{aligned} \underline{b} &= \underline{C}\underline{X}'\underline{Y} \\ \hat{\sigma}^2 &= \frac{1}{n}(\underline{Y}-\underline{X}\underline{b})'(\underline{Y}-\underline{X}\underline{b}) \end{aligned} \tag{13.21}$$

where $\underline{C} = (\underline{X}'\underline{X})^{-1}$. Therefore, the maximum-likelihood estimator of $\underline{\beta}$ is the same as the least-squares estimator under normality assumption. The estimator $\hat{\sigma}^2$ is a biased estimator just like the maximum-likelihood estimator of σ^2 in $N(\mu,\sigma^2)$. An unbiased estimator of σ^2 is

$$S^2 = \frac{1}{n - p}(\underline{Y} - \underline{X}\underline{b})'(\underline{Y} - \underline{X}\underline{b}) \tag{13.22}$$

Since $\underline{b}$ is a linear combination of $\underline{Y}$ and $\underline{Y}$ is normally distributed, so $\underline{b}$ has a multivariate normal distribution with mean vector $\underline{\beta}$ and covariance matrix $\sigma^2\underline{C}$ where $C = (\underline{X}'\underline{X})^{-1}$.

EXAMPLE 13.2 Suppose p = 2 and $X_1 \equiv 1$; then $\underline{\beta}' = (\beta_1,\beta_2)$ and $\underline{b}' = (b_1,b_2)$, the regression line is the one studied in Section 13.2 with $\beta_1 = \alpha$, $\beta_2 = \beta$. The matrix $\underline{C}$ is

$$\underline{C} = \begin{pmatrix} n & \Sigma x_i \\ \Sigma x_i & \Sigma x_i^2 \end{pmatrix}^{-1} = \frac{1}{n\Sigma x_i^2 - (\Sigma x_i)^2}\begin{pmatrix} \Sigma x_i^2 & -\Sigma x_i \\ -\Sigma x_i & n \end{pmatrix}$$

Therefore, $\underline{b}$ has a bivariate normal distribution with mean vector $\underline{\beta}$,

$$V(b_1) = \frac{\sigma^2\Sigma x_i^2}{n\Sigma x_i^2 - (\Sigma x_i)^2} \qquad V(b_2) = \frac{n\sigma^2}{n\Sigma x_i^2 - (\Sigma x_i)^2}$$

and the correlation coefficient between b_1 and b_2 is

$$\rho = \frac{-\Sigma x_i}{\sqrt{n\Sigma x_i^2}}$$

13.5 TESTING HYPOTHESES ABOUT $\underline{\beta}$: It is common that the experimenter or investigator may wish to test statistical hypotheses about $\underline{\beta}$. For example if $\underline{\beta}$ is partitioned with $\underline{\beta} = \begin{pmatrix} \underline{\beta}_1 \\ \underline{\beta}_2 \end{pmatrix}$ where $\underline{\beta}_1$ has q elements and $\underline{\beta}_2$ has p - q elements, the null hypothesis may be H_0: $\underline{\beta}_2 = \underline{\beta}_0$ and the alternative hypothesis H_1: $\underline{\beta}_2 \neq \underline{\beta}_0$, where $\underline{\beta}_0$ is a given known vector. In particular, $\underline{\beta}_0$ may be equal to zero, so under the null hypothesis the regression model consists of only $x_1, x_2, \ldots, x_q$ as fixed variables and $x_{q+1}, x_{q+2}, \ldots, x_p$ are not included in the regression. Further when q = 0, the null hypothesis becomes H_0: $\underline{\beta} = \underline{0}$.

Let us derive the likelihood-ratio test for testing H_0: $\underline{\beta} = \underline{0}$ against H_1: $\underline{\beta} \neq \underline{0}$. The maximum of the likelihood function L under the whole parameter space Ω is

$$L(\hat{\Omega}) = (2\pi\hat{\sigma}^2)^{-n/2}e^{-n/2}$$

Under the null hypothesis space ω: $\underline{\beta} = 0$, the maximum-likelihood estimator of σ^2 is

$$\hat{\sigma}_0^2 = \frac{1}{n}\underline{Y}'\underline{Y}$$

Hence we obtain the maximum of the likelihood function under ω:

$$L(\hat{\omega}) = (2\pi\hat{\sigma}_0^2)^{-n/2}e^{-n/2}$$

The likelihood-ratio criterion is

$$\lambda = \frac{L(\hat{\omega})}{L(\hat{\Omega})} = \left(\frac{\hat{\sigma}^2}{\hat{\sigma}_0^2}\right)^{n/2}$$

or

$$\lambda^{2/n} = \frac{\hat{\sigma}^2}{\hat{\sigma}_0^2} \tag{13.23}$$

The likelihood-ratio test says to reject H_0 if $\lambda \leq \lambda_\alpha$, where λ_α is suitably choosen such that $P(\lambda \leq \lambda_\alpha) = \alpha$, and α is the significance level of the test.

Let us express λ into another form so that we may easily determine the critical value. We can write

$$\begin{aligned}\underline{Y}'\underline{Y} &= (\underline{Y} - \underline{\underline{X}}\underline{b} + \underline{\underline{X}}\underline{b})'(\underline{Y} - \underline{\underline{X}}\underline{b} + \underline{\underline{X}}\underline{b}) \\ &= (\underline{Y} - \underline{\underline{X}}\underline{b})'(\underline{Y} - \underline{\underline{X}}\underline{b}) + 2\underline{b}'\underline{\underline{X}}'(\underline{Y} - \underline{\underline{X}}\underline{b}) + \underline{b}'\underline{\underline{X}}'\underline{\underline{X}}\underline{b} \\ &= n\hat{\sigma}^2 + \underline{b}'(\underline{\underline{X}}'\underline{\underline{X}})\underline{b}\end{aligned}$$

The cross-product term equals zero because $\underline{X}'\underline{Y} = \underline{X}'\underline{X}\underline{b}$. Here the sum of squares $\underline{Y}'\underline{Y}$ is partitioned into two components; one is the error sum of squares and the other, $\underline{b}'\underline{X}'\underline{X}\underline{b} = \underline{b}'\underline{X}'\underline{Y}$, is called the reduction in sum of squares due to regression which is denoted by $R(x_1,\ldots,x_p)$. Now $\lambda^{2/n}$ can be written as

$$\lambda^{2/n} = \frac{\hat{\sigma}^2}{\hat{\sigma}^2 + b'(\underline{X}'\underline{X})b/n}$$

It will be shown later that the ratio $(n - p)[\underline{b}'(\underline{X}'\underline{X})\underline{b}]/pn\hat{\sigma}^2$ has an F distribution. Now

$$\lambda^{2/n} = \frac{1}{1 + [p/(n - p)]F}$$

where

$$F = \frac{(n - p)[\underline{b}'(\underline{X}'\underline{X})\underline{b}]}{p(\underline{Y} - \underline{X}\underline{b})'(\underline{Y} - \underline{X}\underline{b})} \tag{13.24}$$

Since F is a monotonic function of λ, the likelihood-ratio test is equivalent to the test with critical region $F > k$. This test is reasonable because if $\underline{\beta}$ is not equal to zero, then its estimate $\underline{b}$ will also tend to be different from 0. Since $\underline{X}'\underline{X}$ is positive definite, $\underline{b}'\underline{X}'\underline{X}\underline{b}$ will be larger than zero and hence lead to the rejection of H_0.

We now consider testing H_0: $\underline{\beta}_2 = 0$ against H_1: $\underline{\beta}_2 \neq 0$, where $\underline{\beta}_2$ is the last $p - q$ elements of $\underline{\beta}$. Under the null hypothesis, the model is

$$\underline{Y} = \underline{X}_1\underline{\beta}_1 + \underline{\varepsilon}$$

where $\underline{X}_1$ is the first q columns of $\underline{X}$, or we may partition $\underline{X} = (\underline{X}_1,\underline{X}_2)$. The maximum-likelihood estimators of $\underline{\beta}_1$ and σ^2 under H_0 are now

$$\underline{b}_0 = (\underline{X}_1'\underline{X}_1)^{-1}\underline{X}_1'\underline{Y}$$

$$\hat{\sigma}_0^2 = \frac{1}{n}(\underline{Y} - \underline{X}_1\underline{b}_0)'(\underline{Y} - \underline{X}_1\underline{b}_0)$$

The reduction in sum of squares due to $x_1, \ldots, x_q$ is

$$R(x_1, x_2, \ldots, x_q) = \underline{b}_0'\underline{X}_1'\underline{Y} \tag{13.25}$$

The likelihood-ratio test can be shown to be equivalent to the test with critical region F > k, where

$$F = \frac{(n - p)[R(x_1, \ldots, x_p) - R(x_1, \ldots, x_q)]}{(p - q)(\underline{Y} - \underline{\underline{X}}\underline{b})'(\underline{Y} - \underline{\underline{X}}\underline{b})}$$
$$= \frac{\underline{b}'(\underline{X}'\underline{X})\underline{b} - \underline{b}_0(\underline{X}_1'\underline{X}_1)\underline{b}_0}{(p - q)S^2} \tag{13.26}$$

The difference, $R(x_1, \ldots, x_p) - R(x_1, \ldots, x_q)$, may be viewed as the contribution from $x_{q+1}, \ldots, x_p$. If the contribution is large, one should reject the null hypothesis that $\underline{\beta}_2 = \underline{0}$.

In order to find the critical value k, we shall show that the statistic F has an F(p - q, n - p) distribution under H_0. By assumption, $\underline{\varepsilon}$ is distributed as $N(\underline{0}, \sigma^2\underline{I})$. Let $\underline{L}$ be an n × n orthogonal matrix such that $\underline{L}'\underline{L} = \underline{\underline{L}}\underline{L}' = \underline{I}$; then $\underline{U} = \underline{\underline{L}}\underline{\varepsilon}$ is distributed as $N(\underline{0}, \sigma^2\underline{I})$. This is the property of an orthogonal transformation. If we partition $\underline{U} = \begin{pmatrix}\underline{U}_1\\ \underline{U}_2\end{pmatrix}$, where $\underline{U}_1$ has r elements and $\underline{U}_2$ has n - r elements, then $\underline{U}_1$ is distributed as r-dimensional $N(\underline{0}, \sigma^2\underline{I})$, $\underline{U}_2$ is distributed as (n - r)-dimensional $N(\underline{0}, \sigma^2\underline{I})$, and $\underline{U}_1$ and $\underline{U}_2$ are independent. Further, $\underline{U}_1'\underline{U}_1/\sigma^2$ is distributed as $\sigma^2\chi_r^2$, and $\underline{U}_2'\underline{U}_2 = \underline{\varepsilon}'\underline{\varepsilon} - \underline{U}_1'\underline{U}_1$ is distributed as $\sigma^2\chi_{n-r}^2$ (since $\underline{\varepsilon}'\underline{\varepsilon} = \underline{U}'\underline{U}$). These results will be used when we obtain the distribution of the statistic F.

The matrix $\underline{X}$ is a matrix of fixed variables. We may transform the set of x variables into a set of z variables such that the sum of squares of the z's is unity and cross-products are zeros. In particular, let

$$z_{i1} = a_{11}x_{i1}$$
$$z_{i2} = a_{21}x_{i1} + a_{22}x_{i2}$$
$$z_{i3} = a_{31}x_{i1} + a_{32}x_{i2} + a_{33}x_{i3}$$
$$\vdots$$
$$z_{ip} = z_{p1}x_{i1} + a_{p2}x_{i2} + a_{p3}x_{i3} + \cdots + a_{pp}x_{ip}$$

for i = 1, 2, ..., n, where $\Sigma_{i=1}^{n} z_{ij}^2 = 1$ and $\Sigma_{i=1}^{n} z_{ij}z_{ik} = 0$ for j,k = 1, ..., p and $j \neq k$.

Since we assume that $\underline{X}$ is of rank p, so the transformation is nonsingular, and we can solve this transformation for the x variables in terms of the z variable. Replacing the x variables by the z variables in the original model and collecting terms we have

$$\begin{aligned} Y_i &= \beta_1 x_{i1} + \beta_2 x_{i2} + \cdots + \beta_p x_{ip} + \varepsilon_i \\ &= \beta_1^* z_{i1} + \beta_2^* z_{i2} + \cdots + \beta_p^* z_{ip} + \varepsilon_i \end{aligned}$$

where $(\beta_1^*,\ldots,\beta_p^*)' = \underline{\beta}^*$ are the new regression coefficients for the z variables. Denote the least-squares estimator of $\underline{\beta}^*$ by $\underline{b}^*$; then by the orthogonality of the z's, we have

$$b_j^* = \sum_{i=1}^{n} z_{ij} Y_i = \beta_j^* + \sum_{i=1}^{n} z_{ij}\varepsilon_i$$

or

$$b_j^* - \beta_j^* = \sum_{i=1}^{n} z_{ij}\varepsilon_i \qquad j = 1, 2, \ldots, p$$

These give the first p members of an orthogonal transformation of the ε's. Completing the orthogonal transformation by adding the remaining n - p members and using the results of orthogonal transformation of variables (the z's play the role of $\underline{L}$), we have $(b_j^* - \beta_j^*)^2$ is distributed as $\sigma^2\chi_1^2$ for j = 1, ..., p and $\underline{\varepsilon}'\underline{\varepsilon} - \Sigma_{j=1}^{p} (b_j^* - \beta_j^*)^2$ is distributed as $\sigma^2\chi_{n-p}^2$ and they are all mutually independent.

The transformation from x's to the z's is a triangular transformation; so is the inverse transformation from the z's to the x's. From this, it follows that $\beta_{q+1}, \ldots, \beta_p$ are homogeneous linear functions of $\beta_{q+1}^*, \ldots, \beta_p^*$. So the null hypothesis that $\underline{\beta}_2 = \underline{0}$ implies $\underline{\beta}_2^* = \underline{0}$ where $\underline{\beta}_2^* = (\beta_{q+1}^*,\ldots,\beta_p^*)'$. Therefore, under H_0, $\Sigma_{j=q+1}^{p} b_j^{*2}$ is distributed as $\sigma^2\chi_{p-q}^2$. But by the uniqueness of the least-squares solution,

$$R(x_1,\ldots,x_p) = R(z_1,\ldots,z_p) = \sum_{j=1}^{p} b_j^{*2}$$

and by the triangular nature of the transformation,

$$R(x_1,\ldots,x_q) = R(z_1,\ldots,z_q) = \sum_{j=1}^{q} b_j^{*2}$$

Therefore,

$$R(x_1,\ldots,x_p) - R(x_1,\ldots,x_q) = \sum_{j=q+1}^{p} b_j^{*2}$$

which is distributed as $\sigma^2\chi^2_{p-q}$. Now

$$\underline{\varepsilon}'\underline{\varepsilon} = \sum_{i=1}^{n} (Y_i - \beta_1^* z_{i1} - \cdots - \beta_p^* z_{ip})^2$$

$$= \sum_{i=1}^{n} [(Y_i - b_1^* z_{i1} - \cdots - b_p^* z_{ip}) + \sum_{j=1}^{p} (b_j^* - \beta_j^*) z_{ij}]^2$$

$$= \sum_i (Y_i - b_1^* z_{i1} - \cdots - b_p^* z_{ip})^2 + \sum_{j=1}^{p} (b_j^* - \beta_j^*)^2$$

The cross-product term is equal to zero since the z's are orthogonal and

$$\sum_i (Y_i - b_1^* z_{i1} - \cdots - b_p^* z_{ip}) z_{ij} = 0$$

Therefore,

$$\underline{\varepsilon}'\underline{\varepsilon} - \sum_{j=1}^{p} (b_j^* - \beta_j^*)^2 = \Sigma_i Y_i^2 - \sum_{j=1}^{p} b_j^{*2}$$

$$= \underline{Y}'\underline{Y} - R(x_1, \ldots, x_p)$$

$$= (\underline{Y} - \underline{\underline{X}}\underline{b})'(\underline{Y} - \underline{\underline{X}}\underline{b})$$

which is distributed as $\sigma^2\chi^2_{n-p}$. Hence

$$F = \frac{R(x_1,\ldots,x_p) - R(x_1,\ldots,x_q)}{(p - q)S^2} \tag{13.27}$$

is distributed as an $F(p - q, n - p)$ distribution under H_0. We can summarize the testing procedure in an analysis-of-variance table (Table 13.2).

TABLE 13.2 Analysis of Variance

Source of variation	Degrees of freedom (df)	Sum of Squares (SS)	Mean squares (MS)
Due to fitting $x_1, \ldots, x_p$	p	$R(x_1,\ldots,x_p)$	
Due to fitting $x_1, \ldots, x_q$	q	$R(x_1,\ldots,x_q)$	
Difference	p - q	$R(x_{q+1},\ldots,x_p \mid x_1,\ldots,x_q)$	MS_d
Error	n - p	$(\underline{Y} - \underline{X}\underline{b})'(\underline{Y} - \underline{X}\underline{b})$	s^2
Total:	n	$\underline{Y}'\underline{Y}$	

In the analysis-of-variance table,

$$R(x_{q+1},\ldots,x_p \mid x_1,\ldots,x_q) = R(x_1,\ldots,x_p) - R(x_1,\ldots,x_q)$$

and

$$MS_d = \frac{R(x_{q+1},\ldots,x_p \mid x_1,\ldots,x_q)}{p - q}$$

The F test is the ratio of the two mean squares

$$F = \frac{MS_d}{s^2} \tag{13.28}$$

When q = 0, the test reduces to testing $\underline{\beta} = \underline{0}$ and the statistic F has an F(p, n - p) distribution. When q = p - 1, we are testing a single $\beta_j = 0$ and p - q = 1. The F test is equivalent to a t test. Hence for testing H_0: $\beta_j = \beta_{j0}$ against H_1: $\beta_j \neq \beta_{j0}$, where β_{j0} is a known constant, we may use the statistic

$$t = \frac{b_j - \beta_{j0}}{\sqrt{c_{jj}s^2}} \tag{13.29}$$

which has a t distribution with n - p degrees of freedom.

13.6 CONFIDENCE INTERVALS: To obtain a confidence interval for the mean value of Y, i.e., $\underline{X}_k\underline{\beta}$, where $\underline{X}_k = [x_{k1}, x_{k2}, \ldots, x_{kp}]$ is given, we consider its unbiased estimator $\underline{X}_k\underline{b}$. Since $\underline{b}$ is distributed as $N(\underline{\beta}, \sigma^2\underline{C})$, hence $\underline{X}_k\underline{b}$ is distributed as $N(\underline{X}_k\underline{\beta}, \sigma^2\underline{X}_k\underline{C}\underline{X}_k')$ and is independent of S^2. So

$$\frac{\underline{X}_k\underline{b} - \underline{X}_k\underline{\beta}}{S\sqrt{\underline{X}_k\underline{C}\underline{X}_k'}}$$

is a pivotal quantity and it has a t distribution with n - p degrees of freedom. Therefore a 100(1 - α)% confidence interval for $\underline{X}_k\underline{\beta}$ is

$$\underline{X}_k\underline{b} - t_{(\alpha/2,n-p)} S\sqrt{\underline{X}_k\underline{C}\underline{X}_k'} < \underline{X}_k\underline{\beta} < \underline{X}_k\underline{b} + t_{(\alpha/2,n-p)} S\sqrt{\underline{X}_k\underline{C}\underline{X}_k'} \quad (13.30)$$

where $t_{(\alpha/2,n-p)}$ is the 100(1 - α/2)% point of the t distribution with n - p degrees of freedom.

If it is desired to find a prediction interval for any future observation, Y_k, we must consider the difference of the predicted value $\hat{Y}_k = \underline{X}_k\underline{b}$ and Y_k, and it is necessary to estimate the variance of the difference. The difference is $e = \hat{Y}_k - Y_k$ with variance

$$(1 + \underline{X}_k\underline{C}\underline{X}_k')\sigma^2$$

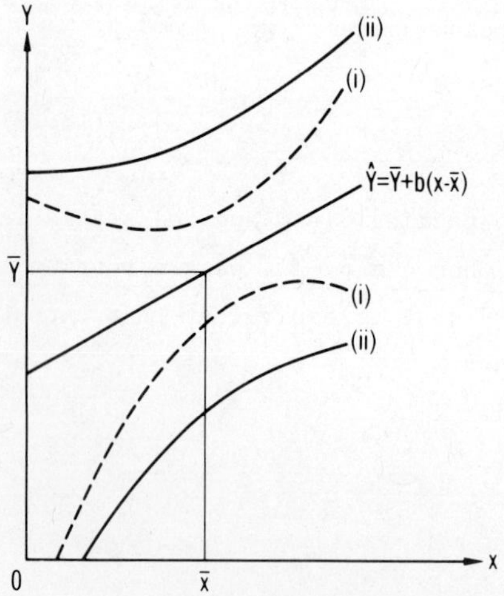

FIG. 13.2 Regression line and (i) confidence band for the mean value, (ii) prediction band for a single Y.

So the prediction interval for Y_k is

$$\hat{Y}_k - t_{(\alpha/2,n-p)} S\sqrt{1 + \underline{X}_k \underline{C} \underline{X}'_k} < Y_k < \hat{Y}_k + t_{(\alpha/2,n-p)} S\sqrt{1 + \underline{X}_1 \underline{C} \underline{X}'_k} \quad (13.31)$$

It should be noted that $S^2(e) = S^2(\underline{X}_k \underline{b}) + S^2$, where $S^2(e)$ and $S^2(\underline{X}_k \underline{b})$ indicate the estimators of the variances. Therefore, the prediction interval for Y_k is wider than the confidence interval for the mean value $\underline{X}_k \underline{\beta}$. These results are illustrated in Fig. 13.2 for one x variable. The reader will note that (1) the confidence bands for $\underline{X}_k \underline{\beta}$; and (2) the prediction bands for Y_k form a hyperbola, with the curvature being much greater for the confidence bands than for the prediction bands.

Confidence intervals for any single β or any linear combination $\underline{L}'\underline{\beta}$, where $\underline{L}$ is a constant $p \times 1$ vector, can be obtained similarly. For example, a $100(1 - \alpha)\%$ confidence interval for β_j is

$$b_j - t_{(\alpha/2,n-p)} S\sqrt{c_{jj}} < \beta_j < b_j + t_{(\alpha/2,n-p)} S\sqrt{c_{jj}} \quad (13.32)$$

13.7 SELECTION OF VARIABLES: In practice, the researcher may have many fixed variables in a regression analysis. The problem arises as to whether the number of variables can be reduced to a smaller number and still yield a usable regression equation. There are many different procedures for selecting such a subset among the fixed variables. Generally speaking, these procedures fall into one of two categories: One is to start with the full model with all variables in and sequentially delete one or more nonessential variables at each time; the other is to start with a basic set of variables and add one or more essential variables at each time. We shall call the former the *sequential deletion procedure* and the latter *forward selection procedure*. For each procedure, statistical tests are employed to determine when to stop the deletion or selection and determine the final model. These statistical tests are preliminary tests in nature and their effects in subsequent statistical analysis should be taken into account. Hence the problem of model building in regression is one of the general class of problems called problems of incompletely specified models involving the use of preliminary tests. We shall consider one sequential deletion procedure and one forward selection procedure which are studied as incompletely specified models by Kennedy and Bancroft (1971). These inference procedures are in the area of inference based on conditional specification discussed in Chapter 12.

Consider the regression model

$$Y = \beta_1 x_1 + \beta_2 x_2 + \cdots + \beta_p x_p + \varepsilon$$

It is assumed that the experimenter has sufficient knowledge in the area of application to allow him or her to select the "basic set" of r fixed variables (which could be a null set) and to designate an "order of importance" for the remaining $p - r$ variables with x_{r+1} being most important and x_p least important. In many cases such knowledge is gained from theoretical considerations or from a substantial amount of experience in the applied area. One important situation is a polynomial regression where a natural order is given. A polynomial regression of Y on a fixed variable z has the form

$$Y = \alpha + \beta_1 z + \beta_2 z^2 + \cdots + \beta_p z^p + \varepsilon$$

Defining $x_i = z^i$, the above model is then reduced to the usual regression model. Here z^p is considered to be the least important variable.

The experimenter has n observations and wants to predict Y. The sequential deletion procedure is described as follows. One first tests the hypothesis that $\beta_p = 0$ (i.e., x_p is not needed in the equation). If one accepts this hypothesis, one deletes x_p and tests that $\beta_{p-1} = 0$. If one accepts the second hypothesis, one deletes x_{p-1} from the regression equation and tests $\beta_{p-2} = 0$, etc. One continues deleting variables in this manner until one rejects a hypothesis that a coefficient is zero, or until one reaches the coefficient of x_r $(r < p)$, then one retains in the regression equation the variable corresponding to that coefficient and all other variables whose coefficients one has not yet tested.

For the forward selection procedure, the experimenter assumes that the first r of the p fixed variables are necessary for prediction of y. One first tests the hypothesis that $\beta_{r+1} = 0$. If one rejects this hypothesis, one adds x_{r+1} to the list of necessary variables and tests that $\beta_{r+2} = 0$. If one rejects this second hypothesis, one adds x_{r+2} to the regression equation and tests that $\beta_{r+3} = 0$, etc. One continues adding variables to the regression equation in this manner until one arrives at a variable whose coefficient does not differ significantly from zero, at which point one does not add that variable to the equation, nor does one add the variables whose coefficients one has not yet tested.

The test for the hypothesis that $\beta_j = 0$ in both procedures is a t test as given in Section 13.5. The t tests are usually made at the same level of significance α. It is apparent that the final prediction equation depends on the results of the t tests which in turn depends on the selection of α level. Since the experimenter is interested in the prediction of Y, he or she would want to select the α level such that the bias and mean square error (= variance + bias^2) of the predicted value are as small as possible. The derivations of the bias and mean square error are complicated and hence omitted here. After considerable study, Kennedy and Bancroft (1971) found that the best level α for the sequential deletion procedure is near .10 and that for the forward selection procedure is near .25. Further, it is found that the sequential deletion procedure is more efficient than the forward selection procedure.

13.8 POOLING REGRESSIONS IN PREDICTION: As indicated in Section 12.2, if two samples are from the same population, it is advantageous to pool the two samples to make statistical inference. In the present context, the experimenter or investigator may have two or more regression equations. If these regression equations are estimates of the same population regression, it would be advantageous to pool them and form a single regression equation as the estimate. When the experimenter is uncertain if the regression equations are the same, he or she may test the hypothesis that they are equal and decide whether to pool or not based on the outcome of the preliminary test. For simplicity we shall consider a pooling methodology for two simple linear regressions and leave the more complicated situations to the papers by Bock et al. (1973) and Han and Bancroft (1978).

Given two regression models

$$Y_{ij} = \beta_{i0} + \beta_{i1}(x_{ij} - \bar{x}_i) + \varepsilon_{ij}$$
$$i = 1, 2;\ j = 1, 2, \ldots, n_i \qquad (13.33)$$

where the x's are fixed known constants and $\{\varepsilon_{ij}\}$ are NID$(0,\sigma^2)$, we are interested in estimating the line

$$Y_{1j} = \beta_{10} + \beta_{11}(x_{1j} - \bar{x}_1)$$

If it is certain that $\beta_{10} = \beta_{20}$ and $\beta_{11} = \beta_{21}$, the coefficients from the two regressions should be pooled. Let b_{10}, b_{20}, b_{11}, and b_{21} be the least-squares estimators of β_{10}, β_{20}, β_{11}, and β_{21}, respectively; then the pooled regression line is

$$Y_{ij} = \bar{b}_0 + \bar{b}_1(x_{1j} - \bar{x}_1)$$

where

$$\bar{b}_0 = \frac{n_1 b_{10} + n_2 b_{20}}{n_1 + n_2}$$

$$\bar{b}_1 = \frac{c_1 b_{11} + c_2 b_{22}}{c_1 + c_2}$$

and

$$c_i = \Sigma(x_{ij} - \bar{x}_i)^2 \qquad i = 1, 2$$

In many practical situations, the experimenter is uncertain whether $\beta_{10} = \beta_{20}$ and $\beta_{11} = \beta_{21}$. In order to resolve the uncertainty he may test the hypothesis H_0: $\beta_{10} = \beta_{21}$ and $\beta_{11} = \beta_{21}$ against H_1: $\beta_{10} \neq \beta_{21}$ or $\beta_{11} \neq \beta_{21}$, or both. A test statistic for H_0 is

$$F = \frac{(b_{10} - b_{20})^2 n_1 n_2/(n_1 + n_2) + (b_{11} - b_{21})^2 c_1 c_2/(c_1 + c_2)}{2S^2} \tag{13.34}$$

where S^2 is the pooled estimator of σ^2. Let S_1^2 and S_2^2 be the unbiased estimators of σ^2 under the two models; then

$$S^2 = \frac{(n_1 - 2)S_1^2 + (n_2 - 2)S_2^2}{n_1 + n_2 - 4}$$

Under H_0, F has a $F(2, n_1 + n_2 - 4)$ distribution.

The estimator of the regression line, referred to as the sometimes-pool predictor, is

$$Y_{1j}^* = \begin{cases} \bar{b}_0 + \bar{b}_1(x_{1j} - \bar{x}_1) & \text{if } F \leq F_\alpha \\ b_{10} + b_{11}(x_{1j} - \bar{x}_1) & \text{if } F > F_\alpha \end{cases} \tag{13.35}$$

where F_α is the $100(1 - \alpha)$ percentage point of the $F(2, n_1 + n_2 - 4)$ distribution. The bias and mean square error of the sometimes-pool predictor are given in Johnson et al. (1977).

EXAMPLE 13.3 Consider Example 13.1; similar data for the year 1969 are given in Table 13.3. The fitted regression line for 1969 is

$$Y_2 = 1.5087 + .1745(x - 4.5)$$

which is very similar to the 1970 regression line

$$Y_1 = 1.5017 + .1732(x - 4.5)$$

The calculated F value is .2 with 2 and 16 degrees of freedom and this is clearly not significant at the .40 level. (The significance level is selected at .40 because this is the optimal level to use in the present case. The reader is referred to the paper by Johnson et al., 1977, for the determination of an optimal significance level in other cases.) Hence the rule of procedure tells us to pool the 1969 line with the 1970 line to provide a "better" prediction. The resulting prediction line is $Y^* = 1.5052 + .1739(x - 4.5)$.

TABLE 13.3 Death Rates for Males (per 1000 Population) by Age--United States, 1969

Age	Coded age, x	Male death rate	$\log_{10}$ (death rate), y
40-44	0	4.92	.692
45-49	1	7.57	.879
50-54	2	11.91	1.076
55-59	3	18.67	1.271
60-64	4	28.54	1.455
65-69	5	41.50	1.618
70-74	6	62.77	1.798
75-79	7	85.14	1.930
80-84	8	119.31	2.077
≥85	9	195.55	2.291

Source: United States Department of Health, Education, and Welfare, Public Health Service, *Vital Statistics of the United States, 1969,* Vol. 2, *Mortality,* Part A (Washington, D.C., 1974).

EXERCISES

13.1 In order to use the usual regression analysis, constant variance is assumed. When the variances are unequal, we may use a transformation $z = f(y)$, to stabilize the variance. In particular if the standard deviation of a random variable Y can be expressed as a function of the mean, say $\sigma_Y = \phi(\mu)$, show that the following transformation approximately stabilizes the variance

$$f(y) = c \int \frac{dy}{\phi(y)}$$

where c is a constant. *Hint:* Use the approximate formula $\sigma_Z = \sigma_Y f'(\mu)$.

13.2 In Exercise 13.1, let Y be distributed as a Poisson distribution. Show that $f(y) = \sqrt{y}$. This is the square root transformation which is used when $\sigma_Y = c\sqrt{\mu}$.

13.3 In Exercise 13.1, let Y be distributed as a binomial distribution. Show that $f(y) = \sin^{-1}\sqrt{y/n}$ which is the angular transformation, where y/n is the observed proportion of successes and the arcsine is in radians.

13.4 In Exercise 13.1, let Y be distributed as a gamma distribution with density function

$$\frac{1}{\Gamma(n)\theta^n} y^{n-1} e^{-y/\theta} \qquad y > 0,\ \theta > 0, \text{ and } 0 \text{ otherwise}$$

where n is known. Show that $f(y) = \log y$. This is the logarithm transformation which is used when $\sigma_Y = c\mu$.

13.5 In the regression model $Y_i = \alpha + \beta x_i + \varepsilon_i$, $i = 1, \ldots, n$, suppose the variance of Y_i is proportional to x_i^2, i.e. $V(Y_i) = \sigma^2 x_i^2$. (a) Find a transformation to stabilize the variance. (b) Obtain the BLUE a and b for α and β. (c) Find the variances and covariance for a and b.

13.6 Use the confidence limits for the mean value, $E(Y') = \alpha + \beta X'$, for a future X' to show that the confidence limits of X' for a future Y' are

$$\bar{X} + \frac{b(Y' - \bar{Y})}{\lambda} \pm \frac{t_{\alpha/2}S}{\lambda}\sqrt{\frac{\lambda}{n} + \frac{(Y' - \bar{Y})^2}{\Sigma(X_i' - \bar{X}')^2}}$$

where $\lambda = b^2 - t_{\alpha/2}^2 S^2/[\Sigma(X_i' - \bar{X}')^2]$ and X, Y, b, S, and $\Sigma(X_i' - \bar{X}')^2$ are based on the original sample of n.

13.7 What changes would be made in the confidence limits in Exercise 13.6 if Y' were the average of k observations?

13.8 Girshick and Haavelmo (1947) have made an analysis of the demand for food in the United States for the years 1922 to 1941. One equation in their analysis involved the relationship between disposable income adjusted for the cost of living (Y) and investment per capita adjusted for the cost of living (x). The values of Y and x are shown in the following table:

Y	x	Y	x	Y	x
87.4	92.9	107.8	142.9	103.1	114.3
97.6	142.9	96.6	92.9	105.1	121.4
96.7	100.0	88.9	97.6	96.4	78.6
98.2	123.8	75.1	52.4	104.4	109.5
99.8	111.9	76.9	40.5	110.7	128.6
100.5	121.4	84.6	64.3	127.1	238.1
103.2	107.1	90.6	78.6	1950.7	2159.7

(a) Set up a simple linear regression of Y on x, and determine the constants in the regression equation. (b) Set up the analysis of variance. (c) Make a test of significance of the usefulness of the regression equation. Are there any aspects of these data which might invalidate this test? (d) Plot the data (rounded to nearest integer), and draw in the regression line and the 95% confidence lines for $E(Y|x)$. From the nature of the residuals from the regression line, would you suggest any changes in the form of the regression equation?

13.9 In Section 13.2 suppose Y_i is normally distributed. (a) Show that $\bar{Y}$ and b are the maximum-likelihood estimators of μ and β, respectively. (b) What is the ML estimator of σ^2? Is this estimator unbiased?

13.10 Suppose a random sample of size n_1 is used to estimate the parameters in the equation $Y_1 = \mu_1 + \beta_1 x_1 + \varepsilon_1$, $\varepsilon_1 \sim N(0,\sigma^2)$, and an independent random sample of size n_2 for the equation $Y_2 = \mu_2 + \beta_2 x_2 + \varepsilon_2$, $\varepsilon_2 \sim N(0,\sigma^2)$. Derive the likelihood-ratio test for testing H_0: $\beta_1 = \beta_2$ against H_1: $\beta_1 \neq \beta_2$.

13.11 Show that the likelihood-ratio criterion in Exercise 13.10 is equivalent to a t test.

13.12 R. A. Fisher (1947) has compared the body weights (in kilograms) with the heart weights (in grams) of 47 female and 97 male cats. The sums of squares and products were as follows:

	Degrees of freedom	$(\text{Body})^2$	(Body × heart)	$(\text{Heart})^2$
Females:				
Total	47	265.13	1,029.62	4,064.71
Correction for mean	1	261.677	1,020.516	3,979.92
Difference	46	3.453	9.104	84.79
Males:				
Total	97	836.75	3,275.55	13,056.17
Correction	1	815.77	3,185.07	12,435.70
Difference	96	20.98	90.48	620.47

(a) Determine the regression of heart weight on body weight for both males and females. (b) Are these two regression coefficients different from one another? Let $\alpha = .05$. (c) Are the two error variances essentially the same? Let $\alpha = .05$.

13.13 In a study of lobster population, D. B. DeLury (1947) presents the following data on the catch per unit of effort for the time interval t, C(t), and the total catch up to t, K(t), in thousands of pounds:

t	1	2	3	4	5	6	7	8	9	10	11	12	13	14	15	16	17
C	.82	.75	.94	.80	.83	.89	.70	.58	.64	.55	.52	.45	.45	.49	.45	.48	.43
K	0	7	13	16	22	25	32	37	40	45	50	53	54	55	57	60	62

(a) A linear equation of the form $C = a + bK + e$ was set up. Determine the values of a and b and their standard errors. (b) The total population at time $t = 0$ is estimated by $N_0 = -a/b$. Determine N_0.

13.14 Consider the regression $Y = \alpha + \beta_1 x_1 + \beta_2 x_2 + \varepsilon$. An attempt is sometimes made to compute a regression coefficient b_2 in two simple stages. (a) Compute first the simple regression coefficient b_1' of Y on x_1; then compute the simple regression coefficient of $Y - b_1' x_1$ on x_2. (b) Compute first the simple regression coefficient b_1^* of

x_2 on x_1; then compute the simple regression coefficient of Y on $x_2 - b_1^* x_1$. Show that the procedure in (b) gives the correct answer but the procedure in (a) does not.

13.15 Given the linear model $\underline{Y} = \underline{X}\underline{\beta} + \underline{\varepsilon}$, show that the unbiased estimator S^2 of σ^2 can be written as $\underline{Y}'(\underline{I} - \underline{X}\ \underline{C}\underline{X}')\underline{Y}/(n - p)$.

13.16 A linear model has the form $\underline{Y} = \underline{X}\beta + \underline{\varepsilon}$, where $\underline{Y}$, $\underline{X}$, and $\underline{\varepsilon}$ are $n \times 1$ vectors and $\underline{X} = (1,2,\ldots,n)'$. $[\Sigma x_i^2 = n(n + 1)(2n + 1)/6.]$ Write down the following results and state the additional assumptions about the ε, if any, needed to justify the result: (a) the least-squares estimator $\hat{\beta}$ and its expected value; (b) the variance of $\hat{\beta}$.

13.17 Similar data to those in Examples 13.1 and 13.3--death rates (per 1000 population) by age in the United States--for females are given as follows:

Age	Coded age, x	Female death rate		$\log_{10}$(death rate), y	
		1970	1969	1970	1969
40-44	0	2.75	2.84	.439	.453
45-49	1	4.23	4.17	.626	.620
50-54	2	6.20	6.13	.792	.788
55-59	3	9.07	9.03	.958	.956
60-64	4	13.17	13.43	1.120	1.128
65-69	5	20.42	21.42	1.310	1.331
70-74	6	32.44	34.16	1.511	1.533
75-59	7	53.80	52.91	1.731	1.724
80-84	8	87.72	88.62	1.943	1.947
≥ 85	9	155.18	187.96	2.191	2.274

Use the methodology in Section 13.8 to estimate the 1970 regression line with the 1969 regression as a doubtful line.

REFERENCES

Bock, M. E., Yancey, T. A., and Judge, G. G. (1973). The statistical consequence of preliminary test estimators in regression. J. Amer. Stat. Assoc. 68: 109-116.

DeLury, D. B. (1947). On the estimation of biological populations. Biometric Bull. 3: 145-167.

Fisher, R. A. (1947). The analysis of covariance methods for the relation between a part and the whole. Biometric Bull. 3: 65-68.

Girshick, M. A., and Haavelmo, T. (1947). Statistical analysis of the demand for food. Econometrica 15: 79-110.

Han, C.-P., and Bancroft, T. A. (1978). Estimating regression coefficients under conditional specification. Commun. Stat. A7: 47-56.

Ipsen, J., and Feigl, P. (1970). *Bancroft's Introduction to Biostatistics*, 2nd ed. Harper & Row, New York.

Johnson, J. P., Bancroft, T. A., and Han, C.-P. (1977). A pooling methodology for regression in prediction. Biometrics 33: 57-67.

Kennedy, W. J., and Bancroft, T. A. (1971). Model building for prediction in regression based upon repeated significance tests. Ann. Math. Stat. 42: 1273-1284.

14

ANALYSIS OF VARIANCE

14.1 INTRODUCTION: In Chapter 13, the regression of Y on p fixed variables was written as $\underline{Y} = \underline{X}\underline{\beta} + \underline{\varepsilon}$ where $\underline{X}$ was assumed to have full rank. In this chapter, we consider that $\underline{X}$ may not be of full rank and the elements of $\underline{X}$ have values either 0 or 1. We shall first consider the one-way classification in which the experimenter wishes to compare several treatments, i.e., test the equality of means of several populations. The likelihood-ratio test is derived for the one-way classification. The test can be summarized in an analysis-of-variance table. The analysis of variance will then be extended to two-way classification. The random-effect model and inference based on conditional specification will also be studied.

14.2 TESTING THE EQUALITY OF SEVERAL MEANS: Suppose independent random samples, $Y_{i1}, Y_{i2}, \ldots, Y_{in}$, are taken from $N(\mu_i, \sigma^2)$, $i = 1, 2, \ldots, t$, and it is desired to test $H_0: \mu_1 = \mu_2 = \cdots = \mu_t$ against H_1: at least one equality under H_0 does not hold. We now derive the likelihood-ratio test. Under Ω, the entire parameter space, the likelihood function is

$$L(\Omega) = (2\pi\sigma^2)^{-N/2} \exp\left[-\sum_{i=1}^{t}\sum_{j=1}^{n} \frac{(y_{ij} - \mu_i)^2}{2\sigma^2}\right]$$

where N = tn. The maximum-likelihood estimates of μ_i and σ^2 are

$$\hat{\mu}_i = \frac{1}{n}\Sigma y_{ij} = \bar{y}_{i.} \qquad \hat{\sigma}^2_\Omega = \frac{1}{N}\sum_i\sum_j (y_{ij} - \bar{y}_{i.})^2 \qquad i = 1, \ldots, t$$

respectively, where a dot indicates the summation over the subscript. Hence

$$L(\hat{\Omega}) = (2\pi\hat{\sigma}_{\Omega}^{2})^{-N/2} e^{-N/2}$$

Under ω that $\mu_1 = \mu_2 = \cdots = \mu_t = \mu$, where μ is unknown, the likelihood function is

$$L(\omega) = (2\pi\sigma^2)^{-N/2} \exp\left[-\sum_i\sum_j \frac{(y_{ij} - \mu)^2}{2\sigma^2}\right]$$

The maximum-likelihood estimates of μ and σ^2 are

$$\hat{\mu} = \frac{1}{N}\sum_i\sum_j y_{ij} = \bar{y}_{..} \qquad \text{and} \qquad \hat{\sigma}_{\omega} = \frac{1}{N}\sum_i\sum_j (y_{ij} - \bar{y}_{..})^2$$

respectively. Hence

$$L(\hat{\omega}) = (2\pi\hat{\sigma}_{\omega}^{2})^{-N/2} e^{-N/2}$$

The likelihood ratio is

$$\lambda = \frac{L(\hat{\omega})}{L(\hat{\Omega})} = \left(\frac{\hat{\sigma}_{\Omega}^{2}}{\hat{\sigma}_{\omega}^{2}}\right)^{N/2}$$

Let us rewrite $\hat{\sigma}_{\omega}^{2}$ as

$$\hat{\sigma}_{\omega}^{2} = \frac{1}{N}\sum_i\sum_j [(y_{ij} - \bar{y}_{i.}) + (\bar{y}_{i.} - \bar{y}_{..})]^2$$

$$= \frac{1}{N}\sum_i\sum_j (y_{ij} - \bar{y}_{i.})^2 + \frac{n}{N}\sum_i (\bar{y}_{i.} - \bar{y}_{..})^2$$

The cross-product term is equal to zero since $\Sigma_j\ (y_{ij} - \bar{y}_{i.}) = 0$. Therefore, we may write

$$\lambda^{2/N} = \frac{\hat{\sigma}_{\Omega}^{2}}{\hat{\sigma}_{\Omega}^{2} + \frac{1}{N}\, n\, \Sigma_i\ (\bar{y}_{i.} - \bar{y}_{..})^2} = \left[1 + \frac{n\, \Sigma_i\ (\bar{y}_{i.} - \bar{y}_{..})^2}{\sum_i\sum_j (y_{ij} - \bar{y}_{i.})^2}\right]^{-1}$$

The likelihood-ratio test is to reject H_0 if $\lambda < \lambda_\alpha$. Equivalently we may use the test statistic

$$F = \frac{n \, \Sigma_i \, (\bar{Y}_{i.} - \bar{Y}_{..})^2 \, /(t - 1)}{\sum_i \sum_j (Y_{ij} - \bar{Y}_{i.})^2 \, / \, t(n - 1)} \tag{14.1}$$

The null hypothesis H_0 is rejected if $F > F_\alpha$, where F_α is an appropriately chosen constant, depending on the level of significance α.

The above testing problem may be stated alternatively as follows. Write

$$Y_{ij} = \mu + \alpha_i + \varepsilon_{ij} \qquad \begin{matrix} j = 1, \ldots, n \\ i = 1, \ldots, t \end{matrix} \tag{14.2}$$

where $\mu = (1/t) \, \Sigma_i \, \mu_i$ and $\alpha_i = \mu_i - \mu$; hence $\Sigma_i \, \alpha_i = 0$, $\{\varepsilon_{ij}\}$ are $NID(0,\sigma^2)$. This model is usually referred to as a one-way classification model. The value α_i is called the ith treatment effect because the ith population may be considered corresponding to certain treatment on the experimenter units, e.g., teaching methods, different fertilizers, etc.

The hypothesis that μ_i are all equal is equivalent to H_0: $\alpha_i = 0$ for all i. The above model may be written as a linear model, discussed in Chapter 13. Let us define

$$x_0 = 1 \qquad \text{for all } i \text{ and } j$$

$$\left.\begin{matrix} x_1 = 1 & \text{if } i = 1 \\ x_1 = 0 & \text{if } i \neq 1 \\ x_2 = 1 & \text{if } i = 2 \\ x_2 = 0 & \text{if } i \neq 2 \\ \vdots & \\ x_t = 1 & \text{if } i = t \\ x_t = 0 & \text{if } i \neq t \end{matrix}\right\} \quad \text{for all } j$$

Then the model is

$$Y_{ij} = \mu + \alpha_1 x_{1j} + \alpha_2 x_{2j} + \cdots + \alpha_t x_{tj} + \varepsilon_{ij}$$

which can be written in matrix notation as

$$\underline{Y} = \underline{X}\underline{\beta} + \underline{\varepsilon}$$

where

$$\underline{Y} = \begin{bmatrix} Y_{11} \\ \vdots \\ Y_{1n} \\ Y_{21} \\ \vdots \\ Y_{2n} \\ \vdots \\ Y_{t1} \\ \vdots \\ Y_{tn} \end{bmatrix} \qquad \underline{X} = \begin{bmatrix} 1 & 1 & 0 & \cdots & 0 \\ \vdots & \vdots & \vdots & & \vdots \\ 1 & 1 & 0 & & \vdots \\ 1 & 0 & 1 & & \vdots \\ \vdots & \vdots & \vdots & & \vdots \\ 1 & 0 & 1 & & \vdots \\ \vdots & \vdots & 0 & & 0 \\ 1 & 0 & & & 1 \\ \vdots & \vdots & & & \vdots \\ 1 & 0 & 0 & & 1 \end{bmatrix} \qquad \underline{\beta} = \begin{bmatrix} \mu \\ \alpha_1 \\ \alpha_2 \\ \vdots \\ \alpha_t \end{bmatrix}$$

$\underline{\varepsilon}$ is similarly defined as $\underline{Y}$ and is distributed as $N(0,\sigma^2\underline{I})$.

The normal equations $(\underline{X}'\underline{X})\underline{b} = \underline{X}'\underline{Y}$ are

$$\begin{bmatrix} tn & n & \cdots & n \\ n & n & 0\cdots & 0 \\ \vdots & 0 & & \vdots \\ \vdots & \vdots & & \vdots \\ n & 0 & & n \end{bmatrix} \begin{bmatrix} \hat{\mu} \\ \hat{\alpha}_1 \\ \vdots \\ \hat{\alpha}_t \end{bmatrix} = \begin{bmatrix} \Sigma\Sigma Y_{ij} \\ \Sigma Y_{1j} \\ \vdots \\ \Sigma Y_{tj} \end{bmatrix}$$

We note that $\underline{X}$ is not of full rank and $\underline{X}'\underline{X}$ does not have an inverse. Since we have the restriction $\Sigma_i\ \alpha_i = 0$, so $\alpha_t = -\alpha_1 - \alpha_2 - \cdots - \alpha_{t-1}$. Substituting this into the model, we obtain

$$Y_{ij} = \mu + \alpha_1(x_{1j} - x_{tj}) + \alpha_2(x_{2j} - x_{tj}) + \cdots + \alpha_{t-1}(x_{t-1,j} - x_{tj}) + \varepsilon_{ij}$$

Let

$$\underline{X}^* = \begin{bmatrix} 1 & 1 & 0 & \cdots & 0 \\ \vdots & \vdots & \vdots & & \vdots \\ 1 & 1 & 0 & & \vdots \\ 1 & 0 & 1 & & \vdots \\ \vdots & \vdots & \vdots & & \vdots \\ 1 & 0 & 1 & & 0 \\ \vdots & \vdots & 0 & & 1 \\ \vdots & \vdots & \vdots & & \vdots \\ \vdots & 0 & 0 & & 1 \\ 1 & -1 & -1 & & -1 \\ \vdots & \vdots & \vdots & & \vdots \\ 1 & -1 & -1 & & -1 \end{bmatrix} \qquad \underline{\beta}^* = \begin{bmatrix} \mu \\ \alpha_1 \\ \alpha_2 \\ \vdots \\ \alpha_{t-1} \end{bmatrix}$$

so the model becomes

$$\underline{Y} = \underline{X}^*\underline{\beta}^* + \underline{\varepsilon}$$

and the null hypothesis is H_0: $\alpha_1 = \cdots = \alpha_{t-1} = 0$. In this revised model, $\underline{X}^*$ is of full rank and we know how to test H_0 from the testing procedure given in the preceding chapter. We first fit the full model; the estimator of $\underline{\beta}^*$ is

$$\hat{\underline{\beta}}^* = (\underline{X}^{*\prime}\underline{X}^*)^{-1}\underline{X}^{*\prime}\underline{Y}$$

where

$$\underline{X}^{*\prime}\underline{X}^* = \begin{bmatrix} tn & 0 & 0 & \cdots & 0 \\ 0 & 2n & n & \cdots & n \\ 0 & n & 2n & \cdots & n \\ \vdots & \vdots & \vdots & & \vdots \\ 0 & n & n & \cdots & 2n \end{bmatrix}$$

$$(\underline{X}^{*\prime}\underline{X}^{*})^{-1} = \frac{1}{tn}\begin{bmatrix} 1 & 0 & \cdots & \cdots & 0 \\ 0 & t-1 & -1 & \cdots & -1 \\ 0 & -1 & t-1 & \cdots & -1 \\ \vdots & \vdots & \vdots & & \vdots \\ 0 & -1 & -1 & \cdots & t-1 \end{bmatrix}$$

and

$$\underline{X}^{*\prime}\underline{Y} = \begin{bmatrix} \Sigma\Sigma Y_{ij} \\ \Sigma Y_{1j} - \Sigma Y_{tj} \\ \vdots \\ \Sigma Y_{t-1,j} - \Sigma Y_{tj} \end{bmatrix}$$

Hence the estimators are

$$\hat{\mu} = \frac{1}{tn}\Sigma_i\Sigma_j\, Y_{ij} = \bar{Y}_{..}$$

$$\hat{\alpha}_i = \frac{1}{tn}\ (t-1)(\Sigma_j\, Y_{ij} - \Sigma_j\, Y_{tj}) - \sum_{k\neq i}\Sigma_j\, Y_{kj} + (t-2)\Sigma_j\, Y_{tj}$$

$$= \frac{1}{tn}(t\,\Sigma_j\, Y_{ij} - \Sigma_i\Sigma_j\, Y_{ij})$$

$$= \bar{Y}_{i.} - \bar{Y}_{..} \qquad i = 1, \ldots, t-1 \tag{14.3}$$

The reduction in sum of squares due to fitting x_0, x_1^*, ..., x_{t-1}^* is

$$R(x_0, x_1^*, \ldots, x_{t-1}^*) = \hat{\beta}^{*\prime}X^{*\prime}Y$$

$$= tn\bar{Y}_{..}^2 + n\sum_{i=1}^{t-1}(\bar{Y}_{i.} - \bar{Y}_{..})(\bar{Y}_{i.} - \bar{Y}_{t.})$$

$$= tn\bar{Y}_{..}^2 + n\sum_{i=1}^{t}(\bar{Y}_{i.} - \bar{Y}_{..})(\bar{Y}_{i.} - \bar{Y}_{t.} - \bar{Y}_{..})$$

$$= tn\bar{Y}_{..}^2 + n\sum_{i=1}^{t}(\bar{Y}_{i.} - \bar{Y}_{..})^2 \tag{14.4}$$

Also

$$\Sigma\Sigma Y_{ij}^2 - R(x_0, x_1^*, \ldots, x_{t-1}^*) = \Sigma_i\Sigma_j\,(Y_{ij} - \bar{Y}_{i.})^2 \tag{14.5}$$

Note that because of the restriction $\Sigma\alpha_i = 0$, we have

$$R(x_0, x_1^*, \ldots, x_{t-1}^*) = R(x_0, x_1, \ldots, x_t)$$

We next fit the reduced model. Under the null hypothesis that $\alpha_i = 0$, we only need to estimate μ. The normal equation is $tn\hat{\mu} = \Sigma\Sigma Y_{ij}$; hence the estimator of μ is $\bar{Y}..$. The reduction in sum of squares is

$$R(x_0) = tn\bar{Y}..^2 \tag{14.6}$$

Now the F test is equivalent to computing the ratio

$$\frac{[R(x_0 x_1^*, \ldots, x_{t-1}^*) - R(x_0)]/(t - 1)}{[\Sigma\Sigma Y_{ij}^2 - R(x_0 x_1^*, \ldots, x_{t-1}^*)]/(nt - t)} \tag{14.7}$$

which is

$$F = \frac{n\ \Sigma_i\ (\bar{Y}_{i.} - \bar{Y}..)^2/(t - 1)}{\Sigma_i \Sigma_j\ (Y_{ij} - \bar{Y}_{i.})^2/[t(n - 1)]} \tag{14.8}$$

The statistic F has an F distribution with [t - 1, t(n - 1)] degrees of freedom under H_0. It is readily seen that the F test is equivalent to the likelihood-ratio test in (14.1).

The numerator of F is the variation among the treatment means and the denominator of F is the variation within treatments. The testing procedure may be summarized in the analysis-of-variance (ANOVA) table (Table 14.1). This table corresponds to the model in (14.2) where the effects α_i are fixed unknown parameters. This type of model is referred to as a fixed-effect model or simply a fixed model.

In the ANOVA table (Table 14.1), the first column describes the source of variation. The second column is the degrees of freedom (df) associated with the source of variation. The third column gives the sum of squares (SS). The fourth column is the mean squares (MS), which is simply the SS divided by df. The last column gives the expected value of the mean squares, E(MS). The derivation of E(MS) will not be given here; the reader may refer to Scheffé (1959).

TABLE 14.1 Analysis of Variance of Fixed-effect Model, One-way Classification

Source	df	SS	MS	E(MS)
Among treatments	$t - 1$	$n \Sigma_i (\bar{Y}_{i.} - \bar{Y}_{\cdot\cdot})^2 = SS_A$	$\frac{SS_A}{t - 1} = MS_A$	$\sigma^2 + \frac{n}{t - 1} \Sigma_i \alpha_i^2$
Within treatments	$t(n - 1)$	$\Sigma_i \Sigma_j (Y_{ij} - \bar{Y}_{i.})^2 = SS_e$	$\frac{SS_e}{t(n - 1)} = S^2$	σ^2
Total:	$tn - 1$	$\Sigma_i \Sigma_j (Y_{ij} - \bar{Y}_{\cdot\cdot})^2$		

The F test for testing H_0: $\alpha_i = 0$ for all i is to compute $F = MS_A/S^2$; H_0 is rejected if $F > F_\alpha$, the $100(1 - \alpha)$ percentage point of $F[t - 1, t(n - 1)]$ distribution. This test procedure may also be explained intuitively by examining E(MS). Under H_0 that all α_i values are zero, $E(MS_A) = E(S^2) = \sigma^2$. Hence we would expect that the computed values of MS_A and S^2 are very close and the F value is close to unity. If the F value is very large, say larger than F_α, then the sample indicates that the α_i values are not all zero and we should reject H_0.

For computational purposes, we may use the following identities in the ANOVA table (Table 14.1):

$$SS_A = n\,\Sigma_i\,(\bar{Y}_{i.} - \bar{Y}_{..})^2 = \frac{1}{n}\,\Sigma_i\,(\Sigma_j\,Y_{ij})^2 - \frac{1}{tn}(\Sigma_i\Sigma_j\,Y_{ij})^2$$

$$SS_e = \Sigma_i\Sigma_j\,(Y_{ij} - \bar{Y}_{i.})^2 = \Sigma_i\Sigma_j\,Y_{ij}^2 - \frac{1}{n}\,\Sigma_i(\Sigma_j\,Y_{ij})^2$$

14.3 TWO-WAY CLASSIFICATION WITH ONE OBSERVATION PER CELL: Suppose in an experiment the measurements depend on two factors, say A and B. For example, factor A could be machines in a factory and factor B could be methods in the manufacture process. In agricultural experiments, factor A may be types of soil and factor B may be levels of fertilizer. Let factor A have t levels and factor B have r levels. One observation is made at each combination of the levels of factors A and B. These observations can be arranged into a two-way table with one observation per cell. The (i,j)th cell has a measurement Y_{ij} and the model can be written as

$$Y_{ij} = \mu + \alpha_i + \gamma_j + \varepsilon_{ij} \qquad \begin{array}{l} i = 1, \ldots, t \\ j = 1, \ldots, r \end{array} \tag{14.9}$$

where α_i is the factor A effect at the ith level, γ_j is the factor B effect at the jth level, and ε_{ij} is a random variable; the usual assumption for the error term is that $\{\varepsilon_{ij}\}$ are $NID(0,\sigma^2)$. We may write the above model as a linear model by defining

$$
\begin{array}{llll}
x_0 = 1 & \text{for all } i \text{ and } j & & \\
x_1 = 1 & \text{if } i = 1 & & \\
 & \text{and } x_1 = 0 & \text{if } i \neq 1 & \text{for all } j \\
\vdots & & & \\
x_t = 1 & \text{if } i = t & & \\
 & \text{and } x_t = 0 & \text{if } i \neq t & \text{for all } j \\
x_{t+1} = 1 & \text{if } j = 1 & & \\
 & \text{and } x_{t+1} = 0 & \text{if } j \neq 1 & \text{for all } i \\
\vdots & & & \\
x_{t+r} = 1 & \text{if } j = r & & \\
 & \text{and } x_{t+r} = 0 & \text{if } j \neq r & \text{for all } i
\end{array}
$$

Then $\underline{Y} = \underline{X}\beta + \underline{\varepsilon}$, where $\underline{Y}$ and $\underline{\varepsilon}$ are $tr \times 1$ vectors, $\underline{X}$ is a $tr \times (t + r + 1)$ matrix, and $\underline{\beta}$ is a $(t + r + 1) \times 1$ vector. Here $\underline{X}$ is not of full rank. If we impose the restriction $\Sigma_i \alpha_i = 0$ and $\Sigma_j \gamma_j = 0$, then we can solve the normal equations $(\underline{X}'\underline{X})\hat{\underline{\beta}} = \underline{X}'\underline{Y}$. The solutions

$$
\begin{aligned}
\hat{\mu} &= \bar{Y}_{..} \\
\hat{\alpha}_i &= \bar{Y}_{i.} - \bar{Y}_{..} \\
\hat{\gamma}_j &= \bar{Y}_{.j} - \bar{Y}_{..}
\end{aligned}
\tag{14.10}
$$

where $\bar{Y}_{..}$ is the overall mean; $\bar{Y}_{i.}$ and $\bar{Y}_{.j}$ are the means for factors A and B, respectively. The reduction is obtained as (see Exercise 14.7)

$$
R(x_0, x_1, \ldots, x_{t+r}) = tr\,\bar{Y}_{..}^2 + r \sum_{i=1}^{t} (\bar{Y}_{i.} - \bar{Y}_{..})^2 + t \sum_{j=1}^{r} (\bar{Y}_{.j} - \bar{Y}_{..})^2 \tag{14.11}
$$

Suppose we want to test the hypothesis that there is no factor B effect, i.e., H_0: $\gamma_j = 0$ for all j. Then under the null hypothesis the model becomes

$$
Y_{ij} = \mu + \alpha_i + \varepsilon_{ij}
$$

This is a one-way classification model and the reduction is

$$R(x_0,x_1,\ldots,x_t) = tr\ \bar{Y}_{\cdot\cdot}^2 + r\ \Sigma_i\ (\bar{Y}_{i.} - \bar{Y}_{\cdot\cdot})^2 \tag{14.12}$$

The difference of the reductions is $t\Sigma(\bar{Y}_{.j} - \bar{Y}_{\cdot\cdot})^2$. Hence the F test statistic for testing H_0: $\gamma_j = 0$ is

$$F = \frac{t\ \Sigma_j\ (\bar{Y}_{.j} - \bar{Y}_{\cdot\cdot})^2/(r - 1)}{SS_e/(t - 1)(r - 1)} \tag{14.13}$$

The sum of squares SS_e is

$$\begin{aligned} SS_e &= Y'Y - R(x_0,x_1,\ldots,x_{t+r}) \\ &= \sum_i \sum_j (Y_{ij} - \bar{Y}_{\cdot\cdot})^2 - r \sum_i (\bar{Y}_{i.} - \bar{Y}_{\cdot\cdot})^2 - t \sum_j (\bar{Y}_{.j} - \bar{Y}_{\cdot\cdot})^2 \end{aligned}$$

with $(rt - 1) - (t - 1) - (r - 1) = (t - 1)(r - 1)$ degrees of freedom. Under H_0 the test statistic has an F distribution with $[r - 1, (t - 1)(r - 1)]$ degrees of freedom.

By symmetry, the F test statistic for testing H_0: $\alpha_i = 0$ is

$$F = \frac{r\Sigma(\bar{Y}_{i.} - \bar{Y}_{\cdot\cdot})^2/(t - 1)}{SS_e/(t - 1)(r - 1)} \tag{14.14}$$

Under H_0 the test statistic has an F distribution with $[t - 1, (t - 1)(r - 1)]$ degrees of freedom.

We can write an ANOVA table (Table 14.2) to summarize the above testing procedure. (Note it is assumed that there is no interaction in the model; hence the residual sum of squares may be used as the error sum of squares.) The F test for testing H_0: $\alpha_i = 0$ is to compute MS_A/S^2 and the F test for testing H_0: $\gamma_j = 0$ is to compute MS_B/S^2. Note that these two tests are not independent.

For computational purposes, we may use the following formulas in the ANOVA table (Table 14.2):

TABLE 14.2 Analysis of Variance of Fixed-effect Model, Two-way Classification with One Observation per Cell

Source	df	SS	MS	E(MS)
Factor A	t - 1	$r \Sigma_i (\bar{Y}_{i.} - \bar{Y}_{..})^2 = SS_A$	$\frac{SS_A}{t - 1} = MS_A$	$\sigma^2 + \frac{r}{t - 1} \sum_i \alpha_i^2$
Factor B	r - 1	$t \Sigma_j (\bar{Y}_{.j} - \bar{Y}_{..})^2 = SS_B$	$\frac{SS_B}{r - 1} = MS_B$	$\sigma^2 + \frac{t}{r - 1} \sum_j \gamma_j^2$
Error	(t - 1)(r - 1)	$SS_T - SS_A - SS_B = SS_e$	$\frac{SS_e}{(t - 1)(r - 1)} = S^2$	σ^2
Total:	tr - 1	$\Sigma_i \Sigma_j (Y_{ij} - \bar{Y}_{..})^2 = SS_T$		

$$SS_A = \frac{1}{r} \Sigma_i (\Sigma_j Y_{ij})^2 - C$$

$$SS_B = \frac{1}{t} \Sigma_j (\Sigma_i Y_{ij})^2 - C$$

$$SS_e = \Sigma_i \Sigma_j Y_{ij} - \frac{1}{r} \Sigma_i (\Sigma_j Y_{ij})^2 - \frac{1}{t} \Sigma_j (\Sigma_i Y_{ij})^2 + C$$

where

$$C = \frac{1}{tr} (\Sigma_i \Sigma_j Y_{ij})^2$$

14.4 TWO-WAY CLASSIFICATION WITH n OBSERVATIONS PER CELL: The two-factor experiment discussed in the previous section can be extended to the situation that each cell has more than one observation. We shall only consider the balanced case, that is, where the number of observations for each cell is equal. The two factors are again denoted by A and B. Let Y_{ijk} be the measurement for the kth individual in the (i,j)th cell; then the model can be written as

$$Y_{ijk} = \mu + \alpha_i + \gamma_j + (\alpha\gamma)_{ij} + \varepsilon_{ijk} \tag{14.15}$$

$$i = 1, \ldots, t, \qquad j = 1, \ldots, r, \qquad k = 1, \ldots, n$$

where μ, α_i, γ_j have the same meaning as in the preceding section and the new term $(\alpha\gamma)_{ij}$ is called the interaction between factor A and factor B. It measures the effect due to the combination of the ith level of factor A and jth level of factor B after their main effects are accounted for. When the interaction term is present, the model is called a nonadditive model; otherwise it is called an additive model.

As in the last section, the above model may be written as a linear model with an appropriately defined $\underline{X}$ matrix. After imposing the restrictions,

$$\Sigma_i \alpha_i = 0 \qquad \Sigma_j \gamma_j = 0 \qquad \Sigma_i (\alpha\gamma)_{ij} = 0 \qquad \text{for all } j$$

$$\Sigma_j (\alpha\gamma)_{ij} = 0 \qquad \text{for all } i$$

the solutions of the normal equations give the estimators

$$\hat{\mu} = \bar{Y}... = \frac{1}{trn} \Sigma_i\Sigma_j\Sigma_k\ Y_{ijk}$$

$$\hat{\alpha}_i = \bar{Y}_{i..} - \bar{Y}... = \frac{1}{rn} \Sigma_j\Sigma_k\ Y_{ijk} - \bar{Y}...$$

$$\hat{\gamma}_0 = \bar{Y}_{.j.} - \bar{Y}... = \frac{1}{tn} \Sigma_i\Sigma_k\ Y_{ijk} - \bar{Y}...$$

$$(\hat{\alpha\gamma})_{ij} = \bar{Y}_{ij.} - \bar{Y}_{i..} - \bar{Y}_{.j.} + \bar{Y}...$$

$$= \frac{1}{n} \Sigma_k\ Y_{ijk} - \bar{Y}_{i..} - \bar{Y}_{.j.} - \bar{Y}... \qquad (14.16)$$

The reduction due to fitting the full model is

$$trn\ \bar{Y}...^2 + rn\ \Sigma_i\ (\bar{Y}_{i..} - \bar{Y}...)^2 + tn\ \Sigma_j\ (\bar{Y}_{.j.} - \bar{Y}...)^2$$
$$+ n\ \Sigma_i\Sigma_j\ (\bar{Y}_{ij.} - \bar{Y}_{i..} - \bar{Y}_{.j.} + \bar{Y}...)^2 \qquad (14.17)$$

The ANOVA table can be written similarly to the one given in the preceding section. This is shown in Table 14.3.

To obtain the analysis-of-variance table, the readers should satisfy themselves by deriving the various reductions (see Exercise 14.8). The tests for testing factor A effects, factor B effects, and interaction are made by computing the ratios of the respective MS with S^2. The computational formulas for the sums of squares are

$$SS_A = \frac{1}{rn} \Sigma_i\ (\Sigma_j\Sigma_k\ Y_{ijk})^2 - C$$

$$SS_B = \frac{1}{tn} \Sigma_j\ (\Sigma_i\Sigma_k\ Y_{ijk})^2 - C$$

$$SS_{AB} = \frac{1}{n} \Sigma_i\Sigma_j\ (\Sigma_k\ Y_{ijk})^2 - \frac{1}{rn} \Sigma_i\ (\Sigma_j\Sigma_k\ Y_{ijk})^2$$
$$- \frac{1}{tn} \Sigma_j\ (\Sigma_i\Sigma_k\ Y_{ijk})^2 + C$$

$$SS_e = \Sigma_i\Sigma_j\Sigma_k\ Y_{ijk}^2 - \frac{1}{n} \Sigma_i\Sigma_j\Sigma_k\ Y_{ijk})^2$$

where

$$C = \frac{1}{trn} (\Sigma_i\Sigma_j\Sigma_k\ Y_{ijk})^2$$

TABLE 14.3 Analysis of Variance of Fixed-effect Model, Two-way Classification with n Observations per Cell

Source	df	SS	MS	E(MS)
Factor A	$t - 1$	$rn\Sigma_i(\bar{Y}_{i..} - \bar{Y}_{...})^2 = SS_A$	$\frac{SS_A}{t-1} = MS_A$	$\sigma^2 + \frac{rn}{t-1}\sum_i \alpha_i^2$
Factor B	$r - 1$	$tn\Sigma_j(\bar{Y}_{.j.} - \bar{Y}_{...})^2 = SS_B$	$\frac{SS_B}{r-1} = MS_B$	$\sigma^2 + \frac{tn}{r-1}\sum_j \gamma_j$
Interaction	$(t - 1)(r - 1)$	$n\Sigma_i\Sigma_j(\bar{Y}_{ij.} - \bar{Y}_{i..} - \bar{Y}_{.j.} + \bar{Y}_{...})^2 = SS_{AB}$	$\frac{SS_{AB}}{(t-1)(r-1)} = MS_{AB}$	$\sigma^2 + \frac{n}{(t-1)(r-1)}\sum_{ij}\sum(\alpha\gamma)_{ij}^2$
Error	$tr(n - 1)$	$\Sigma_i\Sigma_j\Sigma_k(Y_{ijk} - \bar{Y}_{ij.})^2 = SS_e$	$\frac{SS_e}{tr(n-1)} = S^2$	σ^2
Total:	$trn - 1$	$\Sigma_i\Sigma_j\Sigma_k(Y_{ijk} - \bar{Y}_{...})^2$		

14.5 RANDOM EFFECT MODELS: In previous sections, the effects were considered to be fixed unknown parameters. The tests are designed to test the effects being zero. Such models are called fixed-effect models or simply fixed models. In other experimental situations, the effects may be a random variable. For example, a sales company wants to know whether there is any effect due to salespersons. The company selects t salespersons and the amount of sales is recorded for a period of time. For each salesperson there is considerable day-to-day variation. The t salespersons in the experiment are considered to be a random sample from a population of salespersons. The population mean is denoted by μ, which is the daily sales of a typical salesperson. The sales of the salesperson constitute a random variable with mean μ and variance of say σ_A^2. The salesperson effect a_i is then a random variable with mean 0 and variance σ_A^2 (the mean is zero because we are considering the salesperson effect and the population mean μ is subtracted). We may write the random effect model for one-way classification as

$$Y_{ij} = \mu + a_i + \varepsilon_{ij} \qquad \begin{array}{l} i = 1, 2, \ldots, t \\ j = 1, 2, \ldots, n \end{array} \tag{14.18}$$

where μ is the grand mean which is a fixed unknown parameter. a_i is the ith salesperson effect and is a random variable; usually it is assumed that $\{a_i\}$ are NID$(0,\sigma_A^2)$ and ε_{ij} is a random variable with the usual assumption that $\{e_{ij}\}$ are NID$(0,\sigma_e^2)$. Further, ε_{ij} are independent of a_i. Therefore, $V(Y_{ij}) = \sigma_A^2 + \sigma_e^2$; σ_A^2 and σ_e^2 are called *variance components*. The salesperson effects a_i are all zero if and only if $\sigma_A^2 = 0$. Therefore, we test the hypothesis H_0: $\sigma_A^2 = 0$ for salesperson effects.

The partitioning of sum of squares and the analysis-of-variance table for the random model are the same as those of the fixed model. However, the distributions of the mean squares are different because the assumptions of the model are different. From Table 14.1,

$$SS_A = n\,\Sigma_i\,(\bar{Y}_{i.} - \bar{Y}_{..})^2$$

$$SS_e = \Sigma_i\Sigma_j\,(Y_{ij} - \bar{Y}_{i.})^2$$

Substituting Y_{ij} and in terms of a and ε, we have

$$SS_A = n\Sigma_i (a_i + \bar{\varepsilon}_{i.} - \bar{a} - \bar{\varepsilon}_{..})^2$$

$$SS_e = \Sigma_i \Sigma_j (\varepsilon_{ij} - \bar{\varepsilon}_{i.})^2$$

where

$$\bar{a} = \frac{1}{t} \sum_i a_i$$

$$\bar{\varepsilon}_{i.} = \frac{1}{n} \sum_j \varepsilon_{ij}$$

$$\bar{\varepsilon}_{..} = \frac{1}{tn} \sum_i \sum_j \varepsilon_{ij}$$

Since a_i and ε_{ij} are normally independently distributed, we may treat $a_i + \bar{\varepsilon}_{i.}$, $i = 1, \ldots, t$, as a random sample from $N(0, \sigma_A^2 + \sigma_e^2/n)$. Hence $\Sigma_i (a_i + \bar{\varepsilon}_{i.} - \bar{a} - \bar{\varepsilon}_{..})^2/(\sigma_A^2 + \sigma_e^2/n)$ is distributed as a χ^2 $(t - 1)$ distribution. So SS_A is distributed as $(\sigma_e^2 + n\sigma_A^2)\ \chi^2(t - 1)$. Similarly, SS_e is distributed as a $\sigma_e^2\chi^2[t(n - 1)]$ distribution. Further, SS_A and SS_e are independently distributed because the covariance between $a_i + \bar{\varepsilon}_{i.} - \bar{a} - \bar{\varepsilon}_{..}$ and $\varepsilon_{ij} - \bar{\varepsilon}_i$ is zero for all i and j; the fact that they are normal random variables guarantees independence. Using these distributions, we obtain the expectations of the mean squares,

$$E(MS_A) = \sigma_e^2 + n\sigma_A^2$$

$$E(S^2) = \sigma_e^2 \tag{14.19}$$

The ANOVA table of the random-effect model for the one-way classification is given in Table 14.4

To test $H_0\colon \sigma_A^2 = 0$, we use the test statistic $F = MS_A/S^2$ which has an $F[t - 1, t(n - 1)]$ distribution under H_0. Note that the null distribution under H_0 is the same as that of the fixed model, but the nonnull distributions for the two models are different. For the random model, the nonnull distribution is a constant times an F distribution; for the fixed model, it is a noncentral F distribution. Noncentral distributions will not be treated in this book.

TABLE 14.4 Analysis of Variance of Random-effect Model, One-Way Classification

Source	df	SS	MS	E(MS)
Among treatments	$t - 1$	$n\, \Sigma_i (Y_{i.} - \bar{Y}_{..})^2 = SS_A$	$\frac{SS_A}{t - 1} = MS_A$	$\sigma_e^2 + \sigma_A^2$
Within treatments	$t(n - 1)$	$\Sigma_i\, \Sigma_j\, (Y_{ij} - \bar{Y}_{i.})^2 = SS_e$	$\frac{SS_e}{t(n - 1)} = S^2$	σ_e^2
Total:	$tn - 1$	$\sum_i \sum_j (Y_{ij} - \bar{Y}_{..})^2$		

When the experimenter wishes to estimate the variance components σ_A^2 and σ_e^2, unbiased estimators are

$$\tilde{\sigma}_A^2 = \frac{1}{n}(MS_A - S^2) \qquad \text{and} \qquad \tilde{\sigma}_e^2 = S^2 \tag{14.20}$$

The estimator $\tilde{\sigma}_A^2$, though unbiased, may be negative, which is clearly embarrassing. A nonnegative estimator is obtained by truncating the unbiased estimator at zero; that is, when the estimate is negative, replace it by zero. This is the restricted maximum-likelihood estimator considered by Thompson (1962), who further estimates σ_e^2 by pooling MS_A with S^2. This is essentially a preliminary test estimator with $F = MS_A/S^2$ being the test statistic for testing H_0: $\sigma_A^2 = 0$. When H_0 is rejected, σ_A^2 is estimated by $(MS_A - S^2)/n$; when H_0 is accepted, σ_A^2 is set equal to zero. In this testing procedure, the critical region is $\{F > 1\}$, which corresponds to a certain α level of the preliminary test. We may arbitrarily select the α level and consider a general preliminary test estimator. Let $\underline{\tau}' = [\sigma_A^2, \sigma_e^2]$; then a preliminary test estimator of $\underline{\tau}'$ is

$$\underline{\tau}' = \begin{cases} [cMS_A - \dfrac{S^2}{n}, S^2] & \text{if } F > F_\alpha \\ [0, S_p^2] & \text{if } F \leq F_\alpha \end{cases} \tag{14.21}$$

where c is a positive constant and

$$S_p^2 = \frac{SS_A + SS_e}{tn - 1}$$

S_p^2 is the pooled estimator of σ_e^2, F_α is the $100(1 - \alpha)$ percentage point of the $F[t - 1, t(n - 1)]$ distribution. When $F_\alpha \geq 1/cn$, the estimate is nonnegative. When $c = 1/n$, the estimator is the estimator considered by Thompson (1962). The preliminary test estimator is studied by Han (1978) and it is proposed to use $c = 1/nF_\alpha$. The appropriate choice of α, hence F_α, is given in Han (1978). Estimation procedure incorporating preliminary tests is in the area of inference based on conditional specification. Models under conditional specification will be discussed in the next section.

Let us now consider the random-effect model in a two-way classification. The model is

$$Y_{ijk} = \mu + a_i + b_j + (ab)_{ij} + \varepsilon_{ijk} \qquad (14.22)$$
$$i = 1, \ldots, t \qquad j = 1, \ldots, r \qquad k = 1, \ldots, n$$

where μ is the fixed grand mean, a_i and b_j are random main effects, $(ab)_{ij}$ are interactions and assumed random, and ε_{ijk} is the random error term. The usual assumptions about the random variables are

$$a_i \sim N(0,\sigma_A^2) \qquad b_j \sim N(0,\sigma_B^2)$$
$$(ab)_{ij} \sim N(0,\sigma_{AB}^2) \qquad \varepsilon_{ijk} \sim N(0,\sigma_e^2) \qquad (14.23)$$

and all the random variables are mutually independent. The ANOVA table is given in Table 14.5.

The ANOVA table is the same as that of Table 14.3 except the column E(MS). As in the random one-way classification model, the four sums of squares are independently distributed as constants time chi-square distributions, i.e.,

$$SS_A \sim (\sigma_e^2 + n\sigma_{AB}^2 + rn\sigma_A^2)\chi^2(t - 1)$$
$$SS_B \sim (\sigma_e^2 + n\sigma_{AB}^2 + tn\sigma_B^2)\chi^2(r - 1)$$
$$SS_{AB} \sim (\sigma_e^2 + n\sigma_{AB}^2)\chi^2[(t - 1)(r - 1)]$$
$$SS_e \sim \sigma_e^2\chi^2[tr(n - 1)] \qquad (14.24)$$

To test the hypothesis H_0: $\sigma_A^2 = 0$, we observe that SS_A and SS_{AB} have the same multiplier when H_0 is true. Therefore, when we take the ratio MS_A/MS_{AB} the constant multipliers of the chi-squares cancel out and the ratio has an F distribution. Hence the appropriate test statistic is $F = MS_A/MS_{AB}$ which has an $F[t - 1, (t - 1)(r - 1)]$ distribution. Similarly, the appropriate test statistic for testing $\sigma_B^2 = 0$ is $F = MS_B/MS_{AB}$. For testing $\sigma_{AB}^2 = 0$, we use $F = MS_{AB}/S^2$ which has an $F[(t - 1)(r - 1), tr(n - 1)]$ distribution.

TABLE 14.5 Analysis of Variance of Random-effect Model, Two-way Classification

Source	df	SS	MS	E(MS)
Factor A	$t - 1$	$rn\Sigma_i(\bar{Y}_{i..} - \bar{Y}_{...})^2 = SS_A$	$\frac{SS_A}{t - 1} = MS_A$	$\sigma_e^2 + n\sigma_{AB}^2 + rn\sigma_A^2$
Factor B	$r - 1$	$tn\Sigma_j(\bar{Y}_{.j.} - \bar{Y}_{...})^2 = SS_B$	$\frac{SS_B}{r - 1} = MS_B$	$\sigma_e^2 + n\sigma_{AB}^2 + tn\sigma_B^2$
Interaction	$(t - 1)(r - 1)$	$n\Sigma_i\Sigma_j(\bar{Y}_{ij.} - \bar{Y}_{i..} - \bar{Y}_{.j.} + \bar{Y}_{...})^2 = SS_{AB}$	$\frac{SS_{AB}}{(t - 1)(r - 1)} = MS_{AB}$	$\sigma_e^2 + n\sigma_{AB}^2$
Error	$tr(n - 1)$	$\Sigma_i\Sigma_j\Sigma_k(Y_{ijk} - \bar{Y}_{ij.})^2 = SS_e$	$\frac{SS_e}{tr(n - 1)} = S^2$	σ_e^2
Total:	$trn - 1$	$\Sigma_i\Sigma_j\Sigma_k(Y_{ijk} - \bar{Y}_{...})^2$		

In the case that there is only one observation per cell ($n = 1$), the error term in Table 14.5 drops out and the interaction becomes the error.

We have discussed the fixed-effect model and the random-effect model. One may mix these two types of effects and let some effects be fixed and some random. In such a case one has a mixed model. A comprehensive treatment of the mixed model may be found in Scheffé (1959) or Searle (1971).

14.6 MODELS UNDER CONDITIONAL SPECIFICATION: In the previous sections, the models are considered as given and the experimenter does not have any doubt concerning the given specification of the model. In many practical situations, the experimenter is not certain about the model--for example, whether the model includes an interaction term or not. A commonly used procedure to meet such a situation is to let the data resolve the doubt. When an interaction term is doubtful, one may test the hypothesis that the interaction is zero. When the hypothesis is accepted, the interaction term would be eliminated from the model; otherwise the interaction term is retained. Such a model is referred to as a model under conditional specification, since the final form of the model is conditional on the outcome of the preliminary test. This model is also called an incompletely specified model.

Let us consider first the random-effect model given in Table 14.5. When the interaction term is included in the model, the test for factor A effect is to compute the ratio MS_A/MS_{AB}. In the sequel, this test procedure is called the never-pool test. If the interaction term does not present in the model, i.e., $\sigma^2_{AB} = 0$, then $E(MS_{AB}) = E(S^2)$. A better test for testing $\sigma^2_A = 0$ is the test statistic MS_A/S^2_p, where

$$S^2_p = \frac{SS_{AB} + SS_e}{(t - 1)(r - 1) + tr(n - 1)}$$

This test statistic has an $F[(t - 1), (t - 1)(r - 1) + tr(n - 1)]$ distribution. We call this the always-pool test because the two sums of squares are pooled together. This test under the model with no interaction is better because the degrees of freedom in the denominator is increased from (t - 1)(r - 1) to (t - 1)(r - 1) + tr(n - 1). Consequently the estimate of error is more precise and the power is increased.

When the experimenter is not certain whether the interaction is zero or not, he may perform a preliminary test for testing $\sigma_{AB}^2 = 0$. The final test statistic for testing $\sigma_A^2 = 0$ is made to depend on the outcome of the preliminary test. The test procedure is then to reject the hypothesis $\sigma_A^2 = 0$ if either

$$\frac{MS_{AB}}{S^2} \geq F_1[(t-1)(r-1),\ tr(n-1)] \qquad \text{and}$$

$$\frac{MS_A}{MS_{AB}} \geq F_2[t-1,\ (t-1)(r-1)]$$

or

$$\frac{MS_{AB}}{S^2} < F_1[(t-1)(r-1),\ tr(n-1)] \qquad \text{and}$$

$$\frac{MS_A}{S_p^2} \geq F_3[t-1,\ (t-1)(r-1) + tr(n-1)] \qquad (14.25)$$

where $F_i(n_1,n_2)$ is the $100(1 - \xi_i)$ percentage point of the $F(n_1,n_2)$ distribution, i = 1, 2, 3. This test is called the sometime-pool test.

The performance of the sometimes-pool test depends on the significance levels ξ_1, ξ_2, and ξ_3 which are at the experimenter's disposal. Usually one would let $\xi_2 = \xi_3$ and set it equal to some nominal level, say .01 or .05. Then one would select ξ_1 to control the size and maximize the power of the overall test procedure. Bozivich et al. (1956) have investigated the performance of the sometimes-pool test for the random model and gave some general recommendations for the significance level ξ_1. If the experimenter is reasonably certain that σ_{AB}^2 is small, then an ξ_1 level of .25 on the preliminary F test gives a final F test which, when made at the 5% level, actually is around the 5% level. Even if σ^2_{AB} is large, the final F test will still be close to the .05 level, but its power will be diminished.

In the above discussion, we use factor A to illustrate the testing procedure. A similar testing procedure may be established for testing factor B effects. Further, the sometimes-pool testing procedure can be extended to other random models.

Let us now turn to the fixed model. We shall use Table 14.3 to illustrate the sometimes-pool test, although it is found that such a test procedure should not be used except in special cases, in the two-way classification fixed model. Again we consider testing all factor A effects are zero, i.e., $\alpha_i = 0$ for all i. If the interaction term is included in the model, the test statistic is MS_A/S^2. This is called the never-pool test. (Note that this test statistic is different from that of the never-pool test under the random model.) If the interaction is zero, a better test is the always-pool test with test statistic MS_A/S_p^2, where S_p^2 is defined similarly to that of the random model. When the experimenter is uncertain about the interaction term, a preliminary test may be used to test all $(\alpha\gamma)_{ij} = 0$ and then he or she may decide the final test statistic for testing all $\alpha_i = 0$. Hence the sometimes-pool test is to reject the hypothesis that all $\alpha_i = 0$ if either

$$\frac{MS_{AB}}{S^2} \geq F_1[(t - 1)(r - 1),\ tr(n - 1)] \qquad \text{and}$$

$$\frac{MS_A}{S^2} \geq F_2[t - 1,\ tr(n - 1)]$$

or

$$\frac{MS_{AB}}{S^2} < F_1[(t - 1)(r - 1),\ tr(n - 1)] \qquad \text{and}$$

$$\frac{MS_A}{S_p^2} \geq F_3[t - 1,\ (t - 1)(r - 1) + tr(n - 1)] \qquad (14.26)$$

The sometimes-pool test for the fixed model was investigated by Mead et al. (1975). They have found that performance of the test is quite different from that of the random-effect model. Consequently the recommendations for the level of the preliminary tests are also different. Generally speaking, if the interaction term in the fixed model is referred to as a doubtful error with n_1 degrees of freedom and the error term has n_2 degrees of freedom, then the recommendation is as follows. When the doubtful error degrees of freedom are considerably larger than the error degrees of freedom, say $n_1 > 2n_2$, the level of the preliminary test should be set about .25; when n_1 and n_2 are about equal, it should be set equal to .50; if n_1 is smaller than n_2 and n_2 is reasonably large, the never-pool test procedure should be used. Consider the two-way classification given in Table 14.3, the interaction (doubtful error) degrees of freedom are smaller than the error degrees of freedom. Therefore, we should not use the sometimes-pool test procedure and the never-pool test procedure should be used, because the increase in degrees of freedom is small and the power gain is at best small, while there may be a loss.

The above recommendation provides a warning as regards the indiscriminate use of preliminary tests in the case of fixed models. In most fixed models, the doubtful error degrees of freedom are smaller than the error degrees of freedom and the latter are usually reasonably large. The sometimes-pool test usually increases the size of the test and it in turn wipes out the power gain resulting in the small increase in degrees of freedom. However, this situation is a reverse for the random models.

EXERCISES

14.1 Show that the F test given in Section 14.2, when there are two samples, is equivalent to the t test in Section 10.8 (let $Y_{ij} = X_{ij}$).

14.2 Let the observations have the following model:

$$Y_{ij} = \mu + \alpha_i + \varepsilon_{ij} \qquad j = 1, \ldots, n_i \qquad i = 1, \ldots, t$$

where $\Sigma_{i=1}^{t} \alpha_i = 0$, $\varepsilon_{ij} \sim \text{NID}(0,\sigma^2)$. Obtain the ANOVA table analogous to Table 14.1 for testing H_0: all α_i are zero. This exercise deals with the case that each treatment has a different number of observations.

14.3 Given the one-way analysis of variance in Section 14.2, a comparison of the means or a contrast is a linear form of the means $\Sigma_{i=1}^{t} a_i\mu_i$, where $\Sigma_{i=1}^{t} a_i = 0$. (a) Obtain an unbiased estimator of $\Sigma_{i=1}^{t} a_i\mu_i$. (b) Construct a t test for testing H_0: $\Sigma_{i=1}^{t} a_i\mu_i = 0$.

14.4 In Exercise 14.3, two contrasts $\Sigma_i a_i\mu_i$ and $\Sigma_i b_i\mu_i$ are said to be orthogonal if $\Sigma_i a_i b_i = 0$. This definition corresponds to the orthogonal linear forms in Section 6.4. How many orthogonal contrasts in an orthogonal set can one construct if there are t populations? In general, any two comparisons may not be orthogonal. Such comparisons are called multiple comparisons. Determine the covariance of the estimators of any two comparisons $\Sigma_i a_i\mu_i$ and $\Sigma_i b_i\mu_i$. (For various multiple-comparison procedures, see, e.g., Bancroft, 1968.)

14.5 To compare three different teaching methods--(a) lecture only, (b) lecture with laboratory but without the use of a computer, (c) lecture with laboratory and computer--30 students are randomly assigned to the 3 methods. After 10 weeks of learning the same material, a test was given to the students and their scores were:

Group a	Group b	Group c
78	62	72
42	58	66
51	91	39
39	74	58
46	46	77
25	55	85
64	82	88
61	75	61
59	71	48
45	68	73

Test the hypothesis that the three methods are equally effective at the 5% level.

14.6 In Exercise 14.5, perform a t test for the contrast $\mu_a - \mu_b = 0$; the means of the first and second methods are equal at the 5% level.

14.7 Given the two-way classification model in Section 14.3, show that

$$R(x_0, x_1, \ldots, x_{t+r}) = tr\,\bar{Y}_{..}^2 + r \sum_{i=1}^{t} (\bar{Y}_{i.} - \bar{Y}_{..})^2$$

$$+ t \sum_{j=1}^{r} (\bar{Y}_{.j} - \bar{Y}_{..})^2$$

14.8 Given the two-way classification model with n observations per cell in Section 14.4, show that the reduction due to fitting the full model is

$$trn\,\bar{Y}_{...}^2 + rn\,\Sigma_i\,(\bar{Y}_{i..} - \bar{Y}_{...})^2 + tn\,\Sigma_j\,(\bar{Y}_{.j.} - \bar{Y}_{...})^2$$

$$+ n\,\Sigma_i\Sigma_j\,(\bar{Y}_{ij.} - \bar{Y}_{i..} - \bar{Y}_{.j.} + \bar{Y}_{...})^2$$

14.9 The computed ANOVA table of the random-effect model for the one-way classification is given as (t = 3, n = 9)

Source	df	SS	MS
Among treatments	2	96	48
Within treatments	24	360	15
Total:	26	456	

(a) Compute unbiased estimators of σ_A^2 and σ_e^2. (b) Compute the preliminary test estimator given in Eq. (14.21) with $c = 1/(nF_\alpha)$ and $\alpha = .10$.

14.10 The following table summarizes the result of an experiment in a factory which studies the worker effect (factor A) and machine effect (factor B). There are 5 workers and 4 machines in the experiment, and n = 3. Suppose the experimenter is uncertain whether there is interaction between the workers and the machines and wishes to use a random-effect model under conditional specification. (a) Complete the ANOVA table. (b) Test the worker effect using the procedure given in Eq. (14.25); use significance level .25 for the preliminary F test.

Source	df	SS	MS
Workers		36	
Machines		18	
Interaction		42	
Error		128	

REFERENCES

Bancroft, T. A. (1968). *Topics in Intermediate Statistical Methods*, Vol. 1. Iowa State University Press, Ames.

Bozivich, H., Bancroft, T. A., and Hartley, H. O. (1956). Power of analysis of variance test procedures for certain incompletely specified models, I. Ann. Math. Stat. 27: 1017-1043.

Han, C.-P. (1978). Nonnegative and preliminary test estimators of variance components. J. Amer. Stat. Assoc. 73: 855-858.

Mead, R., Bancroft, T. A., and Han, C.-P. (1975). Power of analysis of variance test procedures for incompletely specified fixed models. Ann. Stat. 3: 797-808.

Scheffé, H. (1959). *The Analysis of Variance*. Wiley, New York.

Searle, S. R. (1971). *Linear Models*. Wiley, New York.

Thompson, W. A., Jr. (1962). The problem of negative estimates of variance components. Ann. Math. Stat. 33: 273-289.

APPENDIX A

MATRIX ALGEBRA

A.1 MATRICES: A matrix is an array of numbers or functions and may be represented by the symbol

$$\underline{A} = \begin{bmatrix} a_{11} & a_{12} & \cdots & a_{1n} \\ a_{21} & a_{22} & \cdots & a_{2n} \\ \vdots & \vdots & & \vdots \\ a_{m1} & a_{m2} & \cdots & a_{mn} \end{bmatrix} = [a_{ij}]$$

The numbers or functions a_{ij} are called the *elements* of the matrix. The matrix $\underline{A}$ above has m rows and n columns and is called a matrix of order (m,n), or an $m \times n$ matrix. When $m = n$, the matrix is called a *square matrix* of order n. When $\underline{A}$ is a square matrix, the elements $a_{11}, a_{22}, \cdots, a_{nn}$ constitute the *main diagonal*. If $a_{ij} = a_{ji}$ for all i and j, the matrix is called a *symmetric matrix*. A $1 \times n$ matrix is called a row vector and an $m \times 1$ matrix is called a column vector.

Let $\underline{A}$ and $\underline{B}$ be two matrices of the same order; then $\underline{A} = \underline{B}$ if and only if $a_{ij} = b_{ij}$ for all i and j. The sum of $\underline{A}$ and $\underline{B}$, $\underline{A} + \underline{B}$, is the matrix $[(a_{ij} + b_{ij})]$. For example,

$$\begin{bmatrix} a_{11} & a_{12} \\ a_{21} & a_{22} \end{bmatrix} + \begin{bmatrix} b_{11} & b_{12} \\ b_{21} & b_{22} \end{bmatrix} = \begin{bmatrix} a_{11} + b_{11} & a_{12} + a_{12} \\ a_{21} + b_{21} & a_{22} + b_{22} \end{bmatrix}$$

The difference of $\underline{A}$ and $\underline{B}$ is $\underline{A} - \underline{B} = [(a_{ij} - b_{ij})]$. Note that $\underline{A}$ and $\underline{B}$ here are of the same order. Two matrices of the same order are said to be comformable for addition (or subtraction).

A matrix may be multiplied by a scalar or a suitable matrix. Let c be a scalar, than $c\underline{A} = \underline{A}c$ is defined to be $[(ca_{ij})]$, that is, each element of $\underline{A}$ is multiplied by the scalar c. When $\underline{A}$ is postmultiplied by a matrix $\underline{B}$, the number of rows of $\underline{B}$ must equal to the number of columns of $\underline{A}$. If this be the case, A is said to be conformable to B for multiplication. Let $\underline{A} = [a_{ij}]$ be a m × n matrix and $\underline{B} = [b_{ij}]$ be a n × p matrix, then $\underline{AB}$ is of order (m,p) and the (ij)th element of $\underline{AB}$ is equal to $\Sigma_{k=1}^{n} a_{ik}b_{kj}$. For example,

$$\begin{pmatrix} a_{11} & a_{12} & a_{13} \\ a_{21} & a_{22} & a_{23} \end{pmatrix} \begin{pmatrix} b_{11} & b_{12} \\ b_{21} & b_{22} \\ b_{31} & b_{32} \end{pmatrix}$$

$$= \begin{pmatrix} a_{11}b_{11} + a_{12}b_{21} + a_{13}b_{31} & a_{11}b_{12} + a_{12}b_{22} + a_{13}b_{32} \\ a_{21}b_{11} + a_{22}b_{21} + a_{23}b_{31} & a_{21}b_{12} + a_{22}b_{22} + a_{23}b_{32} \end{pmatrix}$$

It is seen that in general $\underline{AB} \neq \underline{BA}$. If $p \neq m$, $\underline{BA}$ is not even defined. However, the associative and distributive laws hold for matrix multiplication. Hence $(\underline{AB})\underline{C} = \underline{A}(\underline{BC})$ and $\underline{A}(\underline{B} + \underline{C}) = \underline{AB} + \underline{AC}$, $(\underline{A} + \underline{B})\underline{C} = \underline{AC} + \underline{BC}$. Here it is assumed that the multiplications are defined (i.e., conformable for multiplication).

The *transpose* of a matrix is obtained by interchanging rows and columns. If $\underline{A}$ is an m × n matrix, then its transpose, denoted by $\underline{A}'$, is an n × m matrix. For example,

$$\underline{A} = \begin{pmatrix} a_{11} & a_{12} & a_{13} \\ a_{21} & a_{22} & a_{23} \end{pmatrix} \quad \text{then} \quad \underline{A}' = \begin{pmatrix} a_{11} & a_{21} \\ a_{12} & a_{22} \\ a_{13} & a_{23} \end{pmatrix}$$

Some rules for the operation of transpose are

1. $(\underline{A}')' = \underline{A}$
2. $(\underline{A} + \underline{B})' = \underline{A}' + \underline{B}'$
3. $(\underline{AB})' = \underline{B}'\underline{A}'$, $(c\underline{A})' = c\underline{A}'$

If $\underline{A}$ is a square matrix, $\underline{A}$ is a symmetric matrix if $\underline{A}' = \underline{A}$. In statistics, the symmetric matrix plays a very important role. For example, in regression analysis, we have an $\underline{X}$ matrix and it can be shown easily that $\underline{X}'\underline{X}$ is a symmetric matrix. Also, all covariance matrices are symmetrical.

A.2 DETERMINANT AND INVERSE: The determinant of a square matrix $\underline{A}$ of order n is denoted by $|\underline{A}|$. The value of $|\underline{A}|$ may be computed by using cofactors of the elements a_{ij} and the expansion of $|\underline{A}|$ either in terms of the elements of the ith row or jth column. The cofactor of a_{ij}, denoted by A_{ij}, is $(-1)^{i+j}$ times the determinant of the submatrix of order n - 1 obtained by deleting the ith row and jth column of $\underline{A}$. Using A_{ij}, we find $|\underline{A}|$ as follows:

$$|\underline{A}| = \sum_{j=1}^{n} a_{ij}A_{ij} \quad \text{or} \quad |\underline{A}| = \sum_{i=1}^{n} a_{ij}A_{ij} \quad \text{for any } i, j$$

For example, if $\underline{A}$ is of order 2, then

$$|\underline{A}| = a_{11}A_{11} + a_{12}A_{12} = a_{11}a_{22} - a_{12}a_{21}$$

If $\underline{A}$ is of order 3, then

$$\begin{aligned} |\underline{A}| &= a_{11}A_{11} + a_{12}A_{12} + a_{13}A_{13} \\ &= a_{11}(a_{22}a_{33} - a_{23}a_{32}) - a_{12}(a_{21}a_{33} - a_{23}a_{31}) \\ &\quad + a_{13}(a_{21}a_{32} - a_{22}a_{31}) \end{aligned}$$

A square matrix $\underline{A}$ is called a nonsingular matrix if $|\underline{A}| \neq 0$; if $|\underline{A}| = 0$, then $\underline{A}$ is called a singular matrix. Some rules of finding the value of $|\underline{A}|$ are as follows.

1. If any column (or row) of $\underline{A}$ consists of all zero, then $|\underline{A}| = 0$.
2. The value of $|\underline{A}|$ is unchanged if a column (row) multiplied by a constant is added to another column (row).
3. The value of $|\underline{A}|$ is multiplied by -1 if any two columns (or rows) are interchanged.

4. If the only nonzero elements of $\underline{A}$ are on the main diagonal, then $|\underline{A}| = a_{11} \; a_{22} \; \cdots \; a_{nn}$. This matrix is called a *diagonal matrix*.
5. $|\underline{A}| = |\underline{A}'|$.
6. $|\underline{AB}| = |\underline{A}| \; |\underline{B}|$, where both $\underline{A}$ and $\underline{B}$ are square matrices.

Once we know how to evaluate $|\underline{A}|$, we can consider the *rank* of a matrix. The *rank* of a matrix is equal to r if there exists at least one submatrix of order r whose determinant is not zero, but there exists no submatrix of order larger than r whose determinant is not zero. A square matrix $\underline{A}$ is said to be of *full rank* if its rank equals its order. Therefore, a nonsingular matrix is of full rank.

In algebra, if a is a nonzero number, then the inverse of a is a^{-1} and $aa^{-1} = a^{-1}a = 1$. In matrix algebra, the inverse matrix of a nonsingular square matrix $\underline{A}$ is similarly defined. Denote the inverse matrix by $\underline{A}^{-1}$; then $\underline{AA}^{-1} = \underline{A}^{-1}\underline{A} = 1$, where $\underline{I}$ is a matrix with unity on the main diagonal and zero everywhere else. $\underline{I}$ is called an *identity matrix*. The procedure to find $\underline{A}^{-1}$ is as follows:

1. Obtain $|\underline{A}|$
2. Replace each element a_{ij} of $\underline{A}$ by its cofactor A_{ij} dividing by $|A|$.
3. Take the transpose.

For example, let

$$\underline{A} = \begin{pmatrix} 2 & 1 & 3 \\ 1 & 2 & 1 \\ 1 & 2 & 4 \end{pmatrix}$$

Then $|\underline{A}| = 9$, and the matrix of cofactors divided by $|\underline{A}|$ is

$$\begin{pmatrix} \frac{6}{9} & -\frac{3}{9} & 0 \\ \frac{2}{9} & \frac{5}{9} & -\frac{3}{9} \\ -\frac{5}{9} & \frac{1}{9} & \frac{3}{9} \end{pmatrix}$$

So

$$\underline{A}^{-1} = \begin{bmatrix} \frac{6}{9} & \frac{2}{9} & -\frac{5}{9} \\ -\frac{3}{9} & \frac{5}{9} & \frac{1}{9} \\ 0 & -\frac{3}{9} & \frac{3}{9} \end{bmatrix}$$

The inverse of a diagonal matrix is obtained by taking the inverse of each diagonal element.

A matrix $\underline{P}$ is called an orthogonal matrix if $\underline{P}^{-1} = \underline{P}'$, i.e., $\underline{P}\underline{P}' = \underline{P}'\underline{P} = I$. Orthogonal matrices are useful in many statistical applications, e.g., transformation of variable.

A.3 SYSTEM OF EQUATIONS: The following system of equations

$$\begin{aligned} a_{11}x_1 + a_{12}x_2 + \cdots + a_{1n}x_n &= b_1 \\ a_{21}x_1 + a_{22}x_2 + \cdots + a_{2n}x_n &= b_2 \\ &\vdots \\ a_{n1}x_1 + a_{n2}x_2 + \cdots + a_{nn}x_n &= b_n \end{aligned}$$

may be written in matrix notation as

$$\underline{A}\underline{x} = \underline{b}$$

where $\underline{A}$ is an $n \times n$ matrix of the coefficients, and $\underline{x}$ and $\underline{b}$ are $n \times 1$ column vectors:

$$\underline{A} = \begin{bmatrix} a_{11} & a_{12} & \cdots & a_{1n} \\ a_{21} & a_{22} & \cdots & a_{2n} \\ \vdots & \vdots & & \vdots \\ a_{n1} & a_{n2} & \cdots & a_{nn} \end{bmatrix} \qquad \underline{x} = \begin{bmatrix} x_1 \\ x_2 \\ \vdots \\ x_n \end{bmatrix} \qquad \underline{b} = \begin{bmatrix} b_1 \\ b_2 \\ \vdots \\ b_n \end{bmatrix}$$

If $\underline{A}^{-1}$ exists, then the solution of $\underline{x}$ is $\underline{x} = \underline{A}^{-1}\underline{b}$.

A.4 MATRIX DIFFERENTIATION: Let $f(\underline{x})$ be a function of $\underline{x} = [x_1 \ x_2 \ \cdots \ x_n]'$; then the derivative of f with respect to $\underline{x}$ is defined as

$$\frac{\partial f}{\partial \underline{x}} = \begin{pmatrix} \frac{\partial f}{\partial x_1} \\ \frac{\partial f}{\partial x_2} \\ \vdots \\ \frac{\partial f}{\partial x_n} \end{pmatrix}$$

For example, if $f(\underline{x}) = \underline{a}'\underline{x}$, then $\frac{\partial f}{\partial \underline{x}} = \underline{a}$.

A.5 QUADRATIC FORMS: Let $\underline{A}$ be a symmetric matrix of order n and $\underline{x}$ be an $n \times 1$ vector; then $Q = \underline{x}'\underline{A}\underline{x}$ is called a *quadratic form.* We may write $Q = \Sigma_{i=1}^{n} \Sigma_{j=1}^{n} a_{ij}x_i x_j$. Therefore, $\frac{\partial Q}{\partial x_i} = 2 \Sigma_{j=1}^{p} a_{ij}x_j$ and by the definition of matrix differentiation, we have

$$\frac{\partial Q}{\partial x} = 2\underline{A}\underline{x}$$

In particular, if $\underline{A} = \underline{I}$, then $Q = \underline{x}'\underline{x}$ and $\frac{\partial Q}{\partial \underline{x}} = 2\underline{x}$.

The quadratic form $\underline{x}'\underline{A}\underline{x}$ is called positive definite if and only if $\underline{x}'\underline{A}\underline{x} > 0$ for all $\underline{x}$ where $\underline{x} \neq \underline{0}$, and the corresponding matrix $\underline{A}$ is called a positive definite matrix. If $\underline{x}'\underline{A}\underline{x} \geq 0$ for all $\underline{x}$ and $\underline{x}'\underline{A}\underline{x} = 0$ for some $\underline{x} \neq 0$, then $\underline{x}'\underline{A}\underline{x}$ is called positive semidefinite; also the corresponding matrix $\underline{A}$ is called a positive semidefinite matrix. Examples of positive definite and positive semidefinite matrices in statistics are covariance matrices. In this book, we only consider positive definite covariance matrices. We now state some results concerning positive definite matrices:

1. If $\underline{B}$ is a nonsingular matrix and $\underline{A}$ is positive definite, then $\underline{BAB}'$ is positive definite.
2. If $\underline{A}$ is positive definite, then there exists a nonsingular matrix $\underline{P}$ such that $\underline{PAP}' = I$.

A.6 PARTITIONED MATRICES: An $m \times n$ matrix $\underline{A}$ may be partitioned as

$$\underline{A} = \begin{pmatrix} \underline{A}_{11} & \underline{A}_{12} \\ \underline{A}_{21} & \underline{A}_{22} \end{pmatrix}$$

where $\underline{A}_{11}$, $\underline{A}_{12}$ have m_1 rows; $\underline{A}_{21}$, $\underline{A}_{22}$ have m_2 rows; $\underline{A}_{11}$, $\underline{A}_{21}$ have n_1 columns; and $\underline{A}_{12}$, $\underline{A}_{22}$ have n_2 columns; $m_1 + m_2 = m$ and $n_1 + n_2 = n$. $\underline{A}_{11}$, $\underline{A}_{12}$, $\underline{A}_{21}$, and $\underline{A}_{22}$ are called *submatrices*. If $\underline{B}$ is an $m \times n$ matrix and partitioned similarly as $\underline{A}$, then

$$\underline{A} + \underline{B} = \begin{pmatrix} \underline{A}_{11} + \underline{B}_{11} & \underline{A}_{12} + \underline{B}_{12} \\ \underline{A}_{21} + \underline{B}_{21} & \underline{A}_{22} + \underline{B}_{22} \end{pmatrix}$$

If $\underline{C}$ is an $n \times p$ matrix and partitioned into submatrices, then

$$\underline{A}\underline{C} = \begin{pmatrix} \underline{A}_{11} & \underline{A}_{12} \\ \underline{A}_{21} & \underline{A}_{22} \end{pmatrix} \begin{pmatrix} \underline{C}_{11} & \underline{C}_{12} \\ \underline{C}_{21} & \underline{C}_{22} \end{pmatrix}$$

$$= \begin{pmatrix} \underline{A}_{11}\underline{C}_{11} + \underline{A}_{12}\underline{C}_{21} & \underline{A}_{11}\underline{C}_{12} + \underline{A}_{12}\underline{C}_{22} \\ \underline{A}_{21}\underline{C}_{11} + \underline{A}_{22}\underline{C}_{21} & \underline{A}_{21}\underline{C}_{12} + \underline{A}_{22}\underline{C}_{22} \end{pmatrix}$$

provided that the submatrices are conformable for multiplication.

APPENDIX B

TABLE OF MAXIMUM AND MINIMUM RELATIVE EFFICIENCIES FOR POOLING MEANS

$n_1 = 4$

α^* \ n_2		4	8	12	16	20
.50	e*	1.05	1.06	1.06	1.06	1.06
	e_0	.95	.93	.93	.93	.93
	(δ)	(1.20)	(1.03)	(.97)	(.93)	(.89)
.40	e*	1.09	1.11	1.12	1.11	1.11
	e_0	.90	.89	.88	.88	.88
	(δ)	(1.24)	(1.05)	(.99)	(.95)	(.90)
.30	e*	1.15	1.19	1.22	1.21	1.19
	e_0	.84	.81	.80	.80	.81
	(δ)	(1.31)	(1.10)	(1.02)	(.97)	(.93)
.20	e*	1.26	1.35	1.41	1.38	1.36
	e_0	.74	.71	.70	.69	.70
	(δ)	(1.41)	(1.19)	(1.10)	(1.05)	(.99)
.10	e*	1.46	1.67	1.83	1.80	1.75
	e_0	.59	.56	.54	.53	.55
	(δ)	(1.63)	(1.34)	(1.20)	(1.16)	(1.11)
.05	e*	1.64	2.03	2.34	2.34	2.28
	e_0	.46	.44	.43	.42	.43
	(δ)	(1.87)	(1.47)	(1.35)	(1.31)	(1.23)

Note: The values in the parentheses are the values of δ corresponding to e_0.

Table (continued)

$n_1 = 8$

α^* \ n_2		4	8	12	16	20
.50	e*	1.03	1.04	1.04	1.04	1.04
	e_0	.96	.95	.95	.95	.95
	(δ)	(1.03)	(.83)	(.76)	(.71)	(.68)
.40	e*	1.05	1.08	1.08	1.08	1.08
	e_0	.94	.92	.90	.90	.90
	(δ)	(1.07)	(.87)	(.79)	(.74)	(.69)
.30	e*	1.09	1.14	1.15	1.15	1.15
	e_0	.90	.86	.84	.84	.84
	(δ)	(1.10)	(.90)	(.82)	(.78)	(.72)
.20	e*	1.15	1.24	1.26	1.26	1.27
	e_0	.83	.78	.75	.75	.75
	(δ)	(1.18)	(.96)	(.88)	(.81)	(.77)
.10	e*	1.25	1.44	1.50	1.52	1.53
	e_0	.71	.64	.60	.60	.61
	(δ)	(1.31)	(1.07)	(.94)	(.90)	(.85)
.05	e*	1.34	1.62	1.75	1.82	1.84
	e_0	.61	.53	.49	.48	.49
	(δ)	(1.45)	(1.19)	(1.06)	(.96)	(.93)

Note: The values in the parentheses are the values of δ corresponding to e_0.

Table (continued)

$n_1 = 12$

α^* \ n_2		4	8	12	16	20
.50	e*	1.02	1.03	1.03	1.04	1.04
	e_0	.97	.96	.96	.96	.96
	(δ)	(.96)	(.75)	(.67)	(.62)	(.60)
.40	e*	1.04	1.05	1.06	1.06	1.07
	e_0	.95	.93	.93	.93	.92
	(δ)	(.99)	(.78)	(.69)	(.64)	(.61)
.30	e*	1.06	1.09	1.11	1.12	1.12
	e_0	.92	.89	.88	.87	.87
	(δ)	(1.03)	(.81)	(.72)	(.66)	(.64)
.20	e*	1.11	1.16	1.19	1.21	1.22
	e_0	.87	.81	.80	.79	.78
	(δ)	(1.10)	(.91)	(.79)	(.71)	(.67)
.10	e*	1.18	1.28	1.35	1.39	1.41
	e_0	.78	.69	.67	.66	.65
	(δ)	(1.25)	(.94)	(.84)	(.78)	(.71)
.05	e*	1.24	1.40	1.51	1.58	1.62
	e_0	.69	.59	.55	.55	.54
	(δ)	(1.36)	(1.05)	(.93)	(.85)	(.81)

Note: The values in the parentheses are the values of δ corresponding to e_0.

Table (continued)

$n_1 = 16$

α^* \ n_2		4	8	12	16	20
.50	e*	1.01	1.02	1.03	1.03	1.03
	e_0	.98	.97	.96	.96	.96
	(δ)	(.93)	(.70)	(.62)	(.59)	(.56)
.40	e*	1.03	1.04	1.05	1.05	1.06
	e_0	.97	.95	.94	.94	.93
	(δ)	(.94)	(.70)	(.63)	(.60)	(.58)
.30	e*	1.04	1.07	1.08	1.10	1.10
	e_0	.94	.91	.90	.89	.89
	(δ)	(.98)	(.75)	(.67)	(.62)	(.59)
.20	e*	1.07	1.12	1.15	1.17	1.18
	e_0	.90	.85	.83	.82	.81
	(δ)	(1.05)	(.82)	(.70)	(.65)	(.61)
.10	e*	1.12	1.21	1.26	1.30	1.32
	e_0	.82	.75	.72	.70	.70
	(δ)	(1.17)	(.83)	(.79)	(.69)	(.64)
.05	e*	1.17	1.29	1.38	1.44	1.49
	e_0	.74	.65	.62	.60	.59
	(δ)	(1.30)	(.96)	(.85)	(.80)	(.71)

Note: The values in the parentheses are the values of δ corresponding to e_0.

Table (continued)

$n_1 = 20$

α^* \ n_2		4	8	12	16	20
.50	e*	1.10	1.02	1.02	1.02	1.03
	e_0	.99	.98	.98	.98	.97
	(δ)	(.90)	(.68)	(.60)	(.56)	(.51)
.40	e*	1.02	1.03	1.02	1.02	1.03
	e_0	.97	.96	.95	.94	.94
	(δ)	(.93)	(.71)	(.62)	(.57)	(.53)
.30	e*	1.03	1.06	1.07	1.08	1.09
	e_0	.95	.93	.92	.91	.90
	(δ)	(.96)	(.74)	(.65)	(.59)	(.55)
.20	e*	1.06	1.09	1.12	1.14	1.15
	e_0	.92	.88	.86	.84	.83
	(δ)	(1.03)	(.79)	(.69)	(.62)	(.58)
.10	e*	1.09	1.16	1.21	1.25	1.28
	e_0	.86	.79	.76	.74	.72
	(δ)	(1.14)	(.85)	(.73)	(.68)	(.64)
.05	e*	1.13	1.22	1.30	1.36	1.40
	e_0	.79	.70	.66	.64	.62
	(δ)	(1.25)	(.93)	(.80)	(.73)	(.68)

Note: The values in the parentheses are the values of δ corresponding to e_0.

APPENDIX C

TABLES OF PROBABILITY DISTRIBUTIONS

C.1 *EXPLANATION OF THE TABLES*

Table C.1.a Ordinates of the Normal Distribution This table gives values of

$$f(x) = \frac{1}{\sqrt{2\pi}} e^{-x^2/2}$$

for values of x between 0 and 4 at intervals of .01. For negative values of x, one uses the relation $f(-x) = f(x)$.

Table C.1.b Area under the Normal Curve, F(z). This table gives values of

$$F(z) = \int_{-\infty}^{z} \frac{1}{\sqrt{2\pi}} e^{-x^2/2}\, dx$$

for values of z between 0 and 3.5 at intervals of .01. For negative values of z, one uses the relation $F(-z) = 1 - F(z)$.

Values of z corresponding to a few selected values of α are presented beneath the main table. These are useful in making tests of significance for a specified significance probability α or in setting up confidence intervals with confidence probability $(1 - \alpha)$.

Table C.2 Percentage Points of the χ^2 Distribution. This table gives values of χ^2_α for selected values of α and for the number of degrees of freedom, ν, ranging from 1 to 30, plus ν = 40, 50, 60, where

$$\alpha = \int_{\chi^2_\alpha}^{\infty} \frac{1}{\Gamma(\nu/2)} \left(\frac{\chi^2}{2}\right)^{(\nu-2)/2} e^{-\chi^2/2}\, d\left(\frac{\chi^2}{2}\right)$$

For larger values of ν, a normal approximation is quite accurate. The variate $\sqrt{2\chi^2} - \sqrt{2\nu - 1}$ is nearly normally distributed with zero mean and unit variance. Hence χ^2_α may be computed as

$$\chi^2_\alpha = \frac{1}{2}\,(z_{2\alpha} + \sqrt{2\nu - 1})^2$$

where we use $z_{2\alpha}$ because the probability for χ^2 corresponds to a single tail of the normal curve. For example, consider $\chi^2_{.05}$ for $\nu = 40$. The exact value is $\chi^2_{.05} = 55.8$, and the approximate value would be

$$\frac{1}{2}(1.645 + \sqrt{79})^2 = 55.5$$

Table C.3 Percentage Points of the t Distribution. This table gives values of $t_{\alpha/2}$ for selected values of α and for the following number of degrees of freedom n: 1, 2, ..., 30, 40, 60, 120, ∞, where

$$\alpha = 2\int_{t_{\alpha/2}}^{\infty} \frac{\Gamma[(n + 1)/2]}{\sqrt{\pi n}\;\Gamma\;(n/2)} \left(1 + \frac{t^2}{n}\right)^{-(n+1)/2} dt$$

If a percentage point is needed for some n not given in the table, linear interpolation can be used between tabulated percentage points, but the reciprocals of n should be used. For example, the value of $t_{.025}$ for 50 degrees of freedom would be

$$2.021 - \frac{1/40 - 1/50}{1/40 - 1/60}\,(.021) = 2.008$$

Table C.4 Percentage Points of the F Distribution. This table gives values of F_α for α = .01 and .05; n_1 = 1 (1) 10, 12, 16, 20, 24, 30, 50, 100, ∞; and n_2 = 1 (1) 20 (2) 28 (4) 40, 60, 100, 200, ∞. F_α is defined as follows:

$$\alpha = \int_{F_\alpha}^{\infty} \frac{(n_1/n_2)^{n_1/2}}{B(n_1/2,\, n_2/2)}\, F^{(1/2)n_1 - 1} \left(1 + \frac{n_1}{n_2}\,F\right)^{-(n_1+n_2)/2} dF$$

$F = s_1^2/s_2^2$ where s_1^2 is a sample variance with n_1 degrees of freedom and S_2^2 a sample variance with n_2 degrees of freedom. The table also can be used to determine $F_{1-\alpha}$ by use of the formula

$$F(1 - \alpha; n_1, n_2) = \frac{1}{F(\alpha; n_2, n_1)}$$

where the first n indicates the number of degrees of freedom in the numerator and the second n the number of degrees of freedom in the denominator. For example,

$$F(.95;3,75) = \frac{1}{F(.05;75,8)} = \frac{1}{8.57}$$

One should interpolate on the reciprocals of n_1 and n_2 as in Table C.3.

C.2 OTHER TABLES:

Odeh, R. E., Owen, D. B., Birnbaum, Z. W., and Fisher, L. (1977). *Pocket Book of Statistical Tables*. Marcel Dekker, New York.

Owen, D. B. (1962). *Handbook of Statistical Tables*. Addison-Wesley, Reading, Mass.

TABLE C.1.a Ordinates of the Normal Distribution

x	.00	.01	.02	.03	.04	.05	.06	.07	.08	.09
.0	.3989	.3989	.3989	.3988	.3986	.3984	.3982	.3980	.3977	.3973
.1	.3970	.3965	.3961	.3956	.3951	.3945	.3939	.3932	.3925	.3918
.2	.3910	.3902	.3894	.3885	.3876	.3867	.3857	.3847	.3836	.3825
.3	.3814	.3802	.3790	.3778	.3765	.3752	.3739	.3725	.3712	.3697
.4	.3683	.3668	.3653	.3637	.3621	.3605	.3589	.3572	.3555	.3538
.5	.3521	.3503	.3485	.3467	.3448	.3429	.3410	.3391	.3372	.3352
.6	.3332	.3312	.3292	.3271	.3251	.3230	.3209	.3187	.3166	.3144
.7	.3123	.3101	.3079	.3056	.3034	.3011	.2989	.2966	.2943	.2920
.8	.2897	.2874	.2850	.2827	.2803	.2780	.2756	.2732	.2709	.2685
.9	.2661	.2637	.2613	.2589	.2565	.2541	.2516	.2492	.2468	.2444
1.0	.2420	.2396	.2371	.2347	.2323	.2299	.2275	.2251	.2227	.2203
1.1	.2179	.2155	.2131	.2107	.2083	.2059	.2036	.2012	.1989	.1965
1.2	.1942	.1919	.1895	.1872	.1849	.1826	.1804	.1781	.1758	.1736
1.3	.1714	.1691	.1669	.1647	.1626	.1604	.1582	.1561	.1539	.1518
1.4	.1497	.1476	.1456	.1435	.1415	.1394	.1374	.1354	.1334	.1315
1.5	.1295	.1276	.1257	.1238	.1219	.1200	.1182	.1163	.1145	.1127
1.6	.1109	.1092	.1074	.1057	.1040	.1023	.1006	.0989	.0973	.0957
1.7	.0940	.0925	.0909	.0893	.0878	.0863	.0848	.0833	.0818	.0804
1.8	.0790	.0775	.0761	.0748	.0734	.0721	.0707	.0694	.0681	.0669
1.9	.0656	.0644	.0632	.0620	.0608	.0596	.0584	.0573	.0562	.0551
2.0	.0540	.0529	.0519	.0508	.0498	.0488	.0478	.0468	.0459	.0449
2.1	.0440	.0431	.0422	.0413	.0404	.0396	.0387	.0379	.0371	.0363
2.2	.0355	.0347	.0339	.0332	.0325	.0317	.0310	.0303	.0297	.0290
2.3	.0283	.0277	.0270	.0264	.0258	.0252	.0246	.0241	.0235	.0229
2.4	.0224	.0219	.0213	.0208	.0203	.0198	.0194	.0189	.0184	.0180
2.5	.0175	.0171	.0167	.0163	.0158	.0154	.0151	.0147	.0143	.0139
2.6	.0136	.0132	.0129	.0126	.0122	.0119	.0116	.0113	.0110	.0107
2.7	.0104	.0101	.0099	.0096	.0093	.0091	.0088	.0086	.0084	.0081
2.8	.0079	.0077	.0075	.0073	.0071	.0069	.0067	.0065	.0063	.0061
2.9	.0060	.0058	.0056	.0055	.0053	.0051	.0050	.0048	.0047	.0046
3.0	.0044	.0043	.0042	.0040	.0039	.0038	.0037	.0036	.0035	.0034
3.1	.0033	.0032	.0031	.0030	.0029	.0028	.0027	.0026	.0025	.0025
3.2	.0024	.0023	.0022	.0022	.0021	.0020	.0020	.0019	.0018	.0018
3.3	.0017	.0017	.0016	.0016	.0015	.0015	.0014	.0014	.0013	.0013
3.4	.0012	.0012	.0012	.0011	.0011	.0010	.0010	.0010	.0009	.0009
3.5	.0009	.0008	.0008	.0008	.0008	.0007	.0007	.0007	.0007	.0006
3.6	.0006	.0006	.0006	.0005	.0005	.0005	.0005	.0005	.0005	.0004
3.7	.0004	.0004	.0004	.0004	.0004	.0004	.0003	.0003	.0003	.0003
3.8	.0003	.0003	.0003	.0003	.0003	.0002	.0002	.0002	.0002	.0002
3.9	.0002	.0002	.0002	.0002	.0002	.0002	.0002	.0002	.0001	.0001

TABLE C.1.b Area under the Normal Curve, F(z)[a]

z	.00	.01	.02	.03	.04	.05	.06	.07	.08	.09
.0	.5000	.5040	.5080	.5120	.5160	.5199	.5239	.5279	.5319	.5359
.1	.5398	.5438	.5478	.5517	.5557	.5596	.5636	.5675	.5714	.5753
.2	.5793	.5832	.5871	.5910	.5948	.5987	.6026	.6064	.6103	.6141
.3	.6179	.6217	.6255	.6293	.6331	.6368	.6406	.6443	.6480	.6517
.4	.6554	.6591	.6628	.6664	.6700	.6736	.6772	.6808	.6844	.6879
.5	.6915	.6950	.6985	.7019	.7054	.7088	.7123	.7157	.7190	.7224
.6	.7257	.7291	.7324	.7357	.7389	.7422	.7454	.7486	.7517	.7549
.7	.7580	.7611	.7642	.7673	.7704	.7734	.7764	.7794	.7823	.7852
.8	.7881	.7910	.7939	.7967	.7995	.8023	.8051	.8078	.8106	.8133
.9	.8159	.8186	.8212	.8238	.8264	.8289	.8315	.8340	.8365	.8389
1.0	.8413	.8438	.8461	.8485	.8508	.8531	.8554	.8577	.8599	.8621
1.1	.8643	.8665	.8686	.8708	.8729	.8749	.8770	.8790	.8810	.8830
1.2	.8849	.8869	.8888	.8907	.8925	.8944	.8962	.8980	.8997	.9015
1.3	.9032	.9049	.9066	.9082	.9099	.9115	.9131	.9147	.9162	.9177
1.4	.9192	.9207	.9222	.9236	.9251	.9265	.9279	.9292	.9306	.9319
1.5	.9332	.9345	.9357	.9370	.9382	.9394	.9406	.9418	.9429	.9441
1.6	.9452	.9463	.9474	.9484	.9495	.9505	.9515	.9525	.9535	.9545
1.7	.9554	.9564	.9573	.9582	.9591	.9599	.9608	.9616	.9625	.9633
1.8	.9641	.9649	.9656	.9664	.9671	.9678	.9686	.9693	.9699	.9706
1.9	.9713	.9719	.9726	.9732	.9738	.9744	.9750	.9756	.9761	.9767
2.0	.9772	.9778	.9783	.9788	.9793	.9798	.9803	.9808	.9812	.9817
2.1	.9821	.9826	.9830	.9834	.9838	.9842	.9846	.9850	.9854	.9857
2.2	.9861	.9864	.9868	.9871	.9875	.9878	.9881	.9884	.9887	.9890
2.3	.9893	.9896	.9898	.9901	.9904	.9906	.9909	.9911	.9913	.9916
2.4	.9918	.9920	.9922	.9925	.9927	.9929	.9931	.9932	.9934	.9936
2.5	.9938	.9940	.9941	.9943	.9945	.9946	.9948	.9949	.9951	.9952
2.6	.9953	.9955	.9956	.9957	.9959	.9960	.9961	.9962	.9963	.9964
2.7	.9965	.9966	.9967	.9968	.9969	.9970	.9971	.9972	.9973	.9974
2.8	.9974	.9975	.9976	.9977	.9977	.9978	.9979	.9979	.9980	.9981
2.9	.9981	.9982	.9982	.9983	.9984	.9984	.9985	.9985	.9986	.9986
3.0	.9987	.9987	.9987	.9988	.9988	.9989	.9989	.9989	.9990	.9990
3.1	.9990	.9991	.9991	.9991	.9992	.9992	.9992	.9992	.9993	.9993
3.2	.9993	.9993	.9994	.9994	.9994	.9994	.9994	.9995	.9995	.9995
3.3	.9995	.9995	.9995	.9996	.9996	.9996	.9996	.9996	.9996	.9997
3.4	.9997	.9997	.9997	.9997	.9997	.9997	.9997	.9997	.9997	.9998

Percentage Points of the Normal Distribution†

$F(z)$	.75	.90	.95	.975	.99	.995	.999	.9995	.99995	.999995
$\alpha = 2[1 - F(z)]$	.50	.20	.10	.05	.02	.01	.002	.001	.0001	.00001
z_α	0.674	1.282	1.645	1.960	2.326	2.576	3.090	3.291	3.891	4.417

[a]F(z) is the area under the normal curve from $-\infty$ to z; α is twice the area from z to ∞ (area from $-\infty$ to $-z$ plus the area from z to ∞).

TABLE C.2 Percentage Points of the χ^2 Distribution

ν \ α	.995	.990	.975	.950	.900	.750	.500	.250	.100	.050	.025	.010	.005
1	$.0^4393$	$.0^3157$	$.0^3982$	$.0^2393$	.0158	.102	.455	1.32	2.71	3.84	5.02	6.63	7.88
2	.0100	.0201	.0506	.103	.211	.575	1.39	2.77	4.61	5.99	7.38	9.21	10.6
3	.0717	.115	.216	.352	.584	1.21	2.37	4.11	6.25	7.81	9.35	11.3	12.8
4	.207	.297	.484	.711	1.06	1.92	3.36	5.39	7.78	9.49	11.1	13.3	14.9
5	.412	.554	.831	1.15	1.61	2.67	4.35	6.63	9.24	11.1	12.8	15.1	16.7
6	.676	.872	1.24	1.64	2.20	3.45	5.35	7.84	10.6	12.6	14.4	16.8	18.5
7	.989	1.24	1.69	2.17	2.83	4.25	6.35	9.04	12.0	14.1	16.0	18.5	20.3
8	1.34	1.65	2.18	2.73	3.49	5.07	7.34	10.2	13.4	15.5	17.5	20.1	22.0
9	1.73	2.09	2.70	3.33	4.17	5.90	8.34	11.4	14.7	16.9	19.0	21.7	23.6
10	2.16	2.56	3.25	3.94	4.87	6.74	9.34	12.5	16.0	18.3	20.5	23.2	25.2
11	2.60	3.05	3.82	4.57	5.58	7.58	10.3	13.7	17.3	19.7	21.9	24.7	26.8
12	3.07	3.57	4.40	5.23	6.30	8.44	11.3	14.8	18.5	21.0	23.3	26.2	28.3
13	3.57	4.11	5.01	5.89	7.04	9.30	12.3	16.0	19.8	22.4	24.7	27.7	29.8
14	4.07	4.66	5.63	6.57	7.79	10.2	13.3	17.1	21.1	23.7	26.1	29.1	31.3
15	4.60	5.23	6.26	7.26	8.55	11.0	14.3	18.2	22.3	25.0	27.5	30.6	32.8
16	5.14	5.81	6.91	7.96	9.31	11.9	15.3	19.4	23.5	26.3	28.8	32.0	34.3
17	5.70	6.41	7.56	8.67	10.1	12.8	16.3	20.5	24.8	27.6	30.2	33.4	35.7
18	6.26	7.01	8.23	9.39	10.9	13.7	17.3	21.6	26.0	28.9	31.5	34.8	37.2
19	6.84	7.63	8.91	10.1	11.7	14.6	18.3	22.7	27.2	30.1	32.9	36.2	38.6
20	7.43	8.26	9.59	10.9	12.4	15.5	19.3	23.8	28.4	31.4	34.2	37.6	40.0
21	8.03	8.90	10.3	11.6	13.2	16.3	20.3	24.9	29.6	32.7	35.5	38.9	41.4
22	8.64	9.54	11.0	12.3	14.0	17.2	21.3	26.0	30.8	33.9	36.8	40.3	42.8
23	9.26	10.2	11.7	13.1	14.8	18.1	22.3	27.1	32.0	35.2	38.1	41.6	44.2
24	9.89	10.9	12.4	13.8	15.7	19.0	23.3	28.2	33.2	36.4	39.4	43.0	45.6
25	10.5	11.5	13.1	14.6	16.5	19.9	24.3	29.3	34.4	37.7	40.6	44.3	46.9
26	11.2	12.2	13.8	15.4	17.3	20.8	25.3	30.4	35.6	38.9	41.9	45.6	48.3
27	11.8	12.9	14.6	16.2	18.1	21.7	26.3	31.5	36.7	40.1	43.2	47.0	49.6
28	12.5	13.6	15.3	16.9	18.9	22.7	27.3	32.6	37.9	41.3	44.5	48.3	51.0
29	13.1	14.3	16.0	17.7	19.8	23.6	28.3	33.7	39.1	42.6	45.7	49.6	52.3
30	13.8	15.0	16.8	18.5	20.6	24.5	29.3	34.8	40.3	43.8	47.0	50.9	53.7
40	20.7	22.2	24.4	26.5	29.1	33.7	39.3	45.6	51.8	55.8	59.3	63.7	66.8
50	28.0	29.7	32.4	34.8	37.7	42.9	49.3	56.3	63.2	67.5	71.4	76.2	79.5
60	35.5	37.5	40.5	43.2	46.5	52.3	59.3	67.0	74.4	79.1	83.3	88.4	92.0

Source: Abridged from Catherine M. Thompson, Tables of percentage points of the χ^2 distribution, Biometrika, 32:188–189 (1941), and is published here with the kind permission of the author and the editor of *Biometrika*. ν is the number of degrees of freedom and α the probability of exceeding the tabular value.

TABLE C.3 Percentage Points of the t Distribution

n \ α	.50	.40	.30	.20	.10	.05	.02	.01	.001
1	1.000	1.376	1.963	3.078	6.314	12.706	31.821	63.657	636.619
2	.816	1.061	1.386	1.886	2.920	4.303	6.965	9.925	31.598
3	.765	.978	1.250	1.638	2.353	3.182	4.541	5.841	12.941
4	.741	.941	1.190	1.533	2.132	2.776	3.747	4.604	8.610
5	.727	.920	1.156	1.476	2.015	2.571	3.365	4.032	6.859
6	.718	.906	1.134	1.440	1.943	2.447	3.143	3.707	5.959
7	.711	.896	1.119	1.415	1.895	2.365	2.998	3.499	5.405
8	.706	.889	1.108	1.397	1.860	2.306	2.896	3.355	5.041
9	.703	.883	1.100	1.383	1.833	2.262	2.821	3.250	4.781
10	.700	.879	1.093	1.372	1.812	2.228	2.764	3.169	4.587
11	.697	.876	1.088	1.363	1.796	2.201	2.718	3.106	4.437
12	.695	.873	1.083	1.356	1.782	2.179	2.681	3.055	4.318
13	.694	.870	1.079	1.350	1.771	2.160	2.650	3.012	4.221
14	.692	.868	1.076	1.345	1.761	2.145	2.624	2.977	4.140
15	.691	.866	1.074	1.341	1.753	2.131	2.602	2.947	4.073
16	.690	.865	1.071	1.337	1.746	2.120	2.583	2.921	4.015
17	.689	.863	1.069	1.333	1.740	2.110	2.567	2.898	3.965
18	.688	.862	1.067	1.330	1.734	2.101	2.552	2.878	3.922
19	.688	.861	1.066	1.328	1.729	2.093	2.539	2.861	3.883
20	.687	.860	1.064	1.325	1.725	2.086	2.528	2.845	3.850
21	.686	.859	1.063	1.323	1.721	2.080	2.518	2.831	3.819
22	.686	.858	1.061	1.321	1.717	2.074	2.508	2.819	3.792
23	.685	.858	1.060	1.319	1.714	2.069	2.500	2.807	3.767
24	.685	.857	1.059	1.318	1.711	2.064	2.492	2.797	3.745
25	.684	.856	1.058	1.316	1.708	2.060	2.485	2.787	3.725
26	.684	.856	1.058	1.315	1.706	2.056	2.479	2.779	3.707
27	.684	.855	1.057	1.314	1.703	2.052	2.473	2.771	3.690
28	.683	.855	1.056	1.313	1.701	2.048	2.467	2.763	3.674
29	.683	.854	1.055	1.311	1.699	2.045	2.462	2.756	3.659
30	.683	.854	1.055	1.310	1.697	2.042	2.457	2.750	3.646
40	.681	.851	1.050	1.303	1.684	2.021	2.423	2.704	3.551
60	.679	.848	1.046	1.296	1.671	2.000	2.390	2.660	3.460
120	.677	.845	1.041	1.289	1.658	1.980	2.358	2.617	3.373
∞	.674	.842	1.036	1.282	1.645	1.960	2.326	2.576	3.291

Source: Abridged from Table III of R. A. Fisher and Frank Yates, *Statistical Tables for Biological, Agricultural, and Medical Research*, 6th ed. Oliver & Boyd, London and Edinburgh, 1963. It is reproduced here with the kind permission of the authors and their publishers. n is the number of degrees of freedom and α twice the probability of exceeding the tabular value (or the probability of being more than the tabular value or less than the negative of the tabular value).

TABLE C.4 Percentage Points of the F Distribution[a]

$\alpha = .05$

n_2 \ n_1	1	2	3	4	5	6	7	8	9	10	12	16	20	24	30	50	100	∞
1	161	200	216	225	230	234	237	239	241	242	244	246	248	249	250	252	253	254
2	18.51	19.00	19.16	19.25	19.30	19.33	19.36	19.37	19.38	19.39	19.41	19.43	19.44	19.45	19.46	19.47	19.49	19.50
3	10.13	9.55	9.28	9.12	9.01	8.94	8.88	8.84	8.81	8.78	8.74	8.69	8.66	8.64	8.62	8.58	8.56	8.53
4	7.71	6.94	6.59	6.39	6.26	6.16	6.09	6.04	6.00	5.96	5.91	5.84	5.80	5.77	5.74	5.70	5.66	5.63
5	6.61	5.79	5.41	5.19	5.05	4.95	4.88	4.82	4.78	4.74	4.68	4.60	4.56	4.53	4.50	4.44	4.40	4.36
6	5.99	5.14	4.76	4.53	4.39	4.28	4.21	4.15	4.10	4.06	4.00	3.92	3.87	3.84	3.81	3.75	3.71	3.67
7	5.59	4.74	4.35	4.12	3.97	3.87	3.79	3.73	3.68	3.63	3.57	3.49	3.44	3.41	3.38	3.32	3.28	3.23
8	5.32	4.46	4.07	3.84	3.69	3.58	3.50	3.44	3.39	3.34	3.28	3.20	3.15	3.12	3.08	3.03	2.98	2.93
9	5.12	4.26	3.86	3.63	3.48	3.37	3.29	3.23	3.18	3.13	3.07	2.98	2.93	2.90	2.86	2.80	2.76	2.71
10	4.96	4.10	3.71	3.48	3.33	3.22	3.14	3.07	3.02	2.97	2.91	2.82	2.77	2.74	2.70	2.64	2.59	2.54
11	4.84	3.98	3.59	3.36	3.20	3.09	3.01	2.95	2.90	2.86	2.79	2.70	2.65	2.61	2.57	2.50	2.45	2.40
12	4.75	3.88	3.49	3.26	3.11	3.00	2.92	2.85	2.80	2.76	2.69	2.60	2.54	2.50	2.46	2.40	2.35	2.30
13	4.67	3.80	3.41	3.18	3.02	2.92	2.84	2.77	2.72	2.67	2.60	2.51	2.46	2.42	2.38	2.32	2.26	2.21
14	4.60	3.74	3.34	3.11	2.96	2.85	2.77	2.70	2.65	2.60	2.53	2.44	2.39	2.35	2.31	2.24	2.19	2.13
15	4.54	3.68	3.29	3.06	2.90	2.79	2.70	2.64	2.59	2.55	2.48	2.39	2.33	2.29	2.25	2.18	2.12	2.07
16	4.49	3.63	3.24	3.01	2.85	2.74	2.66	2.59	2.54	2.49	2.42	2.33	2.28	2.24	2.20	2.13	2.07	2.01
17	4.45	3.59	3.20	2.96	2.81	2.70	2.62	2.55	2.50	2.45	2.38	2.29	2.23	2.19	2.15	2.08	2.02	1.96
18	4.41	3.55	3.16	2.93	2.77	2.66	2.58	2.51	2.46	2.41	2.34	2.25	2.19	2.15	2.11	2.04	1.98	1.92
19	4.38	3.52	3.13	2.90	2.74	2.63	2.55	2.48	2.43	2.38	2.31	2.21	2.15	2.11	2.07	2.00	1.94	1.88
20	4.35	3.49	3.10	2.87	2.71	2.60	2.52	2.45	2.40	2.35	2.28	2.18	2.12	2.08	2.04	1.96	1.90	1.84
22	4.30	3.44	3.05	2.82	2.66	2.55	2.47	2.40	2.35	2.30	2.23	2.13	2.07	2.03	1.98	1.91	1.84	1.78
24	4.26	3.40	3.01	2.78	2.62	2.51	2.43	2.36	2.30	2.26	2.18	2.09	2.02	1.98	1.94	1.86	1.80	1.73
26	4.22	3.37	2.98	2.74	2.59	2.47	2.39	2.32	2.27	2.22	2.15	2.05	1.99	1.95	1.90	1.82	1.76	1.69
28	4.20	3.34	2.95	2.71	2.56	2.44	2.36	2.29	2.24	2.19	2.12	2.02	1.96	1.91	1.87	1.78	1.72	1.65
32	4.15	3.30	2.90	2.67	2.51	2.40	2.32	2.25	2.19	2.14	2.07	1.97	1.91	1.86	1.82	1.74	1.67	1.59
36	4.11	3.26	2.86	2.63	2.48	2.36	2.28	2.21	2.15	2.10	2.03	1.93	1.87	1.82	1.78	1.69	1.62	1.55
40	4.08	3.23	2.84	2.61	2.45	2.34	2.25	2.18	2.12	2.07	2.00	1.90	1.84	1.79	1.74	1.66	1.59	1.51
60	4.00	3.15	2.76	2.52	2.37	2.25	2.17	2.10	2.04	1.99	1.92	1.81	1.75	1.70	1.65	1.56	1.48	1.39
100	3.94	3.09	2.70	2.46	2.30	2.19	2.10	2.03	1.97	1.92	1.85	1.75	1.68	1.63	1.57	1.48	1.39	1.28
200	3.89	3.04	2.65	2.41	2.26	2.14	2.05	1.98	1.92	1.87	1.80	1.69	1.62	1.57	1.52	1.42	1.32	1.19
∞	3.84	2.99	2.60	2.37	2.21	2.09	2.01	1.94	1.88	1.83	1.75	1.64	1.57	1.52	1.46	1.35	1.24	1.00

TABLE C.4 (Continued)

n_2 \ n_1	1	2	3	4	5	6	7	8	9	10	12	16	20	24	30	50	100	∞
1	4,052	4,999	5,403	5,625	5,764	5,859	5,928	5,981	6,022	6,056	6,106	6,169	6,208	6,234	6,258	6,302	6,334	6,366
2	98.49	99.00	99.17	99.25	99.30	99.33	99.34	99.36	99.38	99.40	99.42	99.44	99.45	99.46	99.47	99.48	99.49	99.50
3	34.12	30.82	29.46	28.71	28.24	27.91	27.67	27.49	27.34	27.23	27.05	26.83	26.69	26.60	26.50	26.35	26.23	26.12
4	21.20	18.00	16.69	15.98	15.52	15.21	14.98	14.80	14.66	14.54	14.37	14.15	14.02	13.93	13.83	13.69	13.57	13.46
5	16.26	13.27	12.06	11.39	10.97	10.67	10.45	10.27	10.15	10.05	9.89	9.68	9.55	9.47	9.38	9.24	9.13	9.02
6	13.74	10.92	9.78	9.15	8.75	8.47	8.26	8.10	7.98	7.87	7.72	7.52	7.39	7.31	7.23	7.09	6.99	6.88
7	12.25	9.55	8.45	7.85	7.46	7.19	7.00	6.84	6.71	6.62	6.47	6.27	6.15	6.07	5.98	5.85	5.75	5.65
8	11.26	8.65	7.59	7.01	6.63	6.37	6.19	6.03	5.91	5.82	5.67	5.48	5.36	5.28	5.20	5.06	4.96	4.86
9	10.56	8.02	6.99	6.42	6.06	5.80	5.62	5.47	5.35	5.26	5.11	4.92	4.80	4.73	4.64	4.51	4.41	4.31
10	10.04	7.56	6.55	5.99	5.64	5.39	5.21	5.06	4.95	4.85	4.71	4.52	4.41	4.33	4.25	4.12	4.01	3.91
11	9.65	7.20	6.22	5.67	5.32	5.07	4.88	4.74	4.63	4.54	4.40	4.21	4.10	4.02	3.94	3.80	3.70	3.60
12	9.33	6.93	5.95	5.41	5.06	4.82	4.65	4.50	4.39	4.30	4.16	3.98	3.86	3.78	3.70	3.56	3.46	3.36
13	9.07	6.70	5.74	5.20	4.86	4.62	4.44	4.30	4.19	4.10	3.96	3.78	3.67	3.59	3.51	3.37	3.27	3.16
14	8.86	6.51	5.56	5.03	4.69	4.46	4.28	4.14	4.03	3.94	3.80	3.62	3.51	3.43	3.34	3.21	3.11	3.00
15	8.68	6.36	5.42	4.89	4.56	4.32	4.14	4.00	3.89	3.80	3.67	3.48	3.36	3.29	3.20	3.07	2.97	2.87
16	8.53	6.23	5.29	4.77	4.44	4.20	4.03	3.89	3.78	3.69	3.55	3.37	3.25	3.18	3.10	2.96	2.86	2.75
17	8.40	6.11	5.18	4.67	4.34	4.10	3.93	3.79	3.68	3.59	3.45	3.27	3.16	3.08	3.00	2.86	2.76	2.65
18	8.28	6.01	5.09	4.58	4.25	4.01	3.85	3.71	3.60	3.51	3.37	3.19	3.07	3.00	2.91	2.78	2.68	2.57
19	8.18	5.93	5.01	4.50	4.17	3.94	3.77	3.63	3.52	3.43	3.30	3.12	3.00	2.92	2.84	2.70	2.60	2.49
20	8.10	5.85	4.94	4.43	4.10	3.87	3.71	3.56	3.45	3.37	3.23	3.05	2.94	2.86	2.77	2.63	2.53	2.42
22	7.94	5.72	4.82	4.31	3.99	3.76	3.59	3.45	3.35	3.26	3.12	2.94	2.83	2.75	2.67	2.53	2.42	2.31
24	7.82	5.61	4.72	4.22	3.90	3.67	3.50	3.36	3.25	3.17	3.03	2.85	2.74	2.66	2.58	2.44	2.33	2.21
26	7.72	5.53	4.64	4.14	3.82	3.59	3.42	3.29	3.17	3.09	2.96	2.77	2.66	2.58	2.50	2.36	2.25	2.13
28	7.64	5.45	4.57	4.07	3.76	3.53	3.36	3.23	3.11	3.03	2.90	2.71	2.60	2.52	2.44	2.30	2.18	2.06
32	7.50	5.34	4.46	3.97	3.66	3.42	3.25	3.12	3.01	2.94	2.80	2.62	2.51	2.42	2.34	2.20	2.08	1.96
36	7.39	5.25	4.38	3.89	3.58	3.35	3.18	3.04	2.94	2.86	2.72	2.54	2.43	2.35	2.26	2.12	2.00	1.87
40	7.31	5.18	4.31	3.83	3.51	3.29	3.12	2.99	2.88	2.80	2.66	2.49	2.37	2.29	2.20	2.05	1.94	1.81
60	7.08	4.98	4.13	3.65	3.34	3.12	2.95	2.82	2.72	2.63	2.50	2.32	2.20	2.12	2.03	1.87	1.74	1.60
100	6.90	4.82	3.98	3.51	3.20	2.99	2.82	2.69	2.59	2.51	2.36	2.19	2.06	1.98	1.89	1.73	1.59	1.43
200	6.76	4.71	3.88	3.41	3.11	2.90	2.73	2.60	2.50	2.41	2.28	2.09	1.97	1.88	1.79	1.62	1.48	1.28
∞	6.64	4.60	3.78	3.32	3.02	2.80	2.64	2.51	2.41	2.32	2.18	1.99	1.87	1.79	1.69	1.52	1.36	1.00

Source: Abridged from Table A14, Part I of G. W. Snedecor and W. G. Cochran, *Statistical Methods*, 6th ed., Iowa State University Press, Ames, 1967. It is reproduced here with the kind permission of the author and the publisher. n_1 is the number of degrees of freedom in the numerator and n_2 the number of degrees of freedom in the denominator of F. α is the probability of exceeding the tabular value.

ANSWERS TO SELECTED EXERCISES

1.1. .086 1.2. .0043

2.3. (a) $\{2 \le x \le 3\}$; (b) $\{-1 \le x \le 10\}$; (c) $\{3 < x \le 5\}$;
(d) $\{-\infty < x < -1,\ 2 < x < \infty\}$

2.4. 36 2.5. 12 2.6. 120

2.7. 20 2.8. 50,400 2.9. (a) 420; (b) 120

2.10. 13,326 2.11. 4,943 2.12. 6, 4, 1

2.13. 31 2.15. 16 2.16. 9

2.17. 41 2.18. 1.731×10^{13} 2.19. no

2.20. 8/663 2.21. 34/77 2.23. 2/3

2.24. (a) 1/64; (b) 1/32; (c) 1/64; (d) 5/16

2.28. 3/5 2.29. 8/9

3.2.

x	2	3	4	5	6	7	8	9	10	11	12
f(x)	1/36	2/36	3/36	4/36	5/36	6/36	5/36	4/36	3/36	2/36	1/36

3.3.

x	0	1	2	3	4	5
f(x)	.0778	.2592	.3456	.2304	.0768	.0102

3.4.

x	0	1	2	3	4	5	6
f(x)	.2231	.3347	.2510	.1255	.0471	.0141	.0035

3.5. m = np = 2

x	0	1	2	3	4	5	6	7
f(x)	.1353	.2707	.2707	.1804	.0902	.0361	.0120	.0034

3.6. .2679, .3182, .3188 3.7. $\left(\frac{1}{3}\right)^{x-1}\left(\frac{2}{3}\right)$ x = 1, 2, ...

3.8. $\dfrac{3!x!(9 - x)!(6 - x)}{(3 - x)!9!9}$ x = 0, 1, 2, 3

3.9. $F(x) = \begin{cases} 0 & x \leq 0 \\ x^2 & 0 < x \leq 1 \\ 1 & x > 1 \end{cases}$

3.13. 609/10,000, 81/10,000

3.14. $\log_e 2$

3.15. $f(x) = 2x/(1 + x^2)^2$ $x > 0$ and 0 otherwise

3.16. no

3.17. .5

3.18. e^{-2}

3.19. $1 - (1 - 1/e)^2$

3.20. μ

3.21. (a) .5; (b) .6826; (c) .0228

3.25. 1, 1, 2, 6

3.31. (a) − 6; (b) m = n = 2

4.1. 26

4.2. (a) $\mu = 10/11$, $\sigma^2 = 5/726$; (b) $\mu = 20/30$, $\sigma^2 = 50/9$

4.3. $p_1 = p_2 = .5$

4.4. a = 1, b = 0

4.5 $\mu_3 = \mu_3' - 3\mu\mu_2' + 2\mu^3$, $\mu_4 = \mu_4' - 4\mu\mu_3' + 6\mu^2\mu_2' - 3\mu^4$

4.6. $\mu_1 = 0$, $\mu_2 = \sigma^2$, $\mu_3 = 0$, $\mu_4 = 3\sigma^4$

4.7. $\gamma_1 = \gamma_2 = 0$

4.8. np, npq, npq(q − p), npq(1 − 6pq)

4.10. $m(t) = (e^{2t} - 1)/(2t)$, $\mu = 1$, $\sigma^2 = 1/3$

4.11. (a) 32; (b) 4

4.12. (a) 5; (b) 55/12

4.13. 1.9

4.14. ac/(a + b)

4.15. s(r + 1)/2

4.16. $(e^t - 1)^2/t^2$

4.17. $(e^{\theta t} - 1)/t$

4.18. $\mu_2' = n(n - 1)p^2 + np$

$\mu_3' = n(n - 1)(n - 2)p^3 + 3n(n - 1)p^2 + np$

$\mu_4' = n(n - 1)(n - 2)(n - 3)p^4 + 6n(n - 1)(n - 2)p^3 + 7n(n - 1)p^2 + np$

4.19. $\alpha_3 = (q - p)/\sqrt{npq}$, $\alpha_4 = 3 + (1 - 6pq)/(npq)$

4.20. (a) $\mu = 35/6$, $\sigma^2 = 35/36$, $\alpha_3 = 4/\sqrt{35}$, $\alpha_4 = 111/35$;

(b) $\mu = 12$, $\sigma^2 = 4$, $\alpha_3 = -1/6$, $\alpha_4 = 35/12$

4.23. $\mu = m$, $\mu_2' = m(m + 1)$, $\mu_3' = m(m^2 + 3m + 1)$,
$\mu_4' = m(m^3 + 6m^2 + 7m + 1)$

4.24. $\mu_2 = m$, $\alpha_3 = 1/\sqrt{m}$, $\alpha_4 = (3m + 1)/m$

4.26. 1/12, 5/12, −3/40, 4/15

4.27. (a) $m(t) = (1 - \theta t)^{-2}$; (b) $\kappa_r = 2(r - 1)!\theta^r \quad r = 1, 2, \ldots$

5.3. 1/6

5.4. $f(z) = \binom{n}{z}(p_1 + p_2)^z(1 - p_1 - p_2)^{n-z}$

5.6. (a) $1 - e^{-2}$; (b) 1/2; (c) 2; (d) 2; (e) 0

5.8. (a) $\mu_X = \mu_Y = 1/4, \quad \sigma_X^2 = \sigma_Y^2 = 3/80, \quad \sigma_{XY} = -1/80$;

(b) $\mu_{Y|x} = (1 - x)/2, \ \sigma_{Y.x}^2 = 1/30$

5.9. (a) 120; (b) $g(x) = 20x(1 - x)^3 \quad 0 < x < 1$;

(c) $\mu_{Y|x} = (1 - x)/2$

5.10. $g(u) = 2u(1 + u)^{-3} \quad 0 < u < \infty; \quad P(U \leq 1) = \frac{1}{4}$

5.13. $g(x) = 4x(1 - x^2) \quad 0 < x < 1; \quad h(y) = 4y^3 \quad 0 < y < 1$

5.18. (a) $\mu_X = 4, \quad \mu_Y = 0, \quad \sigma_x^2 = 4, \quad \sigma_Y^2 = 9, \quad \rho = 0$;

(b) $\mu_X = \mu_Y = 0, \quad \sigma_X^2 = \sigma_Y^2 = 1, \quad \rho = -.8$;

(c) $\mu_X = 3, \quad \mu_Y = 1, \quad \sigma_X^2 = 1, \quad \sigma_Y^2 = 4, \quad \rho = .6$

5.19. $N(0, \sigma_X^2 + \sigma_Y^2 + 2\rho\sigma_X\sigma_Y)$

5.22. $Y_1 \sim N[\mu_1 + \mu_2, 2\sigma^2(1 + \rho)], \quad Y_2 \sim N[\mu_1 - \mu_2, 2\sigma^2(1 - \rho)]$

5.23. Poisson (mn)

5.24. $N(\mu, \sigma^2/n)$

5.25.
$$f(u) = \begin{cases} u & 0 < u < 1 \\ 2 - u & 1 \leq u < 2 \\ 0 & \text{otherwise} \end{cases}$$

5.26. $\underline{U} \sim N(\underline{\mu}, \underline{\Sigma})$

$$\underline{\mu} = \begin{pmatrix} 0 \\ 0 \\ 0 \\ 4\mu \end{pmatrix} \qquad \underline{\Sigma} = \sigma^2 \begin{pmatrix} 2 & 1 & 1 & 0 \\ 1 & 2 & 1 & 0 \\ 1 & 1 & 2 & 0 \\ 0 & 0 & 0 & 4 \end{pmatrix}$$

5.27. $Y_1 \sim N\{\sqrt{k}\mu, [1 + (k - 1)\rho]\}$,

$Y_i \sim N[0, \sigma^2(1 - \rho)] \quad i = 2, \ldots, k$

6.1. $L_5 = X_1 - 4X_2 + 6X_3 - 4X_4 + X_5$, n

6.2. $\kappa_{r(\Sigma X)} = N\kappa_r \quad r = 1, 2, \ldots$

6.3. $\kappa_1 = Nnp, \quad \kappa_2 = Nnpq, \quad \kappa_3 = Nnpq(q - p)$

6.4. Poisson, $\kappa_r = \sum_{i=1}^{N} m_i$ for all $r = 1, 2, 3, \ldots$

Binomial, $\kappa_1 = \sum_{i=1}^{N} n_i p_i, \quad \kappa_2 = \sum_{i=1}^{N} n_i p_i q_i,$

$\kappa_3 = \sum_{i=1}^{N} n_i p_i q_i (q_i - p_i)$

6.5. $a_1 = \sigma_2^2/(\sigma_1^2 + \sigma_2^2), \quad a_2 = \sigma_1^2/(\sigma_1^2 + \sigma_2^2)$

6.8. $L_4 = T_0 + T_1 + T_2 + T_3, \quad \sigma_{L_1}^2 = 12r\sigma^2, \quad \sigma_{L_2}^2 = 2r\sigma^2,$

$\sigma_{L_3}^2 = 6r\sigma^2, \quad \sigma_{L_4}^2 = 4r\sigma^2$

6.10. .993

6.11. $f(y) = [2^{\nu/2}\Gamma(\nu/2)]^{-1}y^{-\nu/2}(1 - y)^{-(\nu+2)/2}\exp[-y/2(1-y)]$

$0 < y < 1$

6.12. $$f(w) = B\left(\frac{\nu_1}{2}, \frac{\nu_2}{2}\right)^{-1}(a_1 - a_2)^{-(\nu_1+\nu_2-2)/2}(a_1 - w)^{(\nu_1-2)/2}$$

$$\times (w - a_2)^{(\nu_1-\nu_2)/2} \qquad a_2 < w < a_1$$

6.14. .87 6.16. $1 - e^{-c^2/2}$ 6.17. $x_1^2 + x_2^2 = 9.21$

6.18. $\chi^2\left(\sum_{i=1}^{r} n_i - r\right)$ 6.19. 1% = 6.63, 9.21, 13.3; 5% = 3.84, 5.99, 9.5

6.20. .0916 6.21. .32 6.22. .44

6.23. $\alpha_3 = 2\sqrt{2}/\sqrt{n - 1}, \quad \alpha_4 = 3 + 12/(n - 1)$

6.24. $$m(t) = \left(\frac{2\sigma^2}{\nu}\right)^{t/2}\frac{\Gamma[(\nu + t)/2]}{\Gamma(\nu/2)}$$

6.28. (a) 4.3026; (b) .0955 6.29. $P(|t| \geq 5) < .001$

6.34. $$\mu_i = \left[\Gamma\left(\frac{n_1}{2}\right)\Gamma\left(\frac{n_2}{2}\right)\right]^{-1}\left[\left(\frac{n_2}{n_1}\right)^i\Gamma\left(\frac{n_1}{2} + i\right)\Gamma\left(\frac{n_2}{2} - i\right)\right.$$

$$- i\left(\frac{n_2}{n_1}\right)^{i-1}\left(\frac{n_2}{n_2 - 2}\right)\Gamma\left(\frac{n_1}{2} + i - 1\right)\Gamma\left(\frac{n_2}{2} - i + 1\right)$$

$$\left. + \frac{i(i - 1)}{2}\left(\frac{n_2}{n_1}\right)^{i-2}\left(\frac{n_2}{n_2 - 2}\right)^2\Gamma\left(\frac{n_1}{2} + i - 2\right)\Gamma\left(\frac{n_2}{2} - i + 2\right) - \cdots\right]$$

6.35. (a) 6.9442; (b) 6.3883 6.36. .104

6.37. .208 6.38. .001

7.2. (c) yes; (d) yes; (e) yes 7.3. $\bar{X}$

7.4. (b) yes 7.5. $\hat{\theta} = 4(\bar{X}/3)$ 7.6. (b) $p(1 - p)/n$

7.7. (b) unbiased; (d) consistent 7.8. (d) yes

7.10. bias = $-\theta/(n + 1)$ 7.11. $X_{(1)}$

7.12. (d) yes; (e) yes

7.13. (a) $\hat{\alpha} = \bar{Y}$, $\hat{\beta} = \Sigma(x_i - \bar{x})(Y_i - \bar{Y})/\Sigma(x_i - \bar{x})^2$,

$\hat{\sigma}^2 = \Sigma[Y_i - \bar{Y} - \hat{\beta}(x_i - \bar{x})]^2/n$;

(b) $V(\hat{\alpha}) = \sigma^2/n$; $V(\hat{\beta}) = \sigma^2/\Sigma(x_i - \bar{x})^2$; (c) no

7.14. (a) $\hat{\mu}_X = \bar{X}$, $\hat{\mu}_Y = \bar{Y}$, $\sigma_X^2 = \Sigma(X_i - \bar{X})/n$,

$\hat{\sigma}_Y^2 = \Sigma(Y_i - \bar{Y})^2/n$, $\hat{\rho} = \Sigma(X_i - \bar{X})(Y_i - Y)/[\Sigma(x_i - \bar{X})^2$

$\times \Sigma(Y_i - \bar{Y}^2]^{1/2}$

(c) yes

7.15. $N(\mu_1 - \mu_2, \sigma_1^2 + \sigma_2^2 - 2\sigma_{12})$, $\bar{X}_1 - \bar{X}_2$

7.16. beta distribution with parameters $a + \Sigma x_i$ and $b + nN - \Sigma x_i$;

$\hat{p} = (a + \Sigma x_i)/(a + b + nN)$

7.17. gamma distribution with parameters $\alpha + \Sigma x_i$ and $1/(n + 1/\beta)$;

$\hat{m} = (\alpha + \Sigma X_i)/(n + 1/\beta)$

7.18. (a) 20.55; (b) 15.5; (c) $\bar{X}_{.05} = 17.44$, $\bar{X}_{.10} = 17.56$;

(d) $\bar{X}_{W1} = 17.35$, $\bar{X}_{W2} = 17.75$

8.3. 14,086; 2335×10^3

8.9. (b) $\left(\frac{1}{n_1} - \frac{1}{N_1}\right)S_1^2 + \left(\frac{1}{n_2} - \frac{1}{N_2}\right)S_2^2$;

(c) $n_i = n \dfrac{S_i/\sqrt{c_i}}{S_i/\sqrt{c_i} + S_2/\sqrt{c_2}}$

$$n = \frac{(C - c_0)\Sigma S_i/\sqrt{c_i}}{\Sigma S_i\sqrt{c_i}}$$

8.10. $n_i = n \dfrac{W_iS_i/\sqrt{c_i}}{\Sigma W_iS_i/\sqrt{c_i}}$ $\quad n = (\Sigma W_iS_i\sqrt{c_i})\left(\Sigma\dfrac{W_iS_i}{\sqrt{c_i}}\right)\left(v + \dfrac{\Sigma W_i^2S_i^2}{N_i}\right)^{-1}$

8.11. $V(\hat{p}) = \left(\frac{1}{n} - \frac{1}{N}\right)\frac{1}{N - 1}\Sigma(p_i - p)^2$

8.14. $V(\bar{\bar{y}}_{sy}) = \left(\frac{1}{n} - \frac{1}{N}\right)\frac{1}{N - 1}\sum_{i=1}^{N}(\mu_i - \mu)^2 + \frac{1}{Nnk}\sum_{i=1}^{N}\sum_{j=1}^{k}(\bar{y}_{ij} - \mu_i)^2$

9.2 $\bar{X}_1 - \bar{X}_2 \pm t_{\alpha/2}\, S_p\sqrt{1/n_1 + 1/n_2}$ 9.3. (.90, 5.22)

9.4. (a) (2.787, 46.5); (b) (.451,12.12); (c) (.15,5.14)

9.6. $v_1 = 4.5054$, $v_2 = 28.4733$, (871,5.507)

9.7. (-172.3,-67.7); (-56.35,-13.65)

9.8. $(X_{(n)}, X_{(n)}\, \alpha^{-1/n})$

9.9. $P\{\chi_1^2 < 2n\bar{X}\theta < \chi_2^2\} = 1 - \alpha$, $\chi_1^2/(2n\bar{X}) < \theta < \chi_2^2/(2n\bar{X})$

9.10. (3.58,8.42)

9.11. (a) $\bar{X}_1 - \bar{X}_2 \pm z_{\alpha/2}\ \sqrt{(\sigma_1^2 + \sigma_2^2 - 2\sigma_{12})/n}$;

(b) $\bar{X}_1 - \bar{X}_2 \pm t_{(\alpha/2,n-1)}\ \sqrt{(S_1^2 + S_2^2 - 2S_{12})/n}$

9.12. (a)

τ^2	1	2	3	5	20	50
Upper limit	- .136	- .129	- .126	- .124	- .121	- .120
Lower limit	1.046	1.081	1.094	1.104	1.116	1.118

(b) the confidence interval is (-0.12,1.12).

10.1. (a) .19; (b) .6561 10.2. $\Sigma X_i > k$

10.3. k = 10.08, n = 29

10.7.

a	.4352	.5427	.7519	1.0000	1.4030	2.5686	5.5200
F_a	9.4259	9.5597	5.4559	4.1025	2.9239	1.5972	0.7432

10.8. $\lambda = [m_0/\bar{X}]^{n\bar{X}}\ e^{-n(m_0 - \bar{X})}$

10.9. $\lambda = (n_1 + n_2)^{(n_1 + n_2)/2}\ n_1^{-n_1/2}\ n_2^{-n_2/2}(\Sigma X_{ij}^2)^{n_1/2}$

$\times\ (\Sigma X_{2j}^2)^{n_2/2}(\Sigma X_{1j} + \Sigma X_{2j})^{-(n_1 + n_2)/2}$

10.10 Replace ΣX_{1j}^2 and ΣX_{2j}^2 by $\Sigma(X_{1j} - \bar{X}_1)^2$ and $\Sigma(X_{2j} - \bar{X}_2)^2$, respectively, in Exercise 10.9.

10.13. $\lambda = \prod_{i=1}^{k} (P_{0i}/\hat{p}_i)^{n\hat{p}_i}$ $\hat{p}_i = X_i/n$

10.14. $\lambda = \left(\dfrac{n_1\bar{X} + n_2\bar{Y}}{n_1 + n_2}\right)^{n_1\bar{X} + n_2\bar{Y}} (\bar{X})^{-n_1\bar{X}}\ (\bar{Y})^{-n_2\bar{Y}}$

10.15. $\lambda = \theta_0^{n}\bar{X}^{n}\exp[n(1 - \bar{X})]$

10.16. (a) $\dfrac{\sqrt{n}|\bar{X}_1 - \bar{X}_2 - \mu_0|}{(\sigma_1^2 + \sigma_2^2 - 2\sigma_{12})^{1/2}} > z_{\alpha/2}$

(b) $\dfrac{\sqrt{n}|\bar{X}_1 - \bar{X}_2 - \mu_0|}{(S_1^2 + S_2^2 - 2S_{12})^{1/2}} > t_{(\alpha/2,\ n - 1)}$

10.17. Reject H_0 if $\dfrac{\sqrt{\Sigma(x_i - \bar{x})^2}\ |\hat{\beta} - \beta|}{S} > t_{(\alpha/2, n - 2)}$

$$\hat{\beta} = \frac{\Sigma(x_i - \bar{x})(Y_i - \bar{Y})}{\Sigma(x_i - \bar{x})^2} \qquad S^2 = \Sigma\,\frac{[Y_i - \bar{Y} - \hat{\beta}(x_i - \bar{x})]^2}{n - 2}$$

10.18. $\lambda = (\hat{\theta}_1\hat{\theta}_2)^n/(\hat{\theta})^{2n}$, $\hat{\theta}_1 = \Sigma X_i^2/n$, $\hat{\theta}_2 = \Sigma Y_i^2/n$,

$\hat{\theta} = (\Sigma X_i^2 + \Sigma Y_i^2)/2n$

10.19. $\left[\log \dfrac{p_0(1 - p_1)}{p_1(1 - p_0)}\right]^{-1} \left(\log 19 - n \log \dfrac{1 - p_0}{1 - p_1}\right) <$

$\Sigma X_i < \left[\log \dfrac{p_0(1 - p_1)}{p_1(1 - p_0)}\right]^{-1} \times \left(\log \dfrac{1}{19} - n \log \dfrac{1 - p_0}{1 - p_1}\right)$

10.20. $\dfrac{1}{\theta_0 - \theta_1}\left(n \log \dfrac{\theta_0}{\theta_1} - \log \dfrac{1 - \alpha}{\beta}\right) < \Sigma X_i < \dfrac{1}{\theta_0 - \theta_1}$

$\left(n \log \dfrac{\theta_0}{\theta_1} - \log \dfrac{\alpha}{1 - \beta}\right)$

11.1. .79, .44, .17 11.2. Reject H_0. 11.4. Accept H_0.
11.5. $\chi^2 = 107$ 11.6. $\chi^2 = 36$ 11.7. $\chi^2 = 49.62$
11.8. $p = .026$ 11.9. $\chi^2 = 4.349$ 11.10. $\chi^2 = 11$
11.11. $\chi^2 = 49,33$

12.1. $\bar{X}* = 40.5$ 12.2. not significant 12.3. $\hat{\mu}_1 = 41.6$
12.5. $\hat{\rho} = .393$ 12.6. $\hat{\sigma}_1^2 = 49.09$

13.5. (a) Y_i/x_i;

(b) $$a = \frac{n\Sigma Y_i/x_i^2 - [\Sigma(1/x_i)][\Sigma(Y_i/x_i]}{d}$$

$$d = n\Sigma\frac{1}{x_i^2} - \left(\Sigma\frac{1}{x_i}\right)^2$$

$$b = \frac{[\Sigma(1/x_i^2)][\Sigma(Y_i/x_i)] - [\Sigma(1/x_i)][\Sigma(Y_i/x_i^2)]}{d}$$

(c) $V(a) = n\sigma^2/d$, $\quad V(b) = \dfrac{\Sigma(1/x_i^2)\sigma^2}{d}$ $\quad \text{Cov}(a,b) = \dfrac{-\Sigma(1/x_i)\sigma^2}{d}$

13.7. Add λ/k under the square root sign.

13.8. (a) $\hat{Y} = 69.56 + .2591x$; (b) $s^2 = 27.05$;

(c) $F = 80.3$, correlated error

13.9. (b) $\hat{\sigma}^2 = \dfrac{\Sigma[Y_i - \bar{y} - b(x_i - \bar{x})]^2}{n}$, biased

13.10. $\lambda = [\hat{\sigma}_\Omega^2/\hat{\sigma}_\omega^2]^{(n_1+n_2)/2}$; $\hat{\sigma}_\omega^2$ and $\hat{\sigma}_\Omega$ are the ML estimators of σ^2 under H_0 and H_1, respectively.

13.12. (a) female: $\hat{Y} = 2.981 + 2.637x$; male: $Y = -1.184 + 4.313x$;

(b) $t = 2.35$; (c) $F = 1.79$

13.13. (a) $a = .9285$, $b = -.007984$, $s(a) = .03911$,

$s(b) = .0009381$; (b) $N_0 = 116{,}330$ pounds

13.16. (a) $\hat{\beta} = 6\Sigma Y_i/[n(n + 1)(2n + 1)]$, unbiased;

(b) $V(\hat{\beta}) = 6\sigma^2/[n(n + 1)(2n + 1)$

13.17. $\bar{a} = 1.2688$, $\bar{b} = .1935$

14.3. $\Sigma a_i\bar{Y}_i$; (b) $T = \dfrac{\sqrt{n}\Sigma a_i\bar{Y}_i}{\sqrt{S^2\Sigma a_i^2}}$

14.4. $t - 1$, $\sum_{i=1}^{t} a_ib_i\sigma^2$ 14.5. $F = 4.21$ 14.6. $t = 2.62$

14.9. (a) $\tilde{\sigma}_A^2 = 2/3$, $S^2 = 42$; (b) $\hat{\sigma}_A^2 = .43$, $S^2 = 42$

14.10. pool error with interaction; $F = 2.75$

AUTHOR INDEX

SUBJECT INDEX